MODERN EARTH BUILDING HANDBOOK

Modern Earth Building Handbook

Editor:

Dr. Tanjina Nur, Ph.D.

ArclerPress

www.arclerpress.com

Modern Earth Building Handbook

Editor: Dr. Tanjina Nur, Ph.D.

708 3rd Avenue, 6th Floor
New York
NY 10017
United States of America
www.arclerpress.com

ISBN: 978-1-68094-378-8

Library of Congress Control Number: 2016950960

About the Editor

Dr. Tanjina Nur, Ph.D.

Tanjina finished her PhD in Civil and Environmental Engineering in 2014 from University of Technology Sydney (UTS). Now she is working as Post-Doctoral Researcher in the Centre for Technology in Water and Wastewater (CTWW) and published about eight International journal papers with 80 citations. Her research interest is wastewater treatment technology using adsorption process.

Contents

List of Contributors

Judy Ueda
MU Design Inc., 176-0026, Tokyo, Japan

Minoru Ueda
MU Design Inc., 176-0026, Tokyo, Japan

Apa rna Das
School of Architecture M.S.Ramaiah Institute of Technology India-560012

S. K. Fasogbon
Mechanical Engineering Department, Faculty of Technology, University of Ibadan, Ibadan, Nigeria

A. B. Wahaab
Building Department, Faculty of EDM, Obafemi Awolowo University, Ile-Ife, Nigeria

M. O. Oyewola
Mechanical Engineering Department, Faculty of Technology, University of Ibadan, Ibadan, Nigeria

Gabriela M. Medero
School of the Built Environment, Heriot-Watt University, EH14 4AS, Edinburgh, UK

Justin H. Kennedy
School of the Built Environment, Heriot-Watt University, EH14 4AS, Edinburgh, UK

Peter K. Woodward
School of the Built Environment, Heriot-Watt University, EH14 4AS, Edinburgh, UK

Meysam Banimahd
Technip, Westhill Industrial Estate, Westhill, Aberdeenshire, AB32 6TQ, UK

Farraj Al-Ajmi
Civil Engineering Department, College of Technological Studies, The Public Authority for Applied Education & Training, Kuwait

Hany Abdalla
Civil Engineering Department, College of Technological Studies, The Public Authority for Applied Education & Training, Kuwait

Magdi Abdelghaffar
Civil Engineering Department, College of Technological Studies, The Public Authority for Applied Education & Training, Kuwait

Jamal Almatawah
Civil Engineering Department, College of Technological Studies, The Public Authority for Applied Education & Training, Kuwait

T. López-Lara
Postgraduate Division, Faculty of Engineering, Autonomous University of Queretaro, Cerro de las Campanas S/N, Colonia Niños Héroes, 76010 Querétaro, QRO, Mexico

Juan Bosco Hernandez-Zaragoza
Postgraduate Division, Faculty of Engineering, Autonomous University of Queretaro, Cerro de las Campanas S/N, Colonia Niños Héroes, 76010 Querétaro, QRO, Mexico

Jaime Horta
Postgraduate Division, Faculty of Engineering, Autonomous University of Queretaro, Cerro de las Campanas S/N, Colonia Niños Héroes, 76010 Querétaro, QRO, Mexico

Eduardo Rojas Gonzalez
Postgraduate Division, Faculty of Engineering, Autonomous University of Queretaro, Cerro de las Campanas S/N, Colonia Niños Héroes, 76010 Querétaro, QRO, Mexico

Carlos Lopez-Cajun
Postgraduate Division, Faculty of Engineering, Autonomous University of Queretaro, Cerro de las Campanas S/N, Colonia Niños Héroes, 76010 Querétaro, QRO, Mexico

Gerson Ramirez
Postgraduate Division, Faculty of Engineering, Autonomous University of Queretaro, Cerro de las Campanas S/N, Colonia Niños Héroes, 76010 Querétaro, QRO, Mexico

Vandna Sharma
Department of Architecture, National Institute of Technology, Hamirpur 177005, Himachal Pradesh, India

Hemant K. Vinayak
Department of Civil Engineering, National Institute of Technology, Hamirpur 177005, Himachal Pradesh, India

Bhanu M. Marwaha
Department of Architecture, National Institute of Technology, Hamirpur 177005, Himachal Pradesh, India

Suchita Rai
Jawaharlal Nehru Aluminium Research Development and Design Centre, Wadi, Amravati Road, Nagpur, India
Advanced Separations and Analytical Laboratory, Department of Chemical Engineering, Visvesvaraya National Institute of Technology (VNIT), Nagpur, Maharashtra 440011, India

Dilip H. Lataye
Department of Civil Engineering, Visvesvaraya National Institute of Technology (VNIT), Nagpur, Maharashtra 440011, India

M. J. Chaddha
Jawaharlal Nehru Aluminium Research Development and Design Centre, Wadi, Amravati Road, Nagpur, India

R. S. Mishra
Jawaharlal Nehru Aluminium Research Development and Design Centre, Wadi, Amravati Road, Nagpur, India

P. Mahendiran
Jawaharlal Nehru Aluminium Research Development and Design Centre, Wadi, Amravati Road, Nagpur, India

J. Mukhopadhyay
Indian Institute of Technology, Gandhinagar, Gujarat, India

ChangKyoo Yoo
Environmental Management & Systems Engineering Lab (EMSEL), Department of Environmental Science and Engineering, College of Engineering, Kyung Hee University, Seocheon-dong 1, Giheung-gu, Gyeonggi-do, Yongin-si 446-701, Republic of Korea

Kailas L. Wasewar
Advanced Separations and Analytical Laboratory, Department of Chemical Engineering, Visvesvaraya National Institute of Technology (VNIT), Nagpur, Maharashtra 440011, India
Environmental Management & Systems Engineering Lab (EMSEL), Department of Environmental Science and Engineering, College of Engineering, Kyung Hee University, Seocheon-dong 1, Giheung-gu, Gyeonggi-do, Yongin-si 446-701, Republic of Korea

Elhadj Benachour
Laboratory of Energy in Arid Regions, University of Tahri Mohamed - Bechar, P.O. P 417. 08000 , Road Kenedza , Bechar, Algeria

Belkacem Draoui
Laboratory of Energy in Arid Regions, University of Tahri Mohamed - Bechar, P.O. P 417. 08000 , Road Kenedza , Bechar, Algeria

Baachir Imine
Department of Mechanical Engineering, University of Sciences and Technology of Oran-Mohamed Boudiaf, El-Mnaouar, Oran, Algeria

Mohammed Hasnat
Laboratory of Energy in Arid Regions, University of Tahri Mohamed - Bechar, P.O. P 417. 08000 , Road Kenedza , Bechar, Algeria

Lakhdar Rahmani
Laboratory of Energy in Arid Regions, University of Tahri Mohamed - Bechar, P.O. P 417. 08000 , Road Kenedza , Bechar, Algeria

Marwan Mostafa
Department of Civil, Construction and Environmental Engineering, University of Alabama at Birmingham, Birmingham, AL 35294, USA

Nasim Uddin
Department of Civil, Construction and Environmental Engineering, University of Alabama at Birmingham, Birmingham, AL 35294, USA

Anna Laura Pisello
CIRIAF–Interuniversity Research Center on Pollution Control and Environment Protection "Mauro Felli", University of Perugia, Via Duranti 67, Perugia 06125, Italy

Franco Cotana
CIRIAF–Interuniversity Research Center on Pollution Control and Environment Protection "Mauro Felli", University of Perugia, Via Duranti 67, Perugia 06125, Italy

Andrea Nicolini
CIRIAF–Interuniversity Research Center on Pollution Control and Environment Protection "Mauro Felli", University of Perugia, Via Duranti 67, Perugia 06125, Italy

Lucia Brinchi
CIRIAF–Interuniversity Research Center on Pollution Control and Environment Protection "Mauro Felli", University of Perugia, Via Duranti 67, Perugia 06125, Italy

Gnanli Landrou
Institute of Construction and Infrastructure Management, Chair of Sustainable Construction, Swiss Federal Institute of Technology (ETH Zurich), Stefano-Franscini-Platz 5, Zürich 8093, Switzerland

Coralie Brumaud
Institute of Construction and Infrastructure Management, Chair of Sustainable Construction, Swiss Federal Institute of Technology (ETH Zurich), Stefano-Franscini-Platz 5, Zürich 8093, Switzerland

Frank Winnefeld
Laboratory for Concrete/Construction Chemistry, Empa, Swiss Federal Laboratories for Material Science and Technology, Überlandstrasse 129, Dübendorf 8600, Switzerland

Robert J. Flatt
Institute for Building Materials, Physical Chemistry of Building Materials, Swiss Federal Institute of Technology (ETH Zurich), Stefano-Franscini-Platz 3, Zürich 8093, Switzerland

Guillaume Habert
Institute of Construction and Infrastructure Management, Chair of Sustainable Construction, Swiss Federal Institute of Technology (ETH Zurich), Stefano-Franscini-Platz 5, Zürich 8093, Switzerland

Roberto Fioretti
Department of Industrial Engineering and Mathematical Sciences, Università Politecnica delle Marche, Via Brecce Bianche 12, 60131 Ancona, Italy

Paolo Principi
Department of Industrial Engineering and Mathematical Sciences, Università Politecnica delle Marche, Via Brecce Bianche 12, 60131 Ancona, Italy

Ivana Miličević
Department for Materials and Structures, Faculty of Civil Engineering, University of Osijek, Osijek 31000, Croatia

Nina Štirmer
Department for Materials, Faculty of Civil Engineering, University of Zagreb, Zagreb 10000, Croatia

Ivana Banjad Pečur
Department for Materials, Faculty of Civil Engineering, University of Zagreb, Zagreb 10000, Croatia

Carmen Galan-Marin
Institute of Architecture and Construction Science, University of Seville, Seville 41012, Spain

Carlos Rivera-Gomez
Institute of Architecture and Construction Science, University of Seville, Seville 41012, Spain

Antonio Garcia-Martinez
Institute of Architecture and Construction Science, University of Seville, Seville 41012, Spain

Marinella Fossetti
Facoltà di Ingegneria e Architettura, University of Enna "Kore", 94100 Enna, Italy

Giovanni Minafò
Dipartimento di Ingegneria Civile, Ambientale, Aerospaziale, dei Materiali (DICAM), University of Palermo, 90128 Palermo, Italy

Preface

Earth is an ancient building material that has been used in many different ways around the world for thousands of years. Building with earth materials can be a way of helping with sustainable management of the Earth's resources. This book covers a wide description of Modern earth buildings and provides an essential exploration of the materials and techniques key to the design, development and construction of such buildings. The first part of the book describes sustainable technology research and demonstration and a sustainable habitat for earth structures. Then review of the architecture of mud and thermal comfort characteristics of some selected building materials in the regional setting of Ile-Ife, Nigeria are presented. Then we try to find out flooding effect on earth walls and strength behaviour of mud brick in building construction.

The second part of the book describes sustainable use of tepetate composite in earthen structure and enhancing sustainability of rural adobe houses of hills by addition of vernacular fiber reinforcement. We also focus on red mud sintering using fly Ash via Taguchi's methodology and effect of the insulation by the mud on the convection in building in the Sahara of Algeria. Then we present the effect of banana fibers on the compressive and flexural strength of compressed earth blocks.

The last part of the book presents Lime as an anti-plasticizer for self-compacting clay concrete and thermal performance of hollow clay brick with low emissivity treatment in surface enclosures. Then residual mechanical properties of concrete made with crushed clay bricks and roof tiles aggregate after exposure to high temperatures and use of natural-fiber bio-composites in Construction are described. At the end, an experimental study of strengthening of masonry columns with BFRCM or with Steel Wires is also presented

Editor
Dr. Tanjina Nur, Ph.D.

CHAPTER 1

SUSTAINABLE TECHNOLOGY RESEARCH AND DEMONSTRATION CENTER FOR EARTH STRUCTURES

Judy Ueda and Minoru Ueda

MU Design Inc., 176-0026, Tokyo, Japan

ABSTRACT

This is a discussion paper that the authors presented at the International Workshop on Rammed Earth Materials and Sustainable Structures and Hakka Tulou Forum 2011: Structures of Sustainability, 28–31 October 2011, Xiamen University, China. A Sustainable Technology Research and Demonstration Center for Earth Structures is proposed to study, preserve, advance, promote, and implement rammed earth structures. The Center concept including the objectives, scope of activities and benefits of the proposed center are outlined. The Center for Alternative Technology in Wales, UK has been examined as a good base model along with a few successful environmental sustainability initiatives in China. The funding options to establish the proposed center have been discussed. The breadth of activities ultimately depends on funding capability. It is believed that the proposed center development will require significant government support at the initial stage but once corporate sponsorships are in place, the proposed center will potentially become self-supporting. The strategies, for the establishment of the proposed center are also addressed.

INTRODUCTION

The purpose of this discussion paper is to identify options for developing a framework for a research center focused on rammed earth construction methods, sustainability and the research, development and promotion of vernacular Tulou and rammed earth construction methods. There are several papers focusing on research activities related to rammed earth construction [1]. This paper takes current research activities into account and looks at research organizations and funding options which could be considered in establishing an institute to advance research, promote rammed earth construction methods and identify potential areas for collaboration.

The International Workshop on Rammed Earth Materials and Sustainable Structures, in conjunction with the International Symposium on Innovation and Sustainability of Structures in Civil Engineering hosted by Xiamen University (XMU), took place from 28 to 31 October 2011. This workshop was the 2nd event of the Hakka Tulou Forum Series. The first Hakka Tulou Forum: Lessons to Be Learned, Past, Present and Future took place on 24 June 2009 at Xiamen University, China where the International Hakka Tulou Alliance was also launched [1,2,3]. The objective of the 2009 Forum was to demonstrate how the sustainability of Hakka village architecture built hundreds of years ago and still in-use today, would bridge the past, present and future, with lessons for our modern world.

Liang [2] notes in his introduction to the Workshop that Hakka rammed earth buildings in Fujian Province of China

Citation: Judy Ueda and Minoru Ueda, Sustainable Technology Research and Demonstration Center for Earth Structures, doi:10.3390/su4102639.

reflect the importance of historical precedents, universal evolution, emerging innovation and advancement in the science and engineering of rammed earth construction, from the 8th to 20th century. They are considered as "Eco-villages" of best practices for planet earth's sustainability in their planning, design, construction, lifestyle, resource management, micro industries, renewable energy, recycling of human and animal waste, and a low ecological footprint [4]. The world heritage Hakka Tulou was recognized with the first "History Made for Tomorrow" award by US History Channel [5]. "History Made for Tomorrow" is a new community outreach initiative to show case historic places where lessons can be learned to build a sustainable 21st century.

Because of its high thermal mass, structural performance and inherent recyclability, rammed earth is considered as a sustainable construction material with significant environmental benefits [6]. The Building Research Establishment (BRE) Environmental Assessment Method (BREEAM) now recognizes rammed earth as a sustainable construction material (Element No. 1006220010–1006220013) with an A^+ rating [7]. Detailed information and research activities related to rammed earth construction and its inherent sustainability can be found at Paul Jaquin's web site Historical Rammed Earth [8].

There are challenges to preserve, restore and modernize Hakka Tulou complexes, and to implement rammed earth technologies in modern constructions. The Workshop at XMU brings together experts from Australia, Canada, China, Japan, UK and USA to examine the research potential of rammed earth materials and sustainable structures. Topics include: (1) Rammed earth as a structural material and construction technique for sustainable structures; (2) Status and issues related to construction specifications and standards; (3) Challenges and strategies for incorporating (advancing) earth based structural materials in modern architecture; and (4) Challenges to preserve, maintain and/or modernize the Hakka complexes while maintaining their inherent environmental sustainability.

An important outcome of this workshop is to: (1) Establish a network of professionals to catalyze collaborative research, development and implementation including international partnerships and (2) Develop joint R&D programs emphasizing the commercial application of rammed earth construction with its inherent sustainability.

Much of the value of this workshop is the recognition of the potential viability of using rammed earth structures to meet the needs for sustainable affordable rural communities and to fuse historical techniques to advanced building science to meet today's building and environmental standards.

MOVING TOWARD TECHNOLOGY TRANSFER IMPLEMENTATION

As noted, this paper outlines the benefits and issues related to establishing a research and demonstration center focused on the transfer and implementation of sustainable solutions to preserve, maintain, advance, and promote rammed earth structures.

There are rammed earth structures throughout the world. However, the concentration of massive structures which have been continually inhabited for centuries is rare. The Hakka rammed earth Tulou complexes of Fujian province in China and the surrounding regions offer a unique opportunity to research and monitor existing structures and examine ancient Hakka rammed earth building techniques for use in modern construction. There is also a need to develop prototypes for environmentally efficient solutions to upgrade and modernize many of these complexes to current living and environmental standards.

There is a vast area of Tulou construction in Fujian, Jiangxi and Guangdong provinces in southern China with more than 20,000 Tulou in Yongding County of Fujian province alone. 46 Tulou in Fujian were inscribed and protected as a UNECSO world heritage site in 2008. However, many of the Tulou have been poorly maintained or are uninhabited. The last Tulou was built over 40 years ago while new buildings in this region are predominantly brick and concrete strucutures. Therefore these traditional construction methods are beings lost. By recognizing the sustainable features of rammed earth and promoting rammed earth construction methods and research to modernize these structures many of the Tulou could be preserved and retrofitted. While some villages would require more restricted preservation and restoration for historical reasons others may be candidates for the fusion of traditional and new technology and could incorporate cutting edge design.

Focusing on earth based construction is an important environmental decision but to create viable sustainable buildings and communities equal emphasis must be placed on total energy efficiency, water and sewage systems and regional environmental management. For this a comprehensive approach is required to select appropriate existing technologies and develop and adapt technology specific to maximizing building and community sustainability.

The proposed center would encourage the engineering community to further examine the advantages of rammed earth construction, bring to their attention what rammed earth is capable of, and promote new research opportunities that can further advance our knowledge on earth materials for modern construction.

By combining what we know of rammed earth from the Hakka people with science and technology of today, the efficiencies of rammed earth can be further expanded and used in modern construction. Such construction methods would reduce the need for concrete and thus help reduce greenhouse gas emissions.

CENTER CONCEPT

The Center for Earth Structures concept is to establish a research and demonstration site at or near an existing Tulou complex, and build the facilities to monitor and demonstrate sustainable technology at the site. The site would be open to researchers and visitors and have a significant demonstration and educational component. Activities would include monitoring existing structures and prototype installations, identifying sustainable solutions for local development, demonstrating Hakka sustainable technologies on site, and determining the adaptability of existing technology to contemporary rammed earth construction. The proximity to Xiamen University, a leading school of architectural and engineering research can significantly augment activities at the center.

Creating a center of excellence for sustainable rammed earth building techniques, in the heart of China's World Heritage Site of Historic Hakka Architecture will increase awareness of the center. At the same time the activities of the research/demonstration center can address important regional issues. Benefits include: (1) Capitalize on the environmental sustainability of these historic structures, and increase their energy efficiency and comfort with advanced technology; (2) Preserve and restore Hakka villages; (3) Retrofit villages to current living standards; (4) Promote the creation of new buildings combining traditional building techniques with new technology and cutting edge designs; (5) Develop a training, education and visitor center which demonstrates environmental stewardship and promotes practical environmentally sustainable solutions; (6) Provide information to support regional development policy and influence decision-makers; (7) Provide a living laboratory for advanced environmental research; (8) Take advantage of government plans for model eco cities, and rural development; (9) Promote heritage-led regeneration and access to the historic environment; (10) Provide training in rural areas to create employment opportunities in sustainable development.

The breadth of research activities to be conducted at the proposed center depends on funding capability. Activities may include: (1) Ongoing temperature, humidity, and energy usage monitoring of inhabited complexes; (2) Maintenance, monitoring and restoration of existing historical buildings; (3) Research in building science and engineering related to earth structures, sustainable communities, for both retrofit and new projects; (4) Research related to codes and standards; (5) Sustainable retrofit design and implementation; (6) Adaptation of rammed earth construction to modern building designs; (7) Technical training to maximize local opportunities to maintain and preserve Hakka complexes.

Efforts to preserve existing complexes in the Fujian region and upgrade these to modern sustainable structures can demonstrate the viability of sustainable rammed earth construction in remote and/or developing regions in Asia, South America and Africa in a practical living environment. As new standards and building codes on rammed earth construction are developed there is increased potential for new rammed earth construction in urban areas. The Wales Institute for Sustainable Education in Wales [9] and River Green Center in Durham, England [10] are excellent examples of modern earth structures which are well suited to commercial and urban applications. This type of demonstration site is an effective tool to promote sustainable earth architecture.

THE MODEL

The Center for Alternative Technology

The authors have recently visited The Centre for Alternative Technology (CAT) in Wales, UK. Given the priorities of this workshop and intent of moving toward joint research, technology transfer and implementation of rammed earth techniques, it is worth looking at CAT. Their comprehensive approach to sustainability and hands on demonstration of practical sustainable technological solutions is an instructive model.

Detailed information on The Centre for Alternative Technology can be found on their web site [11]. CAT established in 1973 as a small environmental technology demonstration site continually expanded over the years. Today it houses the Wales Institute for Sustainable Education (WISE) with a new rammed earth lecture theater, teaching and conference facilities and research laboratories. From its inception CAT focused on development, demonstration and practical application of sustainable technology—an experimental "Eco-village" prototype. Today the site includes a graduate school with a range of postgraduate degrees in environmental architecture and renewable energy.

CAT is an education and visitor center demonstrating practical solutions for sustainability covering all aspects of green living: environmental building, eco-sanitation, woodland management, renewable energy, energy efficiency and organic growing. The visitor center with 7 acres of hands on interactive displays and interpretive tours is open daily and has 65,000 visitors per year.

The site, with 90 permanent staff and 60 volunteers (summer high season), is a unique and practical demonstration center, a living laboratory with a broad range of real life examples of installed sustainable solutions and a test-bed for new ideas and technologies.

Along with their fee based consulting and training services they offer a free information service answering enquiries on all aspects of sustainable living. Day to day operation is funded primarily by income generation through courses, an eco-store, visitor center restaurant, *etc*. External grants and funding were required to build the WISE facilities and for large capital expenditures.

CAT participates in national and international projects including: (1) Zero Carbon Britain—a series of research reports addressing scenarios for reducing the UK's emissions to zero in 20 years; (2) Coed Gwern—15 acres of sustainable managed woodland, which is home to a range of courses, as well as being monitored and managed for biodiversity; and (3) Dyfi Biosphere—the only UNESCO Biosphere in Wales.

CAT provides a good base model for a research demonstration center particularly because of its dual role as visitor and interactive educational center. Access to a demonstration or pilot facility at an inhabited site in Fujian near an existing Tulou complex would be extremely beneficial for training and working with local contractors, residents and decision makers to preserve and promote local Tulou construction methods and promote sustainable solutions for local development.

Environmental Sustainability Programs in China

Examples of established international joint initiatives in China are also useful with reference to the types of activities which could be undertaken by the proposed center. For example, the JUCCCE training program at the National Training Center for Mayors of China is a very interesting initiative. A similar type of regionally focused program undertaken at a research and demonstration center in Fujian could be particularly effective. Three environmental initiatives are briefly described below. Detailed information is available on their web sites.

The EcoPartnership Program [12]: This is a high level organization which encourages U.S. and Chinese governmental and non-governmental stakeholders to share best practices, foster innovation, advance energy security, promote economic growth, and work towards environmental sustainability.

Joint U.S.-China Collaboration on Clean Energy (JUCCCE) [13]: JUCCCE develops training programs at the National Training Center for Mayors of China on how to build more sustainable urban centers, educating city mayors and China's State Owned Enterprises. Their objective is to empower China's decision makers and provides top local decision makers with case studies, experts and resources for developing energy smart cities. The Green School program is also one of their initiatives.

China U.S. Centre for Sustainable Development [14]. This organization is an international public private partnership (PPP) with a joint board of councilors representing business, government, science and non-governmental organizations. They demonstrate the commercial, social and environmental advantages of sustaining enterprises and are committed to all sectors of society: business, governments, universities, non-governmental organizations and people in their communities.

Other International Earth Institutes

We should also consider international organizations whose activities relate in part or in whole with the activities of the proposed center to identify areas for potential collaboration. Three international organizations are highlighted here. Again detailed information can be found on the web sites.

The Auroville Earth Institute in India [15] is Asian representative and UNESCO Chair "Earthen Architecture, Constructive Cultures and Sustainable Development". The institute undertakes training and technology transfer in a number of countries including Egypt, Nigeria and Tanzania. They plan to develop a curriculum for students of various education levels and grant diplomas in earth based construction and design. The Auroville Earth Institute participated in the Post Earthquake Assessment of Vaulted structures in Bam Iran.

Research expertise at the International Vernacular Architecture Unit—Oxford Institute for Sustainable Development [16] includes: (1) Interdisciplinary, cross-cultural and comparative study of vernacular architecture traditions; (2) Transmission of indigenous, traditional and vernacular skills, expertise and knowledge; (3) Conservation, regeneration and sustainability of vernacular building heritage worldwide; (4) Historic towns, tourism and heritage management; (5) Thermal comfort in vernacular architecture; and (6) Design in a vernacular context.

Institute for Historic Building Conservation (IHBC) [17]. IHBC stated on their website that historic places attract people, activity and investment, giving new life to their communities and helps make places more competitive. IHBC plays a central role in regenerating towns, cities and rural areas in the UK. This organization maintains a database of authorized conservation specialists.

FUNDING OPTIONS

Access to funding and funding alternatives is highly correlated to the activities of an organization. The proposed center is a hybrid, including practical solutions to regional issues and the transfer and implementation of environmentally sustainable technologies. Additionally, there is an emphasis on promotion and dissemination of technical information to the public and private sectors. Public private partnerships (PPP) are well established in North America and Europe. This type of funding is particularly attractive for a center focused on applying and demonstrating technology.

Realistically, at the initial stage this center will require significant government support but it is important to consider corporate sponsorship from the beginning with the view that the center becomes partially self-supporting. As research activities progress and emphasis is increasingly placed on the transfer, implementation and promotion of technologies developed at the center, it becomes more feasible to attract private funding. At this later stage PPP's should necessarily play a greater role. However, from the beginning we must understand the reasons corporations fund projects and their incentives for participating in public private partnerships.

Corporate sponsorship should focus on companies with obvious links to the activities and objectives of the center. One example would be the solar industry. Chinese companies are leaders in solar power. Six of China's largest solar companies are listed on the New York Stock Exchange (see Table 1). These companies highlight social responsibility and benefit from international exposure. More information on their activities and priorities are available on their web sites.

Table 1: China's Solar Companies for Potential Corporate Sponsorship

Name of company	Website
SUNTECH, Wuxi, China	[18]
Trina Solar, Changzhou, China	[19]
Yingli Green Energy, Baoding, China	[20]
JA Solar Holdings, Shanghai, China	[21]
LDK Solar Co. Xinyu, Jiangxi, China	[22]
China Sunenergy, Nanjing, Jiangsu, China	[23]

The web sites of these world class companies confirm their international status and commitment to environmental stewardship. Last year Yingli Green Energy joined the list of 2010 FIFA World Cup sponsors by providing solar panels and recently Trina Solar announced official sponsorship of the Lotus Renault GP Formula 1.

The tourist industry is another source of sponsorship. Increasingly large hotel chains are restoring and reforming historic sites. The Hyatt Regency incorporated an historic site in their convention complex in Calgary, Canada. The Four Seasons Hotels recently opened the Hangzhou at West Lake [24]. The Four Seasons Hotel Istanbul [25] at Sultanahmet in the historic center of Istanbul is created from a century-old neoclassic Turkish prison steps from the Blue Mosque and Topkapi Palace.

The center's location either within, or near the World Heritage site is the type of positive exposure attractive to both regional and large international corporations. Large international enterprises consider corporate sponsorship for both commercial opportunities and social responsibility. Smaller regional companies are more likely to participate as corporate members, commercial opportunities are critical for them. Other considerations include: (1) Links between companies and research activities; (2) Awareness of products and services local and regional suppliers could supply; (3) A strong networking environment to exchange research and promote implementation opportunities; and (4) Database for member companies.

The CAT model discussed previously relies significantly on volunteers but in the rural communities of Fujian providing training and employment opportunities for local inhabitants would have greater impact. A training component is equally beneficial if and when the center participates in regional or international development projects. Governments, in general, are more willing to fund programs which promote training and employment opportunities. Rural development and sustainability are also important government objectives. China in particular is moving in this direction.

At an eco-forum in Guiyang, on 17 July 2011, Li Ganjie China's Vice Environment Minister stated that 14 Chinese provinces are striving to build themselves into "eco-provinces", where local governments seek to promote sustainable development with respect to environmental protection. So far, more than 1000 cities and counties have worked out blueprints and timetables to achieve eco-civilization which features harmonious relations between people and nature, as well as environment-friendly consumption and lifestyles. Over the next few years, China will set up indices and build an evaluation system to further speed up the process of an eco-civilization [26].

Government support as noted earlier is crucial. Private sector involvement depends on the commercial opportunities and goodwill associated with participation in the center. Financing of a private public partnership depends on the structure and activities of the center. However, a useful guideline of how financial resources could be broken down is shown Table 2.

Table 2: Financial guideline for the proposed center

Center income 20% from activities such as:

- training programs
- visitor admission fees
- seminars

Membership fees 10% (corporate, academic, government agencies)

Private Sector sponsorship 25% (award programs, project sponsorship)

Government sponsorship 45%

The CAT model is worth consideration in developing a research demonstration facility in Fujian to promote the preservation of these spectacular rammed earth structures and to develop sustainable solutions to modernize the facilities and infrastructure of rural villages of the region, and to apply this experience nationally and internationally. Establishing a center that is "hands on", interactive and educational will promote the development of rammed earth structures and ensure sustainable choices are understood and promoted. Participation, joint research and information exchanges with relevant organizations will enhance the capabilities and success of this program.

STRATEGY

The first step is to establish a committee of government, academic, and corporate representatives to identify potential partners. Representatives from each sector must work together to determine and understand the objectives for establishing a research/demonstration center. At the initial stage it is critical that government objectives and guidelines are clearly understood and that government organizations understand the intent of the initiative, how it fits into government policy and how it benefits the region. All levels of government should be consulted. The academic role at this stage is crucial in identifying a clear range of potential research activities and their relevance to the goals of the center. The purpose at this stage is to clarify intent and ensure that it is evident to all potential parties what the goals and role of the centre will be.

Private sector involvement is best initiated at the second stage, after activities and goals of the center are defined and government organizations are on side. At this point, a market survey is extremely useful in identifying corporate interest and potential sponsors. A survey is multifunctional; it should introduce companies to the concept, gage their level of interest, help in identifying private sector priorities and finally provide ample opportunity for input. A well designed survey is an invaluable marketing tool to design an effective membership campaign. Interviews with local business associations are also crucial at this stage. Organizations and sectors to be considered include: (1) Environmental technology companies in areas such as, solar, water management, energy efficiency,*etc.* (2) Structural, civil and consulting engineers; (3) Architecture and design companies; (4) Urban and community design; (5) Landscape architects; (6) Tourist organizations; and (7) Local tradesmen.

These preliminary stages determine the level of interest in the research/demonstration center, provide concrete goals and objectives for each sector and hopefully identify organizations and companies willing to actively pursue and participate in the project. Without consensus and one or more champions it would be difficult to proceed with this type of project. However these two steps are essential in determining the direction a project should take and the type of structure that would ultimately be acceptable.

Finally, if there is sufficient interest in establishing a Sustainable Technology Research and Demonstration Center for the Preservation and Development of Earth Structures a major issue will be selecting an appropriate site. One attractive solution is to hold a competition for villages to bid on the project. This type of selection process offers a high degree of community participation and helps inform communities in the region of the intention and potential for working with the center once it is established.

CONCLUSIONS

Given the large number of Tulou complexes in the three provinces Fujian, Jiangxi and Guangdong in southern China, this area would be appropriate for a research and development center based on rammed earth construction and sustainability. Establishing a Sustainable Technology Research and Demonstration Center for Earth Structures either within or near the world heritage site of Fujian Tulou of China would not only help preserve traditional construction methods but also introduce and promote new sustainable building practices in this region. The proposed center would be a base to help preserve and revitalize Hakka rammed earth buildings for our common World Heritage. The centre could also promote Hakka sustainable technologies for design, construction and operation of green structures, and advance and promote the use of rammed earth as a viable sustainable building material. It is hoped that the participants of the workshop, those interested in the preservation of Hakka Tulou and the community of rammed earth research and construction professionals, can identify areas for collaborative research and the promotion of rammed earth construction. A Center for Earth Structures would benefit the region and would be applicable to researchers advancing rammed earth construction internationally.

REFERENCES

1. International Hakka Tulou Alliance (IHTA). Available online: http://www2.cemr.wvu.edu/~rliang/ihta/ihta.htm (accessed on 7 July 2011).
2. Liang, R. International Workshop on Rammed Earth Materials and Sustainable Structures. Available online: http://www2.cemr.wvu.edu/~rliang/ihta/Workshopflyer.pdf (accessed on 8 July 2011).
3. Liang, R.; Hota, G.; Ostrowski, J.; Lei, Y. Applying Hakka Wisdom to Future Sustainable Structures. In Hakka Tulou Forum 2009: Lessons to Be Learned, Past, Present and Future, Xiamen University, Xiamen, China, 24 June 2009.
4. Ostrowski, J.; Ostrowski, H.; Li, X.Y.; Ueda, M. *Hakka Eco-Village Study Tour Report; Internal Report*; ASH-Autonomous & Sustainable Housing Inc.: Calgary, AB, Canada, 2007 September 28.
5. AETN. First History Made for Tomorrow Award Goes to Fujian Hakka Tulou. In Press Conference, Beijing, China, 21 October 2010.
6. Ueda, M. Environmental Assessment and Preservation for Fujian Hakka Villages. In Proceedings of International Workshop on Rammed Earth Materials and Sustainable Structures and Hakka Tulou Forum 2011: Structures of Sustainability, Xiamen University, Xiamen, China, 28–31 October 2011.
7. Building Research Establishment (BRE). Available online: http://www.bre.co.uk (accessed on 15 August 2012).
8. Historical Rammed Earth. Available online: http://www.historicrammedearth.co.uk (accessed on 8 July 2011).
9. Wales Institute for Sustainable Education (WISE). Available online: http://www2.cat.org.uk/wise/ (accessed on 7 July 2011).
10. River Green Center, Durham, England. Available online: http://www.rivergreen.co.uk/accessible/rivergreen_centre_new.html (accessed on 8 July 2011).
11. The Centre for Alternative Technology (CAT). Available online: http://www.cat.org.uk/ (accessed on 7 July 2011).
12. EcoPartnership Program. Available online: http://www.ecopartnerships.gov/ (accessed on 18 July 2011).
13. Joint U.S.-China Collaboration on Clean Energy (JUCCCE). Available online: http://www.juccce.com/ (accessed on 18 July 2011).
14. China U.S. Centre for Sustainable Development. Available online: http://www.chinauscenter.org/ (accessed on 19 July 2011).
15. Auroville Earth Institute (India). Available online: http://www.earth-auroville.com/ (accessed on 20 July 2011).
16. International Vernacular Architecture Unit-Oxford Institute for Sustainable Development. Available online: http://www.brookes.ac.uk/schools/be/oisd/architecture/ivau/index.html(accessed on 22 July 2011).
17. Institute for Historic Building Conservation (IHBC). Available online: http://www.ihbc.org.uk/(accessed on 22 July 2011).
18. SUNTECH. Available online: http://www.suntech-power.com/ (accessed on 20 July 2011).
19. Trina Solar. Available online: http://www.trinasolar.com/index.php (accessed on 20 July 2011).
20. Yingli Green Energy. Available online: http://www.yinglisolar.com/ (accessed on 20 July 2011).
21. JA Solar Holdings. Available online: http://www.jasolar.com/ (accessed on 20 July 2011).

22. LDK Solar Company. Available online: http://www.ldksolar.com/ (accessed on 20 July 2011).
23. China Sunenergy. Available online: http://www.ldksolar.com/ (accessed on 20 July 2011).
24. Four Seasons Hotel Hangzhou at West Lake. Available online: http://www.fourseasons.com/hangzhou/ (accessed on 14 July 2011).
25. Four Seasons Hotel Istanbul at Sultanahmet. Available online: http://www.fourseasons.com/istanbul/ (accessed on 14 July 2011).
26. Deng, S. China Aims for "Ecological Civilization". Available online: http://news.xinhuanet.com/english2010/china/2011-07/17/c_13990879.htm (accessed on 20 July 2011).

CHAPTER **2**

"EARTHEN CONSTRUCTIONS" - TOWARDS CREATING A SUSTAINABLE HABITAT BY MINIMISING THE ECOLOGICAL FOOTPRINT

Apa rna Das

School of Architecture M.S.Ramaiah Institute of Technology India-560012

ABSTRACT

Sustenance of the human race has put an immense pressure on our planet Earth in terms of sustainability of natural resources. The greenhouse effect and the ozone hole are the two most threatening effects of pollution. Constructions of buildings as well as materials contribute to a large percentage to this pollution. Again every material used in the building industry has its source in the Earth. In general the low energy materials will be least polluting. The conventionally used building materials like bricks, cement, steel, timber, plastics, glass etc. usually involve huge transportation costs and also manufacturing processes which are detriment al to the environment. On the other hand the demand for new buildings as well as the cost of building construction is growing a tremendous pace. We have to search for alternative materials which are energy efficient, environment friendly and economical like our traditional building materials - mud walls and thatch roofs. Of (Ill the alternatives available to us which lead the way to sustainability, building with earth has been an ancient and accepted practice among communities all over the world. It is estimated that the construction and the operation of buildings is res possible for around half of all glob al C02 emissions, thereby contributing the largest single source attributable to climate change. Earthen construction has been, is and will continue to be a reality. Stabilized rammed earth walls can be used as a building integrated source of passive cooling technique. A huge population in Indi a lives in the rural areas where there has been a growing trend in shifting towards brick and concrete constructions in search for social status. Even a small percentage can lead to massive increase in global C02 emissions if the trend is not checked at this point. This papers looks into the current scenario and hence the corresponding responsibility on architects, planners and policy makers to bring in technology reforms; the green way of thinking' which will see a better and safer tomorrow.

INTRODUCTION

United Nations World Commission on Environment Conference in 1986, chaired by the then Prime Minister of Norway GroHarlem, Bruudtland, established the concept of sustainability, which it defines as, ' those paths of social, economic an d political progress that meet the needs of the present without compromising the ability of future gene rations to meet their own needs' [Steele 19971J. Sustainability is the moral imperative of the age of architects as said by the President of the Royal institute of British Architects (RIBA). At the beginning of the 21st century architects are expected to be ever more creative, while being mindful of the impact of their building designs on the ecological systems of the planet.

Citation: Apa rna Das, "Earthen constructions" - towards creating a sustainable habitat by minimising the ecological footprint, https://doaj.org/article/df8f889bec6d4fe5a6da39a31562f9f5

Figure 1: Ramesseum vaults

Earthen Construction

Next to food, water and air, shelter is man's most basic need. Earth ha s been the most basic building material since the dawn of man. To be sure, the earliest known kinds of earth construction were very crude by our standards today. Gradually man learned that some kinds of earth made better houses than others. Some of the best would last his entire lifetime. Rammed earth construction was first recorded by the Babylonians in 5000 B.C. The 1500 mile long Great Wall of China is the only man-made structure that can be seen from the moon with the naked eye. The wall was built around 300 B.C. and parts of the western portion are built of rammed earth. From there rammed earth migrated to India, over the Indian Ocean to Madagascar, across the continent of Africa, to Morocco, and then with Hannibal to Spain, the Pyrenees Mountains (where he built a string of watch towers, made from rammed earth, within eye sight of each other, many still standing), through the Alps, down into Italy (218-201 B.C., Hannibal introduced rammed earth to the Romans during the Second Punic War), then the Romans transported the idea to France and the lowland countries of Europe (http:/ / webs. ashlandctc.org / jnapora/ hum-faculty / syllabi/ tradhtml.]. The Kanuri Nubians, faced with the challenge of a harsh, treeless environment and the problem of roofing over spaces without wood for beams or supporting members for centering, have perpetuated a building technique that has been used in Egypt since at least the 13th century BC, as in the Ramcsseum, the mortuary complex of Ramses II. This method of construction

Figure 2: Vikas Community: Houses designed by the people of Auroville itself as well as architects. Photo Acknowledgements: 'Earthen Architecture and Stabilized earth techniques in Auroville', Satprern maini, 2007.

(Figure 1) shows a great understanding of the laws of statics. The vaults that form the basis of the system are made in a parabolic shape that will keep the material in compression only. Earth construction exists in two formats today: the traditional mode which relies essentially on the use of unsterilized earth and the modern format often using stabilized earth. A majority of the modernizing societies tend to reject the traditional approach, with the possible exception of the use of adobe in certain parts of the world. With the addition of cement or cement and lime, earth acquires the strength and durability comparable to brick and is often accepted on that account. However, there are many societies which consider stabilized earth to be closely related to traditional unsterilized earth and prefer to use brick and concrete whose performance is never in question [Jagadish 20073]. The use of earth on the site as a building material saves cost, time, energy, and transportation. Buildings with earth has been an ancient and accepted practice amongst communities all over the world. It is estimated that the construction and the operation of buildings is responsible for around half of all global $C0_2$ emissions, thereby contributing the single largest source attributable to climate change. "Earthen Constructions" Stabilized rammed earth (SRE) materials have low embodied energy content because approximately 95% of the raw materials are unfired and are locally available, thus minimizing transportation. Recently, the use of crushed recycled aggregates (e.g. demolition waste from bricks, concrete or SRE itself) is increasingly used instead of virgin sub-soils. This offers the advantage of reducing landfill and converting onsite waste materials into a high quality product for new-build processes. Earthen architecture in Auroville shows a wide range of buildings: from houses to apartments and public buildings. The development of Vikas community showed that earth can be used for building a progressive and harmonious architecture up to 4 floors high. Compressed stabilised earth blocks (CSEB) are mostly used in Au roville and many stabilised earth technologies have been developed, from foundations to roofs and for disaster resistance.

BUILDING TECHNOLOGIES BASED ON EARTH

Compressed Stabilized Earth Block (CSEB)

A wide range of equipment for building with earth, Auram, has been researched and developed from the very onset by the Autryville Earth Institute. The press 3000 is today being sold worldwide. The Auram press 3000 as shown in the figure can fit 17moulds on it, for producing about 75 different types of blocks, with various shapes and thicknesses. The figures below indicate the wide variety of compressed stabilized earth blocks produces by the Auram press. SEB made in Auroville with 5% cement, have an average dry compressive crushing strength of 5MPa and a wet compressive crushing strength of 2.5MPa. The water absorption is around 10%. Country fired bricks have around 35 kg/ em' for the dry compressive strength and 12% water absorption. The cost of CSEB walls (at Aurovillc) is 25.9 % cheaper than country fired bricks.

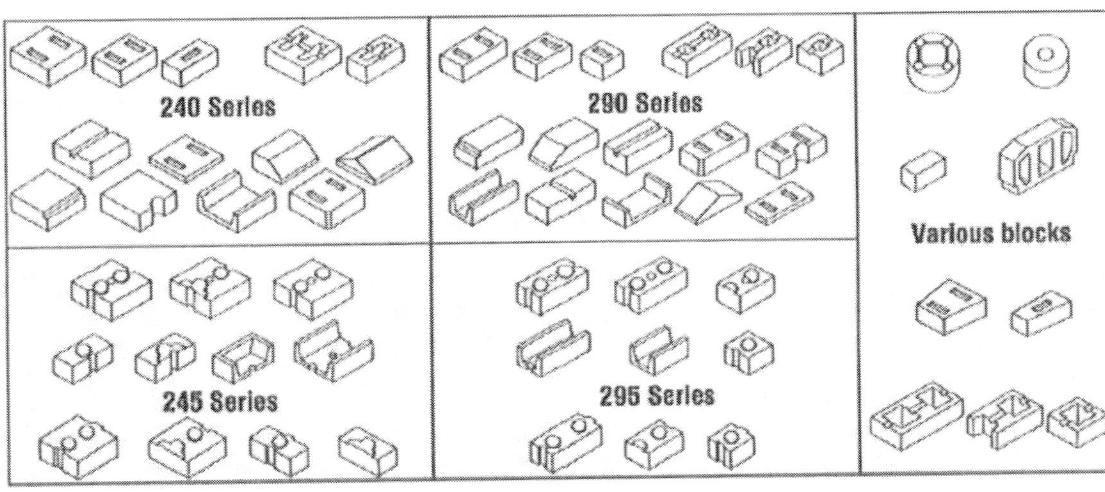

Figure 3: Wide variety of compressed stabilized earth blocks, by the Auram pre ss 3000.

Stabilized Rammed Earth (SRE)

The soil is mixed with sand and stabilized with 5% cement. The mix is rammed by hand. Foundations are rammed directly in the trench. Walls are rammed in between Ironworks. Rammed earth walls are45.3 %cheaper than country fired bricks.

Composite Basement and Plinth Beam

Basements are made with CSEB stabilized with 5% cement. The plinth beam is cast into a U shaped CSEB. Reinforced concrete is cast in the U shape blocks. This is a very neat and efficient way to do a plinth beam.

Composite Columns

Round hollow CSEB arc reinforced with cement concrete. Reinforcements vary but the rod diameter cannot exceed 10 nun for the round blocks 290 (290 mm diameter) and 12 mm for the round blocks 240 (240 nun diameter).

Composite Beams and Lintels

Reinforced cement concrete is cast in U shaped CSEB. The bottom part of the beam is precast in a reversed position on the ground. Once cured, it is lifted and the middle and top parts are built on it. Three different heights of beam are done, according to the span (see Figure 32) and the blocks are used as lost shuttering in all cases:

- Single height beams: up to 1.25 m span with 2 bars of 12 rnm. U blocks also help in this case the compressive strength of the beam, as they are on top of the beam.
- Double height beams: up to 2 m span with 2 bottom bars of 12mm and 2 top bars of 8nun.
- Triple height beams: up to 3 m span with 2 bottom bars of 12nun and 2 top bars of 10mm.

a. Stabilized rammed earth foundation and wall

b. Reinforcements and casting

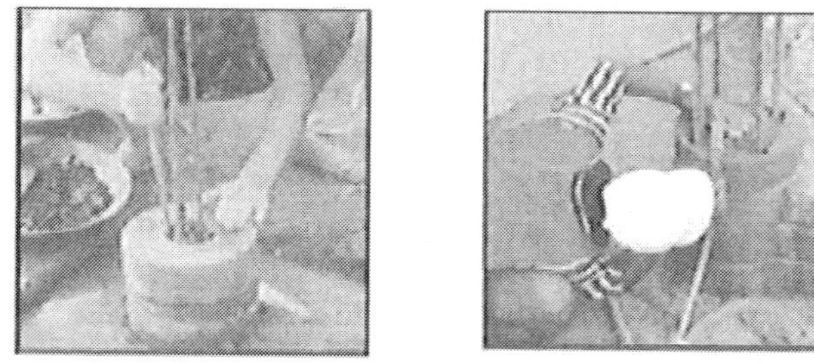

c. Composite columns

Figure 4: Uses of solid rammed earth

Figure 5: Segment al vau lt 10.35m span, 2.20m rise, 6m long, built in 18 days wit h 4 masons

Vaulted Structures

At the Auroville Earth Institute, the Nubian technique of vault construction has been further researched and new ways of building vaults and arches have been worked upon. The free spanning technique allows courses to be laid horizontally, which presents certain advantages compared to the Nubian technique which has vertical courses. Depending on the shape of vaults, structures are built either with horizontal courses, vertical ones or a combination of both. The Auroville Earth Institute has developed an optimization method for calculating the stability of vaulted structures. This method has been especially developed for designing and building vaults without Centering's [Mai'ni Satprem 20074].

Building Process/ Weller Cob - Walling Technique

The most important technical progress in comparison to the traditional house is the introduction of a damp proof course, a brick foundation and mixing of loam with straw. The traditional building technique (where very wet loam is used) was replaced by the "Weller" technique that is quite similar to the traditional one. The 'Wellerba u' is a historical earth building technique. Loam is mixed with straw, which is a kind of reinforcement for the wall. Neither cement is neither added, nor any other supporting structure. The "Weller" - wall (550mm) is constructed in layers. Each layer is built approximately 70 cm high, dries up within 3 - 4 days and is trimmed on the sides with a spade to get the accurate shape (Figure 6). After a second drying period, another layer can be added. The building process is much faster in comparison to the traditional method: the layers can be much higher due to the straw "reinforcement" and the bearing strength is high enough after some days to step on the wall and to add another layer. Even sandy loam is suitable for this technique and fertile sticky soil can be saved for agriculture.

Figure 6: Source, Anna Herringer, 2007. Weller construction : mixing; placing of first layer, trimming with the spade; 2nd layer (Herringer, 2007)

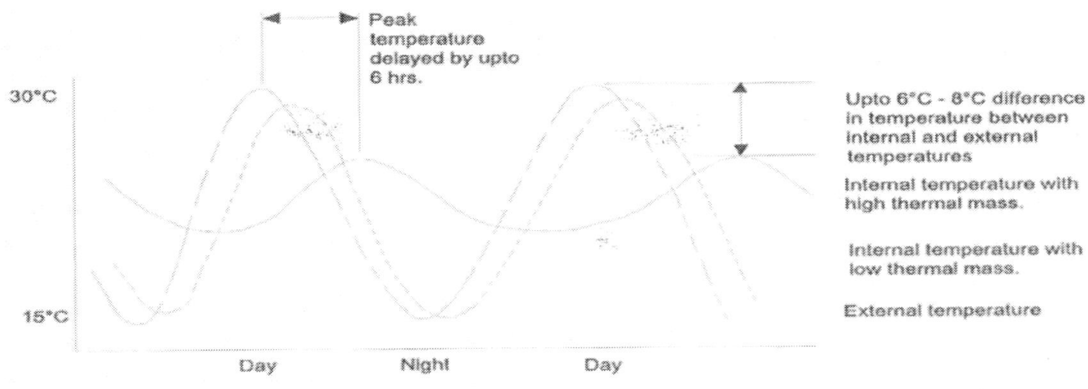

Figure 7: Internal temperature stabilizing effect of thermal mass (de Saulles 2005).

Passive Air Conditioning Potential of SRE

The use of air conditioning to control the indoor thermal environment (temperatures and ieshumidity) of buildings, within comfortable levels for the occupants, is now commonplace. Stabilized rammed earth (SRE) walls can be used as a building integrated source of passive cooling technique.. SRE has a similar heat capacity to concrete and the additional advantage of being hygroscopic while offering a low carbon alternative to the high cement content of solid concrete. Today, building designers are increasingly becoming aware that SRE walls can be used in modern buildings for the combined functions of load-bearing structure, external envelope as well as for passive cooling technique. Some of the row housing approaches in traditional rural India have built-in features which protect earth walls against rain. For instance, the side walls of row houses, which are perpendicular to the street, are never exposed to rain as the space between two walls is too narrow and is further protected by roof projections. The front walls of the house are also often protected by a sloping verandah (or cow shed) roof. The lighting is often provided by small courtyards and the mud walls are completely protected against rain erosion.

The only threat to such mud walls is possible flow of water from leaking roofs which can locally damage the walls if it is not attended to in time [SAllinson 0 & Hall M 2007].

Alternative Specifications for Earth Walls

It is possible to consider a hierarchy of earth walling solutions wherein some of the stringent performance criteria are progressively relaxed retaining some of the essentia l criteria. The following hierarchy of solutions is suggested by the author]agadish and Ullas, 2007 [6]

Figure 8: Rigid cavity fill insulation in composite SRE walls. Photo courtesy of Earth Structures Private ltd

a. When unsterilized earth construction is used it is possible to consider walls which have satisfactory dry compressive strength (-3.0 MPa) and resistance to rain erosion. This is quite cost effective. b. Partly stabilized earth can be considered which has dry compressive strength (-3.0 MPa), resistance to rain erosion and retention of shape of earth block even after soaking in water for 24 hour s although the saturated compressive strength is inadequate. This is moderately cost effective. c. Fully stabilized earth construction will satisfy all the four criteria. Costlier than the earlier options, but favorable with brickwork and concrete block work.

Tolal Cost 01 house Rs.23000 Cosl/sqrt- Rs.t 07.62

Figure 9: Circular Rammed Earth Structure Bhunga at Hodka, Kutch, Cujarat

Use in disaster resistant homes

There was a huge demand for houses to rehabilitate the earthquake affected families in Gujarat state after 2001 Kutch earthquake. Initially Compressed Stabilized Earth Block (CSEB)construction technology was use d but as a variation of the CSEB technology, cement stabilized rammed earth techniques were adopted and developed to suit the various forms of houses include the traditiona lhouses called Bhunga [Jagadish & Ullas 2007]. However, as societies come under the influence of 'Development' earth construction appears to be less attractive. Environmentalists are eloquent about the ecological benefits of earth construction but the poorer sections in developing countries are not impressed.

Future Research Aspects

There is a need to examine afresh the durability aspects of traditional earth constructions since any useful local practices are vanishing from our knowledge systems due to the perceived inferiority of earth construction. Such technologies are likely to spread in areas where entrepreneurs or entrepreneur NCOs are able to provide the technology delivery systems [Abey & Smallcombe 2007].

1. While it is energy efficient to use stabilized block technology instead of bricks, its accessibility to the poorer sections of society does not appeal' to be satisfactory.
2. In terms of new materials, cob is a traditional building material in the South West of England. It is a mixture of subsoil, straw and water and is built with no shuttering. Cob has a real future in mainstream construction - it is sustainable, low environmental impact, versatile, durable and aesthetically pleasing.
3. There have been a number of high profile examples of rammed earth construction in recent years, heavily influenced by the high quality contemporary work on continental Europe by Martin Rauch. In these projects earth construction acts as an emblem of 'green construction', though the walls are rarely used to their structural or environmental potential. Again the work tends to be slow and expensive, and where understanding or quality control is not good, the results can be somewhat disappointing.
4. The high public profile and 'engineered' aesthetic quality of such contemporary earth building has served to dispel the 'mud hut' image of a material associated with hand-made buildings of low quality in dry climates, rather than a one suitable for contemporary UK construction projects. However, because of its expense, rammed earth will not become a common form of contemporary construction and remain in the niche of eco-buildings' special clients.
5. The improved image of earth construction brought about by such projects has coincided with a general increase in the importance of considering environmental impact in the U.K. construction industry. Driven as much by cultural aspiration as by changes in building regulation, there are increasing opportunities for new sustainable products and methods to break out of the green fringe into common use.
6. Criteria for Mainstreaming : The niche of green building has allowed experimentation in a wide variety of 'alternative' construction materials over the last 15 years, but very few have broken into the mainstream of everyday buildings thereby 20 making a significant difference to the environmental impact of the way we create and manage our built environment. Some are too expensive, some are difficult to use, some do not suit commercial construction processes and program and some are difficult to develop as commercial products. Ones that have mainstreamed, such as sheep's wool insulation, have been simple to use as substitutes for conventional, more environmentally damaging products, are not disproportionately more expensive, have a marketable image and could be brought to market with limited investment. In the family of earth construction techniques, only earth masonry fits these criteria in a U.K. context.

It can be affordable and, in a climate of increasing energy costs, will become increasing low cost compared to competing products.

It is easy to use and relates closely to existing common construction practices.

There is a large, established manufacturing industry with excess capacity that could develop commercial products relatively quickly, at low initial investment.

Given basic guidance, building designers could easily incorporate it into current building designs

In design terms, it would have multiple technical benefits and respond to a growing interest in the use of thermal and moisture mass to nonuse energy consumption and improve building health

Clients have a track record of responding favorably to its use.

These credentials mean earth masonry has the ability to be turned, relatively quickly, into a new mass construction material. Investment by the brick industry is key to developing the image of an ecofriendly product of progress and building demand for these products in a market it knows well.

The effect of Housing on Ecological Footprint

The Living Planet Report 2006, [9] (WWF, Institute of zoology, Global Footprint Network, 2006) declares India among the countries having smallest "ecological footprint" at less than 0.9 global hectares per person. In 2003 the footprint per Indian was reported as 0.8 global hectares (gha). In comparison, USA required equivalents of 9.6, China 1.6 and Germany 4.5 global hectares per person (as on 2003). However there is a change in this positive scenario if we look at the National ecological footprint in relation to the nationally available bio capacity. At a total bio-capacity of 0.4 gaper Indian there is a deficit of 0.4 gaper person. With a population of 10667 Million the sum of -0.4 Ha/ person causes a footprint which is more than100% large r than the bio-capacity of the country. There is a crucial need to find strategies to improve this situation. But how? India is one of the developing countries. It is very clear: we must fight poverty in India and find strategies that improve life for each individual person. At the same time we must avoid the increasing of the ecological deficit. This might sound presumptuous taking into account the situation of high-income countries that are causing much more damage to our planet than developing of the ecological deficit. This might sound presumptuous taking into account the situation of high-income countries that are causing much more damage to our planet than developing There is a crucial need to find strategies to improve this situation. But how? India is one of the developing countries. It is very clear: we must fight poverty in India and find strategies that improve life for each individual person . At the same time we mu st avoid the increasing of the ecological deficit. This might sound presumptuous taking into account the situation of high-income countries that are causing much more damage to our planet than developing countries. But at present there is no country in this world that is released from the effort to save our planet. Sustainable development is a commitment to "improving the quality of human life while living within the carrying capacity of supporting ecosystems" (IUeN et al., 1991)[10]. What is the role of architecture in the process of development - in an improvement of living standard and a balanced eco system?

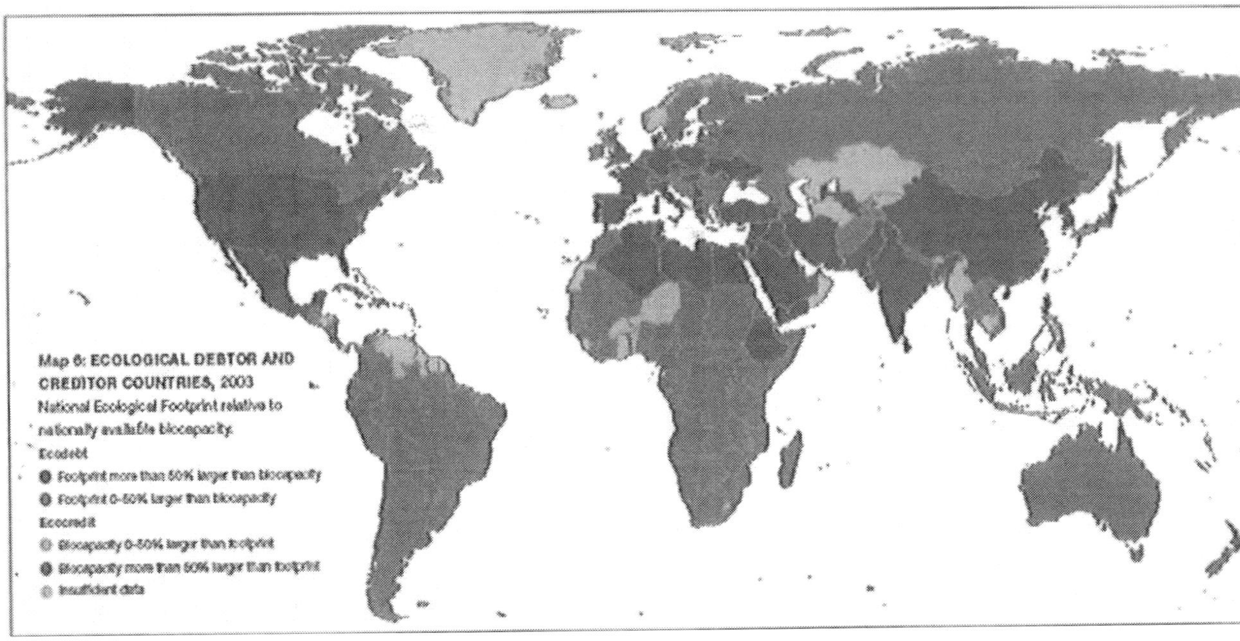

Figure 10: Source: Living Planet Report 2006

A survey of Rural architecture in Indi a (Tobias Hagleitner et al., 2003)shows (Figure 7), that all expansion is happening only is mostly horizontal. In the survey area all with majority of the houses being houses were single story. In addition to that Wealth causes two major effects on architecture and settlement pa tern: It helps transforming settlement patterns from very dense and compact settlement to broader and expatiating ho use holds. Two hamlets with very low income families have for instance 41 where 200 people live on an area of 10 000 sq. rn. and 17 families with 80 inhabitants on 5 500 sqm respectively. But on 5000 sqm only 8 households wit h 35 persons live in "Saha - pa ra" which means "the hamlet of the rich".Theother effect offinancia lcapability according to the survey mentioned above is reflected in the building material. This is visible in the private temples and shrines. They all follow the unwritten law, which prescribes that all temples must be built in the most valuable material that is used in the homestead. The hierarchy is clear to see while walking through a Bengali village: Brick (best), loam (mid dle), bamboo and straw (last). To bring it to a point: All the expansion caused by the demographic development as we ll as financial ability is shown in capturing sp ace and buying material instead of taking the (dirty) mud to one's feet. No t only comparatively " rich" people follow that pattern: also NGOs or public buildings from government do the same, which stand for development and improvement. What will be the consequence if this trend will continue? Back to the concept of the ecological footprint and its respective components: The actual footprint of Indi a due to the built-up land (meaning infrastructure including hydro power) in 2003 was 0.04 gh at person. More specific: the biocapacity of cropland was 0.29, graze land 0.00 and forest 0.02 gh at person. The total bio-capacity of the land is only 0.314 gha per inhabitant whereas the nearby country Myamar with similar climatic Conditions has a bio capacity of 1.04 gh at person. (Living Planet Rep ort, 2006) [Livillg Planet Report, 20069] The decisive difference is the population density. On an area of 1442973190.000 sqkrn a and a population of 1066 Million ([http:/ / www. World a tlas.com / web image / country s / asia / in.htm#facts]. Wikipedia, 2006) Every square meter is vital to cultivate most important ea tables, which is mainly rice, wheat and jute. Of course it is mainly a mailer of agriculture to introduce a cultivation strategy to enhance the bio capacity. But it is a matter of architecture to care for the basic prerequisite, i.e, land. India must get out of poverty and seek to find a strategy to avoid an increasing footprint in the sector built-up land. The strategy is simple: a vertical extension of built-up land to save the precious land for a sustainable cultivation. "Population pressure continues to place as ever burden on productive capacity, creating a food deficit. Foreign assistance and commercial imports fill the gap. Underemployment remains a serious problem. Finding alternative sources of employment will continue to be a daunting problem for future governments, particularly with the increasing numbers of landless peasants who already account for about half the rural labor force." (Pearson Education, 2006)[l'earson Education 200612] As this quotation states, saving land is interlinked with many aspects such as economics increasing debt, health / food, employment facilities and in a wide r perspective - even with demographic changes in larger scale (migration), stabilized democracy and peace [Her ringer 200713]..

Another crucial factor to measure India's footprint in future will be the country's C02 emissions from fossil fuels. And architecture will have a major influence on that. Following article states the present trend: The Brick industry is worldwide constantly growing - Since 1990 the worldwide production of bricks is significantly increasing. The sales volume of that sector is estimated to be 26 million Euros at present. In China the brick production currently samples a vast boom. Now also the brick production in the Arabic countries of North-Africa, which up to now knew mainly handmade bricks is increasing. Globally seen bricks and roof tiles have reached a level of quality and sales quantity that let us look positive into the future (ARCH mastic, 2006).[ARCHmatic 2006 14]. At present ap proximately 20 percent of the houses in rural area are built with bricks. The rest are bamboo/straw or loam buildings. What if India continues the present trend in its rural areas and follows the global model in investing in brick ho uses? This would cause a tremendous rising of the C02 emission due to the coal, which is the energy source of the brickfields. The same would apply for steel instead of bamboo plus the energy consumption for the transport. In 2006 the total population of India was 1067 Million (Wikipedia, 2006),[15]. 75% of the total population) were settled in rural areas. Out of these 80% of; which that means over 80640 Million Indiansi live in houses built in loam or bamboo /straw. If these buildings will be replaced by houses built out of locally fabricated bricks, in brickfield that not even underlie the standards for filtering and reduction of emissions - the rising of C02 emissions would increase in dramatic scale (in worst case about 20150 Million houses). C02 in the atmosphere causes 80% of the global warming. The rising of the sea level would hit Indi a as one of the first countries and 60 million people in Calcutta and Bangladesh would be homeless, the number and power of . destroying floods (due to melting glaciers in the Himalaya) and cyclones (due to the water evaporation caused by higher temperature of the Indian Ocean) as well as a severe and permanent water crisis du e to the drying-up of the river Ganges and Brahmaputra are predicted to be the immediate effects (AlGore, 2006)[16J.That means building with loam is a valuable action against an increasing ecological footprint and it is even affordable for the poorest of a least developed country.

The Role of Architects and Planners

It is our responsibility as planners to support implementing bodies in terms of research in appropriate techniques, training and design. Policy making organizations, government and nongovernment organizations must set good examples through representative public buildings as well as through pilot projects for residential houses that can be multiplied by the villagers. For a real change towards a social and fair economic improvement we need to find ways to strengthen local identity and craftsmanship. Culture could be a possible way to succeed .There is a great example in Indian history of an effective protest against global industrialization and for an independence through focusing on local production: Gandhi's philosophy of homespun, traditional clothes was a strategy to free from imported goods, to free from powerlessness again st social and economic discrimination through self-production. Beside this it was a method to eliminate social differences fortified by the distinction between those who can afford buying prestigious English clothes and those who need to spin it by themselves. Isn't that situation very similar to the one of brick- and loam-buildings? The homespun "khadi" even became a kind of uniform for the Indian independence movement and was a symbol of independence as well as a deliberate return and appreciation of its own culture. (Nanda, 1958)[Rabindranath Tagore 193118]

CONCLUSION

Every man needs a shelter. In today's urban world, every man needs a house that is economic, durable as well as aesthetic. Right from the dawn of the 20th century, mass production of housing using industrial materials has been the only answer to cater to the growing demands. This has lead to large negative impacts on the existing environment, The man's day to day activities which are at par with the modern society has lead him to spend huge amount of energy in order to gain a certain level of comfortable living. However, with the present day trends, the future looks bleak with high carbon emissions, large ozone holes in the atmosphere and degraded human health. The time has come to rethink scientifically new ways of being more innovative and modern but less consumptive. The time has come when we need to give a serious thought as to how we can build upon the existing natural resources rather than unthinkingly deplete the existing ones. The famous Bengali philosopher Tagore describes beauty as "perfect harmony". And can that harmony exclude the elements of sustainability? How could something be beautiful if it is not sustainable? "When our universe is in harmony with man, the eternal, we know it as truth, we feel it as beauty. Beauty is in the ideal of perfect harmony, which is in the universal being." (Rabindranath Tagore, 1931).[Nanda 195817]. As architects we are in the service of beauty - it is on us to realize an architecture that is not a matter of self-infatuation but a trial to reach true harmony on all levels of the entire being. 24 Earth as building material has the potentials to reach a high level of sustainability in all its social, economic and ecological aspects. But there is a surplus in building with earth: it is much more than just a building material: it is an essential element that touches our inner emotions. In every culture there are poems or songs about earth, but I have never heard one about concrete and steel. If we do not only find the right rhymes but also the appropriate architectural language I believe earth will no longer be the "material of the poor" - but the material of the future . This paper has discussed the role of earth in construction of new buildings as a means of sustainable humane living. Some serious efforts on our parts can help us upgrade our lifestyles so that we can optimize our dependency on the nonrenewable energy sources.

REFERENCES

1. Abey, J and Smallcombe, J (2007). Cob in Contemporary Architecture . International Symposium on earthen Structures, Indian Institute of Sciences, Bangalore, India.
2. AlGore, (2006), AnInconvenientTruth,Rodale, Emmaus,PA, USA, pp. 58- 67.
3. Allinson D and Hall M, (2007). Investigating the Optimisation of Stabilized Rammed Earth Materials for Passive Air Conditioning in Buildings. International Sympo siumon earthen Structures, Indian Institute of Sciences, Bangalore, India.
4. ARCHmatic, (2006). Baulinks, Report on Ceramics, Retrieved January 2007.
5. Herringer, A (2007). Ration and Emotion: the ration ofEarth HousesMETI School building inBangaldesh in Weller Cob-Walling Technique.

6. InternationalSymposiumonearthenStructures, Indian Institute of Sciences, Bangalore, India.

7. http: // webs.ashlandctc.org / jnapora /humfaculty/ syllabi/ trad.htm!.

8. hup:l lwww.world atlas.com /web image l countrysI asiaI in.htm#facts [agadish, K.S (2007). Earth construc tion today :Prospec ts and tasks. International Symposiumon earthen Structures, Indian Institute of Sciences, Bangalore, India.

9. Jagadish, K.S and Ullas, S.N (2007). Conceptsin Earth Construction for Durable Housing. .

10. Internationalsymposilim on earthenStructures, Indian Institu te of Sciences, Bangalore, India.

11. Kotak Tejas, (2007). Constructing cement stabilize d rammed earth houses in Gujarat after 2001 Bhuj Earthquake. Intern ational Symposium on earthen Structures, Indian Institute of Sciences, Bangalore, India .

12. Living Planet Report, (2006), WWF, Institute of Zoology, Global Footpri nt Network, pp .14-10.IUCN/ UNEP/WWF, caring for the Earth: A Strategy for Susta inable Living. Gland, Swit zerland.

13. "EarthenConssmaions" Mami Satprem (2007). Earthen Architectureand Stabilized Earth Techniques in Auroville, India. International Symposium on earthen Structures, Indian Institute of Sciences, Bangalore, India.

14. Nanda, B. R. (1958). Mahatma Gandhi, A Biography, Oxford Uni. Press, New Delhi, 147 - 150.

15. Pearson Education (2006). U.S. Depa rtment of State Background Note, Retrieved January 2007. www.infoplease.com /countryI profilesl bangladesh. html Rabindranath Tago re, (1931). Three conversations:Tagore Talks w ith Einstein, with Rolland, and Wells, ASIA3I 1931, p.139-143,196

16. i.,Retrieved January 2007, htt p:/ /www.muktomona .com/ArticlesI einstein_lagoce.htm Steele, J, (1997). Architecture Today. Phaidon Press Limited.

CHAPTER 3

REVIEW OF THE ARCHITECTURE OF MUD AND QUDAD DVDS

Review of *The Architecture of Mud*

This glimpse of the Hadramaut before nightfall showed that, even if in some ways it is sleepy and insular, one activity continues with irrepressible vitality: building. Everywhere, there are unfinished top floors and herringbone stacks of mud bricks. Many of the older houses and turret-cornered little forts are crumbling - in the last stage of disintegration, the structures look like termite mounds. But new buildings rise all around. It is a cycle of dissolution and rebirth.Tim Mackintosh-Smith, *Yemen: Travels in Dictionary Land,* John Murray (London), 1997, p. 175.

INTRODUCTION

This superb documentary by Caterina Borelli is a study of the craftsmanship involved in the construction of the mud brick architecture of the Hadramaut and Do'an valleys of the Yemen, and the cultural aspects of a traditional architecture which incorporates an understanding of buildings which dates back centuries. Expanding the existing knowledge of these earthen heritage properties, examining their behaviour in the local climate and the preservation of traditional craftsmanship as part of a sustainable conservation future are the other prominent concerns of this work.

Figure 1: The valley environment in Yemen.

The traditional heritage of the Hadramaut and Do'an regions in the south east of the Yemen are entirely constructed from loam. Throughout the centuries, the population has developed very sophisticated building techniques, and created a unique architectural environment. Spectacular structures such as ten-story mud brick tower houses rise up from green valleys that are surrounded by arid mountains. As in other parts of the world, with the advent of modern materials such as cement, indigenous construction and conservation practices carried out by craftsmen are rendered intellectually invisible by a process similar to the drawing of a veil. The elimination of these practices equals the erosion of centuries of building and conservation culture. As this heritage is rapidly disappearing, as shown by these obsolete construction techniques, this documentary is an excellent archive for future generations.

Figure 2: A mud brick village in the wadi Do'an.

Figure 3: Some of the mud brick buildings can be up to ten-story high.

The Content

The documentary relies heavily on the information provided by masons, carpenters, and master builders during several interviews (the featured masons had up to 60 years of experience and becoming a master takes up to 30 years).

Figure 4: Some of the masons interviewed in the documentary, showing how moulds are constructed.

In interviews throughout the documentary, the masons describe their working techniques and the challenges they face with the introduction of new, imported materials. The documentary is divided in four sections:

Introduction to the area;

Craftsmanship (types of moulds, mud brick, foundations, walling materials, and roofs);

The production of nurah (lime) and plastering techniques; The advent of modern materials.

The documentary describes the points well and stresses that traditional building techniques are particularly important because they are the result of centuries of development and practice. The recording of craftsmanship is extremely vital, not only because it provides empirical evidence of original practice, but also because it can be directly applied to practical conservation. Employing local materials in construction and repair work has many advantages, not least of which is the fact that the sources for the original materials are close to the site. Furthermore, it is explained that the workmanship necessary for using traditional materials such as mud and lime requires more skill, sensitivity and grounding in traditional culture than what is required for modern materials.

Figure 5: The process of mud brick making.

The documentary ends by explaining that in some villages in these remote valleys in Yemen, emigrant remittances constitute the most important source of income. This money is often used in building, repairing, or enlarging village homes. The construction industry, which employs a wide range of skilled and unskilled workers, and uses a range of materials, has thus become an important part of the village economy. This mechanism of financing construction work also accounts for the half finished state of many buildings, for the work takes place sporadically according to the supply of money sent.

Additional Information

The project was carried out with the consultancy of leading experts on earthen material conservation: Pamela Jerome (Columbia University), Giacomo Chiari (Chief Scientist, Getty Conservation Institute), and Salma Al-Radi (New York University, Institute of Fine Arts).

The work could be of use for those conservators working in the American Southwest, Latin America, Africa, Europe, and Middle East earthen archaeological excavations. Furthermore, it has a strong didactical vocation and, together with another documentary by Caterina Borelli (*Qudad, Re-inventing a Tradition,* DER, 2004) should be on the shelves of every conservation library (see following review). For further information on the mud brick architecture of the Yemen, see the comprehensive book by Salma Samar Damluji, *The Valley of Mud Brick Architecture (Shibam, Tarim, and Wadi Hadramaut),* Garmet Editions, Reading, 1992. See also the technical paper by Jerome P, Chiari G, Borelli C, titled *The Architecture of Mud: Construction and Repair. Technology in the Hadhramaut Region of Yemen,* APT Bulletin, Vol XXX, No 2-3, 1999, pp. 39-48.

Figure 1: The 'Amariya Madrasa' in Rada, while undergoing conservation.

Figure 2: Detail of the some of the conserved elements in the 'Amariya Madrasa'.

REVIEW OF *QUDAD*

Introduction

The architectural heritage of Yemen is the most extensive in the Arabian Peninsula, and certainly from a formal point of view, one of the richest in the world. This outstanding documentary illustrates *qudad*, a lime-based plastering technique that was historically widespread in the Arabian Peninsula for the waterproofing of religious and major buildings. The documentary discusses the sudden decline of ethics which occurred in the conservation of the historic buildings of Yemen during the last three decades, and why *qudad* is today neglected by local craftsmen. The fact that the number of masons who still know this technique is so small makes the work ever so important and necessary. This study of preindustrial materials and manufacturing techniques is documented through an 18 year long project started in 1983 after a collaboration between the Yemeni Government and the Dutch Government.

Both collaborated to preserve a sixteenth century mosque in the town of Rada called the 'Amiriya Madrasa'. As*qudad* is found in most parts of the building (interior and exterior plaster, floor), the conservation team decided to replicate mixes and types of application. In so doing, several masons were trained so that they could impart knowledge and expertise to the new generations of Rada. By explaining the repair work carried out in the Amiriya, suggestions are made for a more appropriate conservation of the architectural heritage of Yemen.

The Content

The documentary is divided into three sections. Part one provides an introduction to the history of the site, on the poor state of conservation of the building before the project started, and on project management. Part two explains the materials and tools employed by craftsmen and masons, and part three outlines the conservation work carried out in the building. The production of *qudad* plaster is time consuming, and the process described in the documentary includes: limestone quarrying (one day), kiln loading (one day), kiln firing (four days), kiln cooling (two days), kiln unloading (one day), lime slaking (the longer the better), mortar mixing (ten weeks). This amounts to a total of over 100 days just to prepare *qudad*. Only local materials were employed throughout the conservation project, and great importance was given to craftsmanship. Materials, tools, and techniques are explained. For instance, a full description of the following plastering materials and tools is included: limestone (nurah), powder limestone (nurah daqiq), pebbles (karri), volcanic cinders (hushash), mortar mix (khaltah), polishing tools (madla kak), pounding tools (marqam), and planer (ma'rij).

Figure 3: Loading the kiln with limestone.

Figure 4: Firing the kiln.

By using these materials and tools the *qudad* plaster survives well in damp conditions, but needs a full year before setting. Furthermore, *qudad* is more expensive than cement because the application of the latter takes one day, whilst the first needs up to three months just to polish. Polishing is a very important aspect of the whole process. Masons explain that "in certain areas it takes up to six or seven months just with water and a polisher. If you do not polish enough, then you start getting cracks". After the last polishing with lime water comes the final phase where you apply "animal fat, from cows for example. It acts as protective layer over the*qudad* so it does not start cracking." The documentary also stresses how the earthen buildings of the Hadramaut valley are repaired nowadays and what are the consequences of that repair.

Figure 5: Polishing the floor.

Figure 6: Repairing a wall crack.

This gives an idea of the lack of understanding during the period when new building materials and types started to be heavily introduced in the area. This can be related to the improvement of transportation facilities and to the consequential drop in the cost of purchasing building materials. The local construction industry has a strong motivation for keeping modern materials and methods on the market as these are profitable. At the same time they have little to no interest in using traditional skills and repair methods because these are too "low key and employ local materials" in other words they are far less profitable.

Figure 7: The group of craftsman from Rada, who are now trained to work with *qudad* as a result of this project.

Additional Information

The work, winner of several film awards, could be of use for archaeologists and conservators working with Middle Eastern archaeological sites and should be in the shelves of every conservation library. For a history of the building and its preservation, see *The Amiriya in Rada* by Selma Al-Radi, Oxford University Press, 1997 and*Amiriya Madrasa: The Conservation of the Mural Paintings* by Selma Al-Radi, Roberto Nardi and Chiara Zizola (published 2005 by the CCA, Centro di Conservazione Archeologica). For further information on the mud brick architecture of the Yemen, see the previous documentary by Caterina Borelli (*The Architecture of Mud,* DER) and the book by Salma Samar Damluji, *The Valley of Mud Brick Architecture* (Shibam, Tarim, and Wadi Hadramaut), Garmet Editions, Reading, 1992. Both describe the traditional processes of production and the traditional methods for repairing earthen houses in the Hadramaut valley.

CHAPTER 4

THERMAL COMFORT CHARACTERISTICS OF SOME SELECTED BUILDING MATERIALS IN THE REGIONAL SETTING OF ILE-IFE, NIGERIA

S. K. Fasogbon[a], A. B. Wahaab[b] and M. O. Oyewola[a]

[a] Mechanical Engineering Department, Faculty of Technology, University of Ibadan, Ibadan, Nigeria

[b] Building Department, Faculty of EDM, Obafemi Awolowo University, Ile-Ife, Nigeria

ABSTRACT

In Ile-Ife, it is very common to see quite a lot of local people engaging in mud bricks moulding, concrete block making, mud house construction and the bricklaying profession in general. These have particular bearing on the economy and the political situation of the people involved as in the profession there is set of rules and regulations guiding the rate and procedure for carrying out any such moulding or construction work. Considering the importance of this, the present study investigated the thermal performance of some selected building materials in the ancient city of Ile-Ife in Nigeria. The work demonstrated how a building envelope responds to outdoor conditions through graphic illustrations. This was followed by constructing three physical building models, with model 1 constructed of mud bricks, model 2 constructed of concrete blocks and model 3 constructed of cast concrete. Each of the models was first roofed with galvanised Iron roofing sheets, later with aluminium roofing sheets and finally with asbestos roofing. Readings were taken by inserting TGP-4500 Data loggers into appropriate positions. The results showed that internal temperatures in the mud-brick and cast concrete buildings remained fairly stable despite external diurnal fluctuations. Humidity data collected also showed that fluctuations in external humidity levels do not affect humidity levels within the mud-brick and cast concrete structures significantly. On the other hand, the structure made of concrete blocks responded more to external fluctuations in temperature and humidity. Moreover, humidity levels in the cast concrete building were the least, followed by those in the mud-brick house; while the concrete block structure had the highest level of humidity. It was also discovered that aluminium roofing gave the highest internal temperature, followed by galvanised iron roofing and the least was seen with asbestos roofing. For external diurnal fluctuations, building model 3 with cast concrete gave the most satisfactory performance in terms of thermal comfort; however, building model 1 with mud bricks also gave a satisfactory performance close to that of model 3. The study concludes that building model 1 with mud bricks roofed with Asbestos is the best building model in terms of thermal comfort and affordability. The study also concluded that developing model 1 (mud bricks) in the regional setting of Ile-Ife will go a long way to developing people's sustainably.

Citation: S. K. Fasogbon, A. B. Wahaab and M. O. Oyewola, Thermal Comfort Characteristics Of Some Selected Building Materials In The Regional Setting Of Ile-Ife, Nigeria, DOI 10.5027/jnrd.v5i0.07

INTRODUCCTION

There are three basic fundamental needs of every man and woman which are food, clothing and shelter. The basic function of building is to provide shelter from harsh climatic conditions (sun and rain, heat and cold), external aggressions (protection from animals and attacks from humans), privacy and storage of possession. Hence, buildings are meant to protect man against harsh climatic conditions [1], [8], [12], [13]. A building cannot be built without a fundamental knowledge of the building materials and the determinant factors (thermal performance, strength, cost etc.) that lead to the choice of one material over another [2], [9], [10], [11]. A building material is any material which is used for a construction purpose. Many naturally occurring substances, such as clay, sand, wood and rocks, even twigs and leaves have been used to construct buildings. Apart from naturally occurring materials, many man-made products are in use, some more and some less synthetic. The manufacture of building materials is now an established industry in Nigeria and the use of these materials is typically segmented into specific specialty trades, such as bricklaying, carpentry, plumbing, roofing and insulation work. Building design and material properties influence the thermal performance and energy consumption for residential and commercial buildings. Different building materials have different thermal performance levels. For example, the embodied energy of mud brick is estimated to be around 0.7MJ/Kg, less than 30% of the embodied energy of clay bricks (2.5MJ/Kg) and less than 20% of the embodied energy of lightweight aerated concrete blocks (3.6MJ/Kg) [3], [4], [14]. According to the Australian institute of Refrigeration, Air-conditioning and Heating the conductivity of 250mm-thick mud bricks with a density of 1540Kg/m3 and specific heat capacity of 1260J/Kg.K is 1.25W/m.K, while concrete with a density of 2240Kg/m3 will have similar conductivity (which is 1.3W/m.K); however the specific heat would be between 800 to 1000J/Kg.K [5]. This means that mud bricks can absorb more heat than concrete even though they is less dense. Thermal performance of building materials is of significant importance in identifying the heating and cooling load within a building and hence the capacity of the mechanical equipment required to handle the load. It is important to know that the thermal performance of any building material is the result of several factors, which are: thermal conductivity "k" of the building material, thermal resistance "R" of the building material, emissivity "ε" of the building material, absorptivity "α" of the building material, reflectivity "ρ" of the building material, specific heat capacity "cp" of the building material, density "ℓ" of the building material. Climate is determinant of how much solar radiation gets to the earth's surface and the temperature within that particular geographical location. This in turn influences the thermal performance of building materials used for construction in a particular region. Two of the most important factors determining an area's climate are temperature and relative humidity (R.H) [6], [7].

Regional management is a system of self-coordination where by a region or unit harmonises its ecological functions with the economic and social needs of its people for sustainable development [15]. In regional management, the importance of a region as a political entity is emphasised. Political responsibility concerning important decisions regarding regional development is left up to regional actors. The territorial principle is being replaced by the functional principle i.e. political responsibility of regions is no longer solely defined by administrative levels and borders (e.g. governmental districts), but rather also by the concept of Regional Governance which understands a region to be a dynamic area of cooperation between actors which is formed due to the density of social relations between regional actors [16]. Through this, the dynamic area of cooperation is able to perform different tasks on which the regional actors mainly decide themselves (for example as an economic, tourist or nature protection region [16]). There is inter sectorial cooperation through weakly institutionalised regional networks and partnerships. The potential of networks and cooperation between regional actors is central to the concept of Regional Governance ([16], [17], [18], [19], [20] and [21]). Cooperation encompasses the horizontal level on the one hand (partnerships on a regional level) and the vertical level on the other (partnerships between the region and higher political levels, which for example can clearly be seen in the EU structural policy's partnership principle). There is also hierarchical steering of incentives through various instruments and forms. The fact that more often than not, regional cooperation in the sense of Regional Governance/Management does not come about naturally, and that it must usually first be initiated, is problematic [16]. Different instruments which may act as potential incentives in building regional cooperation must then come into play. Within this context, Benz discusses "the hierarchical steering of incentives" [16], which he refers to as the always important hierarchical potential of the State when initiating regional cooperation. This kind of incentive steering can be fulfilled through different forms: Higher state levels often use competition as an instrument in order to identify regions as promising for successful development of Regional Governance [22]. Regional management in Nigeria predates the creation of an entity called Nigeria (1914), when various peoples that existed independently had established their own indigenous systems of administration. The recognised political entities then included: Ife Kingdom, Benin Empire, Kanem Bornu Empire, Sokoto Caliphate, and Oyo Empire, to mention a few [23]. The Empires and Kingdoms had established contact with one another and with other peoples through trading activities. On 1st October 1960, Nigeria became self-governing from British colonial rule and was administered at the centre by the federal government and three regional governments in the east, west and north of the country. In 1963, the Midwest Region was carved out of the Western Region making a federation of four Regions. During this First Republic, a parliamentary system of government was in operation and this lasted until January 1966 when the military took over and this lasted reasonably until 1999 when democratically elected government became stable in the country. During this period of regional settings, the eastern region was particularly noted for buying and selling, and this was used to support its economy. The northern region developed its subsistence farming and was able to build a groundnut pyramid. The western region (where Ile-Ife is located) also developed its cash-crop farming such as cocoa, coffee, kola nut and others, and with this, the region saw tremendous development of its education, social services, health services and infrastructural facilities (e.g. cocoa house) [23] and [24]. Figure 1 shows the map of ancient city of Ile-Ife.

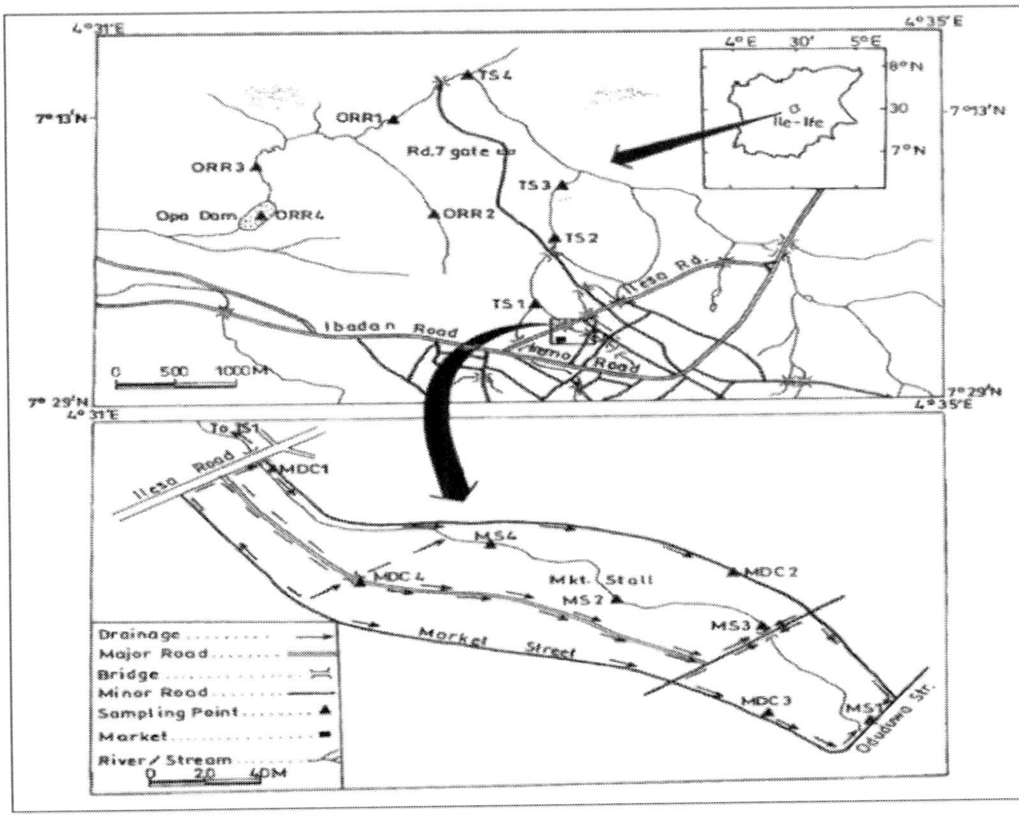

Figure 1: Map showing Ancient City of Ile-Ife. Source: [6]

MATERIALS AND METHODS

Materials

With the service of our qualified builder in the Department of Building at the Obafemi Awolowo University, three building models each with the area of 0.7025m2 were constructed. Model 1 was constructed of mud bricks, model 2 was constructed of concrete blocks and model 3 of cast concrete. Each of the models was first roofed with galvanised Iron roofing sheets, later with aluminium roofing sheets and finally with asbestos roofing. The models are as shown in Figure 2 (Figure 2a, 2b and 2c, respectively).

Figure 2: Developed building models constructed of various materials

Method

The thermal behavior of the aforementioned building materials was studied by taking measurements on site. Temperature and humidity measurements were taken with the help of Tinytag data loggers (TGP-4500). For each set of data, one data logger was placed outside in a shaded area protected from direct sunlight and rain, while the others were placed inside the buildings to record external and internal temperatures concurrently. Data thus collected could help demonstrate how the building envelope performed in terms of its insulation and thermal inertia properties under natural conditions.

RESULTS AND DISCUSION

For the sake of clarity, temperature and humidity measurements for only three days are presented graphically i.e. from 26th to 28th August 2013 for Model 1 (constructed of mud bricks), model 2 (constructed of concrete blocks) and model 3 (constructed of cast concrete). The 24 hours for which the test was conducted was such that the first hour started from 12:00 midnight.

Temperature Situation under Galvanized Iron Roofing Sheet

Figure 3 shows typical behavior of cast concrete, mud-brick and concrete-block buildings under galvanized iron roofs. It was noticed that the average indoor temperature of building model 1 was higher than ambient temperature by 3.5°C, model 2 was higher by 0.5°C, and model 3 was higher by 4°C. It can be seen that model 3 remains superior in terms of temperature variation among the three models for this type of roofing.

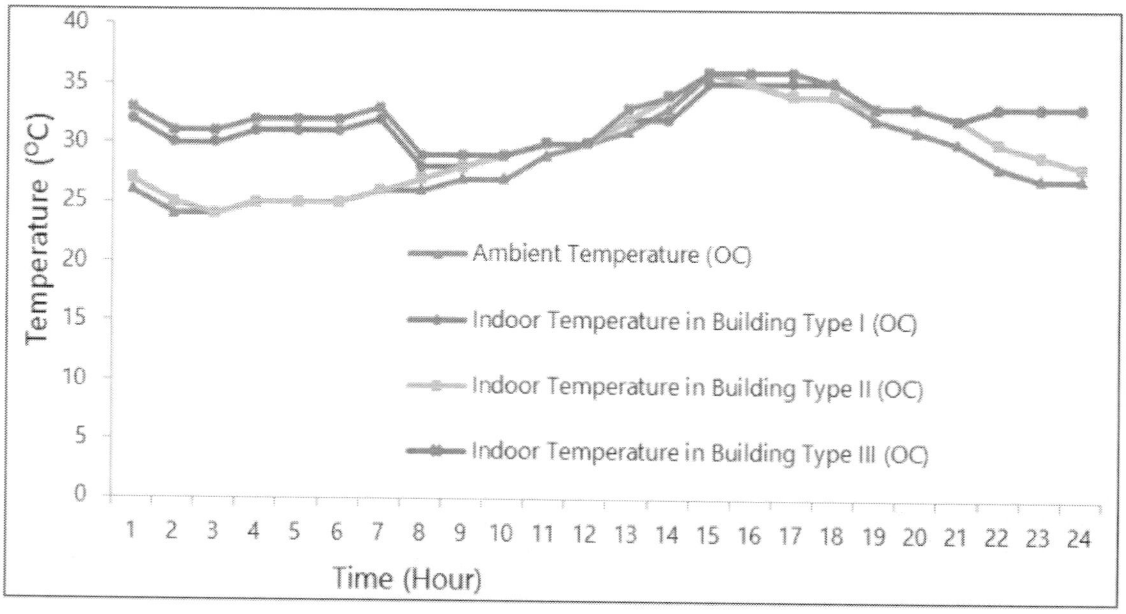

Figure 3: Plot of resulting indoor temperature in the three building types against time of day when roofed with galvanized iron roofing sheets

Temperature Situation under Aluminum Roofing

Figure 4 shows the typical behavior of cast concrete, mud-brick and concrete block buildings under aluminum roofs. It can be seen that the average indoor temperature of building model 1 was higher than ambient temperature by 3°C, while that of model 2 was not higher by any degree, and model 3 was higher by 3.5°C. It can be seen that model 3 still remains superior in terms of temperature variation among the three models for aluminum roofing.

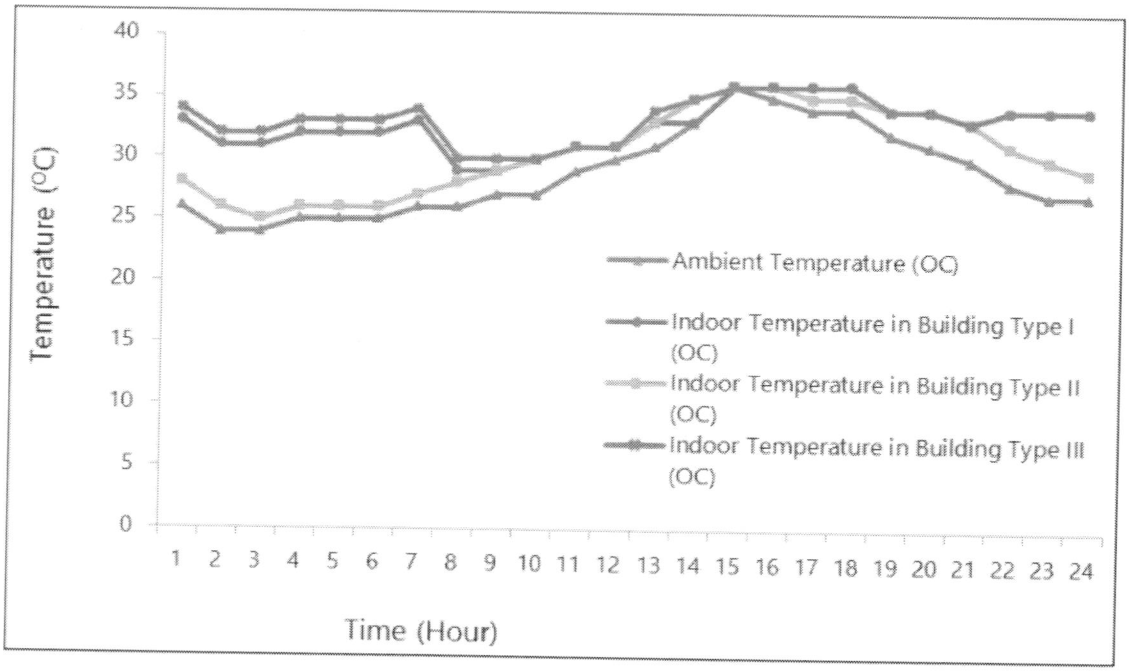

Figure 4: Plot of resulting indoor temperature in the three building types against time of day when roofed with aluminum roofing sheets

Temperature Situation under Asbestos Roofing

In tropical regions, with reference to Nigeria, materials like zinc, aluminum and asbestos are commonly used in the form of sheets for roofing in modern building constructions [25]. Figure 5 shows the behavior of cast concrete, mud-brick and concrete block buildings under asbestos roofs. It can be seen that the average indoor temperature of building model 1 was higher than ambient temperature by 4°C, model 2 was higher by 1°C, and model 3 was higher by 5°C. It can be seen that model 3 is superior in terms of temperature variation among the three models for asbestos roofs.

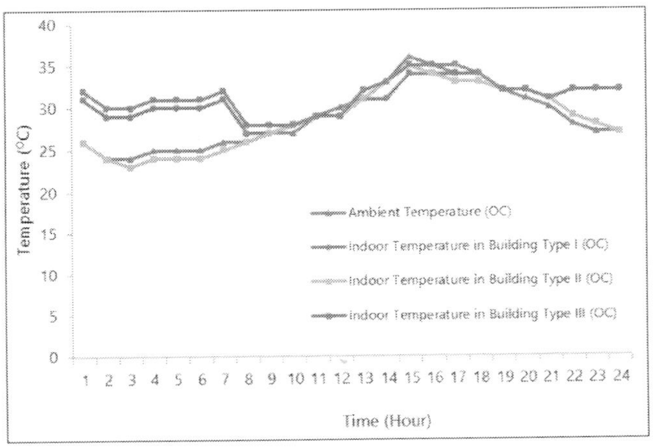

Figure 5: Plot of resulting indoor temperature in the three building types against time of day when roofed with asbestos roofing sheets

Relative Humidity under each Roofing Sheet

Figure 6 shows typical behavior for the different building models under a particular type of roofing. It was observed that relative humidity remains stable for all three types of building models with a particular type of roofing sheet. 67.6% relative humidity was noticed for galvanized iron sheets, 80% for aluminum roofing sheets and 64.4% for asbestos. We suspect that the observation to have the same level of relative humidity in all the three types of building models under the same type of roofing must have been occasioned by the fact that irrespective of the types of building involved, the type of roofing must have informed the same level of room temperature in each case which correspondingly determined the same value of moisture content or relative humidity of the air. In general, the results for internal temperatures in the mud-brick and cast concrete buildings (as shown in Figure 3, Figure 4 and Figure 5) remained fairly stable despite external diurnal fluctuations. The humidity data collected also showed that fluctuations in external humidity levels do not affect humidity levels within the mud-brick and cast concrete structures significantly (Figure 6). We suspect that the good thermal insulation/heavy thermal mass properties of mud bricks and cast concrete are responsible for this observation. It follows that even without providing any air conditioning it is possible to keep internal air temperatures of a building close to thermal comfort levels by preventing extreme temperature conditions and also by avoiding temperature and humidity fluctuations, which cause discomfort to occupants. This can be done by taking into consideration the thermal properties of wall materials during the design stage. On the other hand, the structure made of concrete blocks responded more to external fluctuations in temperature and humidity. Equally, we suspect this to be attributed to the numerous pores in the blocks or the light thermal mass properties of the concrete blocks. It was also discovered that aluminum roofing gave the highest internal temperature, followed by galvanized iron roofing sheets and the lowest values were seen with asbestos roofing. This must be related to the thermal conductivity of each roofing sheet involved.

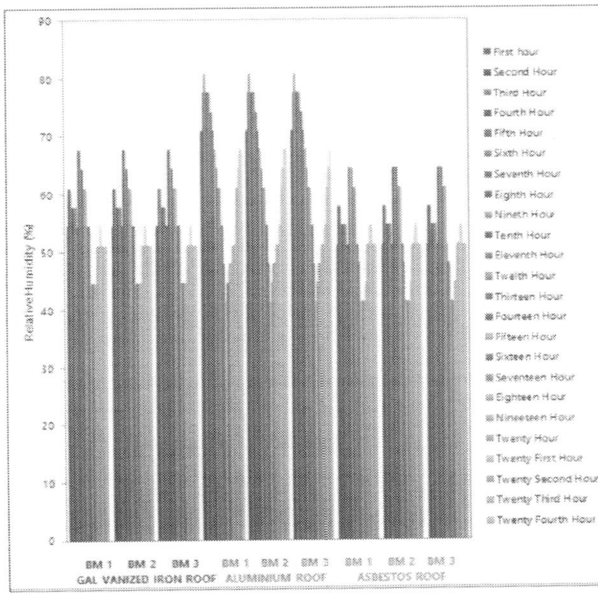

Figure 6: Plot of resulting indoor relative humidity in the three building types against time of day when roofed with the galvanized iron, aluminum and Asbestos roofing sheets

CONCLUSION

This study established that both mud bricks and cast concrete are fairly stable in terms of temperature and humidity variations despite external diurnal fluctuations. By implication, buildings made of mud bricks and cast concrete can be said to be energy efficient as there will be no need to always condition the buildings using extra energy. In addition to these findings, the literature shows that the energy required to produce mud-brick buildings is minimal and the material is completely bio-degradable. Although for the external diurnal fluctuations, building model 3 with cast concrete gave the most satisfactory performance in terms of thermal comfort, building model 1 with mud bricks also gave satisfactory performance that was close to that of model 3. The study concludes that building model 1 with mud bricks roofed with asbestos is the best building model in terms of thermal comfort, affordability and the other aforementioned advantages in Ile-Ife and other similar cities/towns in the tropics. The study also concludes that if the federal government of Nigeria can help develop model 1 (mud bricks) in the region, it will lead to sustainable development for the people.

REFERENCES

1. Otegbulu AC, Osagie JU, and Afe YO. The built environment perspective of climate change- a focus on household activities in Lagos metropolis. Journal of Sustainable Development 2011; 4(5): 174-187. Doi: http://dx.doi.org/10.5539/jsd.v4n5p174

2. Runfolaa D M, Polsky-b M, Nicolsonc C, Giner-b NM, Pontius Jr. b RG, Krahe-b J, Decatur A. A growing concern? Examining the influence of lawn size on residential water use in suburban Boston, MA, USA. Landscape and Urban Planning 2013; 119: 113-123. Doi: http://dx.doi.org/10.1016/j.landurbplan.2013.07.006
3. West S. Improving the sustainable development of building stock by the implementation of energy efficient, climate control technologies. Building and Environment 2001; 36(3): 281-289. Doi: http://dx.doi.org/10.1016/S0360-1323(00)00007-X
4. Carnes TN. Indoor sound criteria according to the American Society of Heating, Refrigerating and Air-Conditioning Engineers (ASHRAE)—An introduction. Journal of the Acoustical Society of America 2005; 118: 1843. Doi: http://dx.doi.org/10.1121/1.4778485
5. Ferguson AJ. Handbook of heating, ventilating and air conditioning: 8th Edition. International Journal of Refrigeration 1983; 6(3): 186-187. Doi: http://dx.doi.org/10.1016/0140-7007(83)90102-0
6. Eludoyin AO, Ofoezie IE, Ogunkoya OO. The effect of oja-titun market effluent on the chemical quality of receiving opa reservoir in Ile-Ife, Nigeria. Journal of environmental management 2004; 72(4): 249–259. Doi: http://dx.doi.org/10.1016/j.jenvman.2004.05.005
7. Delso G, Ziegler S. PET/MR System Design. In I. Carrio and P. Ros, Eds., "PET/MRI". Springer Berlin Heidelberg; 2014, p. 1-19. Doi: http://dx.doi.org/10.1007/978-3-642-40692-8
8. Ajibola K. Design for comfort in Nigeria – a Bioclimatic Approach. Journal of Renewable Energy 2001; 23: 57-76. Doi: http://dx.doi.org/10.1016/s0960-1481(00)00113-0
9. Bodalal A, Zhang JS, Plett EG . A method for measuring internal diffusion and equilibrium partition coefficients of volatile organic compounds for building materials. Building and Environment 2000; 35(2): 101-110. Doi: http://dx.doi.org/10.1016/S0360-1323(99)00005-0
10. Cengel YA. Heat Transfer. In Capehart BL, editor. Encyclopedia of Energy Engineering and Technology-3 volume Set (Print Version). US: Taylor and Francis Group, LLC; 2007, p. 822-829. Doi: http://dx.doi.org/10.1201/9780849338960.ch97
11. Wilkes KE, Rucker JL. Thermal performance of residential attic insulation. Energy and Buildings 1983; 5(4): 263-277. Doi: http://dx.doi.org/10.1016/0378-7788(83)90014-2
12. Glavic P, Lukman R. Review of sustainability terms and their definitions. Journal of clean production 2007; 15 (18): 1875-1885. Doi: http://dx.doi.org/10.1016/j.jclepro.2006.12.006
13. Gagliano A, Patania F, Nocera F, Signorello C. Assessment of the dynamic thermal performance of massive buildings. Energy and Buildings 2014; 72: 361-370. Doi: http://dx.doi.org/10.1016/j.enbuild.2013.12.060
14. A. Davraz, R. Karaguzel, I. I. Soyaslan. The importance of hydrogeological and hydrological investigations in the residential area: a case study in Burdur, Turkey. Environmental Geology 2003; 44(7): 852-861. Doi: http://dx.doi.org/10.1007/s00254-003-0830-1
15. Fürst D, Kilper H. The innovative power of regional policy networks: A comparison of two approaches to political modernization in North Rhine-Westphalia. European Planning Studies 1995; 3(3): 287-304. Doi: http://dx.doi.org/10.1080/09654319508720308
16. Michael Böcher. Regional Governance and Rural Development in Germany: the Implementation of LEADER+. Sociologia Ruralis 2008; 48(4): 372-388. Doi: http://dx.doi.org/10.1111/j.1467-9523.2008.00468.x
17. Benz A. Public Administrative Science in Germany: Problems and Prospects of a Composite Discipline. Public Administration 2005; 83(3): 659-668. Doi: http://dx.doi.org/10.1111/j.0033-3298.2005.00468.x
18. Benz A, Fürst D. Policy Learning in Regional Networks. European Urban and Regional Studies 2002; 9(1): 21-35. Doi: http://dx.doi.org/10.1177/096977640200900102
19. Simon Guy S, Graham S, Marvin S. Privatized Utilities and Regional Governance: The New Regional Managers? Regional Studies 1996; 30(8): 733-739. Doi: http://dx.doi.org/10.1080/00343409612331350018
20. Baldassare M. Regional Variations in Support for Regional Governance. Urban Affairs Review 1994; 30(2): 275-284. Doi: http://dx.doi.org/10.1177/004208169403000205
21. Mangiameli S. The Role of Regional and Local Government in European Governance. In Blanke H, Mangiameli S, editors. Governing Europe under a Constitution. US: Springer Berlin Heidelberg; 2006, p. 457-481. Doi: http://dx.doi.org/10.1007/3-540-31291-9_29
22. Albrecht M, Elbe J, Elbe S, Meyer W. Analyzing and evaluating regional governance networks: Three challenges for applications. Evaluation 2014; 20(1): 58-74. Doi: http://dx.doi.org/10.1177/1356389013518457
23. Frankema E, Waijenburg M. Metropolitan Blueprints of Colonial Taxation? Lessons from Fiscal Capacity Building in British and French Africa, c. 1880–1940. The Journal of African History 2014; 55(3): 371-400. Doi: http://dx.doi.org/10.1017/S002185371400036X
24. Adalemo IA, Balogun OY. The Status Of Cartography In Nigeria. In Festschrift A for Ormeling FJ editor. Cartography Past, Present and Future. US: Elsevier Science Publishers Ltd; 1989, p. 15-26. Doi: http://dx.doi.org/10.1016/B978-1-85166-336-1.50006-0
25. Etuk SE, Akpabio LE, Akpabio KE. Investigation of Raphia hookeri trunk as a potential ceiling material for passively cooled building design. Ghana Journal of Science 2005; 43(1): 3-7. Doi: http://dx.doi.org/10.4314/gjs.v43i1.15894.

CHAPTER 5

FLOODING EFFECT ON EARTH WALLS

Gabriela M. Medero [1], Justin H. Kennedy [1], Peter K. Woodward [1] and Meysam Banimahd[2]

[1]School of the Built Environment, Heriot-Watt University, EH14 4AS, Edinburgh, UK

[2]Technip, Westhill Industrial Estate, Westhill, Aberdeenshire, AB32 6TQ, UK

ABSTRACT

Earth building is a sustainable, environmentally friendly and economical method of construction that has been used worldwide for many centuries. For the past three decades, earth has seen a revival as a building material for a modern construction method due to its benefits in terms of low carbon content, low cost and energy involved during construction, as well as the fact that it is a sustainable technology of building. Climate change is influencing precipitation levels and patterns around the world, and as a consequence, flood risk is increasing rapidly. When flooding occurs, earth buildings are exposed to water by submersion, causing an increase in the degree of saturation of the earth structures and therefore a decrease of the suction between particles. This study investigated the effect of cycles of flooding (consecutive events of flooding followed by dry periods) on earth walls. A series of characterization tests were carried out to obtain the physical and mechanical properties of the studied earth material. In a second stage, Flooding Simulation Tests (FST) were performed to explore the earth walls' response to repeated flooding events. The results obtained for the tested earth wall/samples with reinforced material (straw) reveal hydraulic hysteresis when wall/samples are subject to cycles of wetting and drying.

INTRODUCTION

Earth is still the second most common building material worldwide after bamboo [1], with 40% of the world's population living in earth buildings [2]. In the UK, it is estimated that half a million earth buildings currently exist [3,4]. Since the beginning of 1980s, earth as a building material has seen a revival as a modern construction technique [1,5] due to its low carbon content, sustainable method, low cost and energy involved during construction and inherent recyclability [5,6]. Not only these aspects are driving the come-back of the earth building, but the fact that it is a reasonable and simple building technique and it provides good temperature and moisture exchanges with outdoors. With this growth of earth buildings, there is a need to understand the behavior of these structures under different climate conditions. A serious enemy of earth construction is water, which can be in the form of moisture ingress [1,7] (e.g., rising damp from the ground due to capillary effect) or by flooding of the earth walls [8] (e.g., submerging the earth structures).

Recently, observed increases in precipitation levels in the UK are driving the need for investigation of the effects of flooding on earth walls. Evidence for this is provided by the UK Climate Impacts Program (UKCIP; www.ukcip.org.uk), which concluded that the decade of the 1990s was the wettest in the UK since records began. Some areas received

Citation: Gabriela M. Medero, Justin H. Kennedy, Peter K. Woodward and Meysam Banimahd, Flooding Effect on Earth Walls, doi:10.3390/su3010069.

one month's precipitation in just 24 hours. Furthermore, the predictions by the UKCIP09 indicate that these trends will continue with heavy precipitation becoming more frequent across the UK set to rise further. In addition, climate change is acting as a driver for large increases in flood risk worldwide and in the UK (Figure 1).

(a)

(b)

(c)

Figure 1: Recent flooding occurring in the UK at (a) Catcliffe near Sheffield; (b) the village of Boscastle; and (c) Yorkshire (www.bbc.co.uk).

These processes of extreme precipitation create cycles of wetting/drying (wetting during rain and possible flooding, followed by drying after drainage) on the earth walls, which generate a change in the earth walls structure (fabric and bonding), strength and stiffness. When flooding occurs, the earth walls are exposed to water by submersion as well as by capillary forces, causing an increase of the degree of saturation (amount of water filling the porous of the wall) of the earth structures and therefore decreasing the suction between the particles. At low degrees of saturation (for example, before flooding), the suction is higher and this helps the bonding between particles, increases cohesion, strength and stiffness. When the degree of saturation increases (for example, during flooding), the suction decreases and the structure suffers a drastic change in strength and this can happen together with volumetric variations. After flooding, the earth structures that were exposed to water, start presenting evapotranspiration, decreasing again the degree of saturation of the earth material. Climate change is causing a rise in frequency of flooding, so repeatable flooding (wetting of the earth walls) will generate cycles of wetting/drying on the earth structures. As a consequence, all structures exposed to such weather conditions can be affected by behavioral features due to these cycles of wetting/drying and catastrophic failures can happen.

PREPARATION AND CHARACTERIZATION OF THE WALL/SAMPLES

An experimental program was planned and carried out to investigate the effect of flooding events on the earth walls' response on a reduced scale in the laboratory. The first stage was to characterize the earth materials: see Table 1 and Figure 2 for the used material, particle size distribution and preparation details.

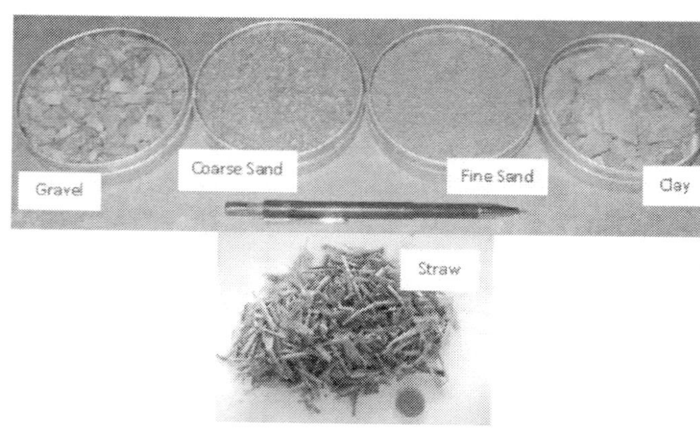

Figure 2: Detail of the material used to mold the wall/samples.

Table 1: Particle size distribution and details of preparation (data from [9])

Material	Percentage required (% by weight)	Details of preparation
Coarse sand	19.7	Dried in oven at 105 °C for 24 h
Gravel	24.4	-
Fine sand	32.5	Dried in oven at 105 °C for 24 h
Clay	20.6	Dried in oven at 105 °C for 24 h
Straw	1.25	Cut into approximate length of 15–50 mm*
Water	1.55	-

* Depending on the aim of the test.

A series of tests were performed to investigate the moisture content at storage of the materials used*in situ* before the mixture process for building. It was observed that these materials were not completely dry and presented an average moisture content of 1.55%. For the experimental program of this study, 1.55% of water content was adopted for the mixture to reproduce the field condition when constructing.

The mixing of the materials was carefully done in the laboratory. Some characterization tests were carried out, such as liquid and plastic limit, specific gravity, compaction and unconfined compression tests. The authors took special care to reproduce—on a small scale—specimens that were representative of the process of building *in situ*. At the same time; attention was taken to prepare repeatable samples to allow the analysis and comparison of the obtained results.

All the wall samples were prepared without using any additives (e.g., starches, lignosulfates, or gypsum) or any coatings (e.g., clay plaster, lime plaster, gypsum plaster, limewash or paint), because it was aimed to study the most critical possible conditions without the benefits that an additive could add to the mixture and without a coating material that will improve the durability of the external face of the walls.

Table 2 presents some of the obtained results. This initial characterization was crucial to enable the second stage of the project in which wall/samples were prepared and tested under flooding conditions.

Table 2: Obtained results from characterization tests

Plastic Limit	14.6%
Liquid Limit	22.5%
Plasticity Index	7.9%
Specific Gravity (without straw)	2.6
Specific Gravity (with straw)	2.5

A series of compaction tests were executed during this investigation to identify the optimum water content and co-related maximum dry density for the different configurations studied (mixture without straw, mixture with 15 mm of straw, mixture with 50 mm of straw). After the mixing process of the materials described in Table 1, the mixture was placed in a steel mold and compacted with the automatic compactor following the BS 1377 [10], using the 2.5 kg rammer method. From the results of this test, it is possible to obtain the optimum moisture content (w_{opt}), at which a maximum value of dry density is obtained. At low values of moisture content, lower than w_{opt}, most soils tend to be stiff and are difficult to compact. As the water content (w) is increased, the soil becomes more workable, facilitating compaction and resulting in higher densities. At high water contents, higher than w_{opt}, however, the dry density decreases with increasing proportion of the soil volume being occupied by water. Comparing the optimum water content and the maximum dry density of all three configurations of soil mixture, it was obtained:

- Soil mixture without straw $w_{opt} = 7.35\%$ and $\rho_d = 2.09$ Mg/m^3
- Cob with Straw cut down to 15 mm $w_{opt} = 9\%$ and $\rho_d = 2.045$ Mg/m^3
- Cob with Straw cut down to 50 mm $w_{opt} = 9.3\%$ and $\rho_d = 2.02$ Mg/m^3

The obtained results show that the addition of straw affects, in terms of compaction, the amount of water that should be added to the mixture to obtain a denser soil structure. This information should be considered when mixing the earth materials for construction in the field. Another important parameter is the moisture content at storage of the materials used, which should always be verified before the amount of water that will be used for construction is decided. The authors studied the effect of straw in the building material mixture motivated by a historical building in Perthshire (Scotland), constructed between 1745 and 1770 [10], which recently started suffering due to flooding events. Part of the building had been flooded (with submergence of the base of the walls) five-times in recent years [11]. In response to these events, isolated repairs were undertaken to the damaged walls using pre-molded and dried earth blocks (clay bat), using straw as reinforcement material [8].

To prepare the samples, the correct amounts of each material was placed in a mixer and blended together as a dry mix until homogeneous. After, water was added in stages. The water was then mixed with the other materials until a homogeneous mixture was obtained. At this stage, the sample was ready to be molded. The molds were hollow cylinders (as shown in Figure 3(a)), that were halved so that the sample could be removed after the blows and were clamped together with Jubilee clips. These molds had marks of 50 mm on the height of the section. These were to mark where the soil layers should reach after each set of blows (see Figure 3(a)). The first layer of the mold was estimated to be filled after ramming. The mixture of the first layer was weighed and the same amount of soil was used in each layer. Each layer received the same number of blows with a manual rammer; this procedure was done by hand, as in cob construction walls and in the pre-molded blocks in Pethshire (Scotland). The authors noticed that the obtained density was dependent on the operator, as is observed in the field. A variation in the obtained density can be observed from operator to operator and should be taken into account when designing an earth structure.

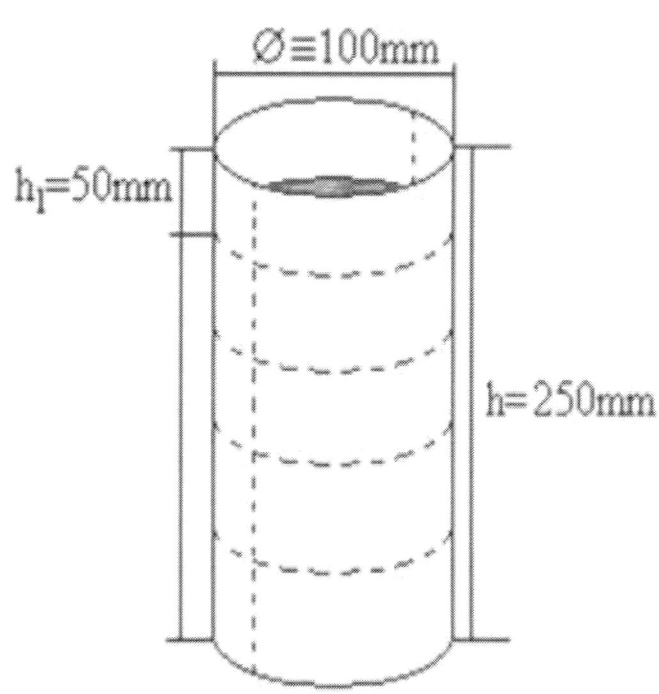

(a)

(b)

Figure 3: (a) Diagram of cylindrical mold. (b) Molded wall/sample (adapted from [9]).

The unconfined compression tests were performed to investigate the behavior of the wall/samples under simple compression; a similar situation would be observed in the field when a vertical load is applied on the top of an earth wall (see Figure 4 for the layout of the test).

(a)

(b)

(c)

Figure 4: (a) Detail of the wall/sample before starting the test. (b) Layout of the wall/sample during testing. (c) Example of a sample presenting failure.

Figure 5 shows the curves obtained from carrying out unconfined compression tests for four samples prepared at different densities (40, 60 and 80 blow per layers), with an initial water content of approximately 7.8%.

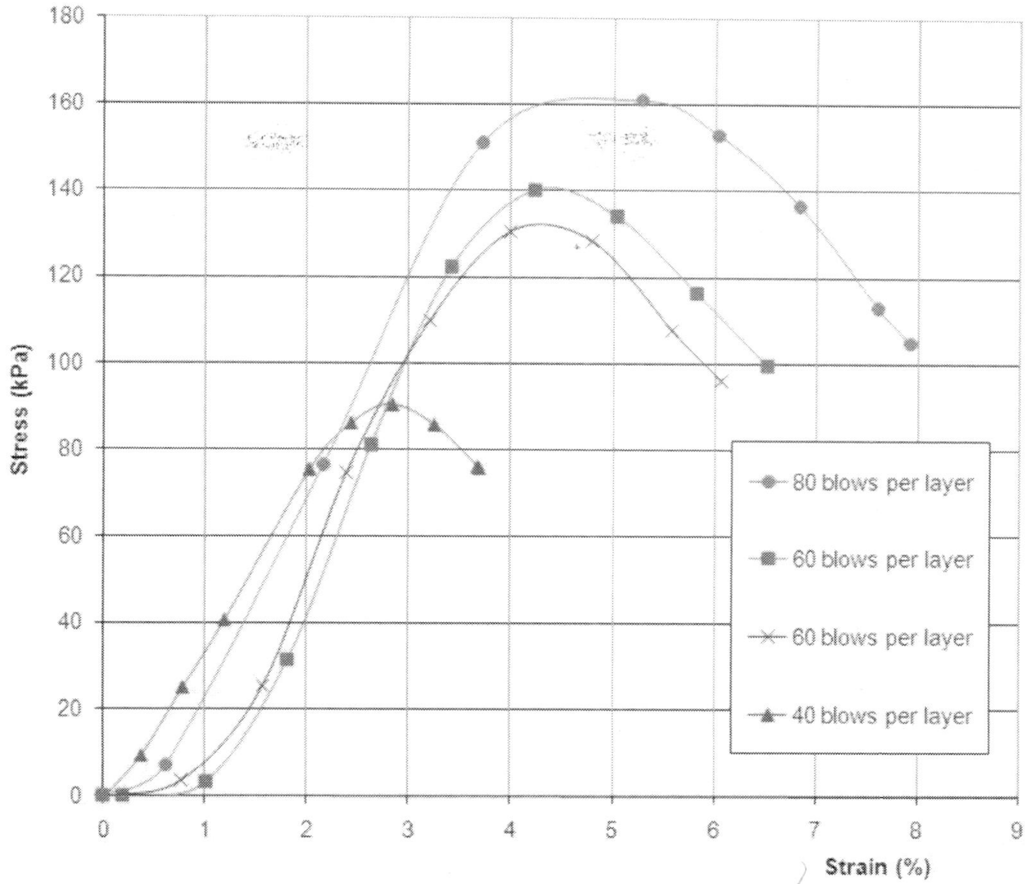

Figure 5: Unconfined compression test results obtained for wall/samples prepared with 40, 60 or 80 blows per layer.

It is evident that the higher the number of blows per layer, *i.e.*, the higher the density of the wall/sample, the higher the gain in simple compression strength. When the range of the number of blows per layer between 40 and 80 was used, the capability of the wall/samples to sustain simple compression was increased: it was 78% between 40 to 80 blows, 50% between 40 and 60 blows and 19% between 60 and 80 blows.

This trend in behavior is important to demonstrate the relevance of the number of blows per layer that should be used in the field. Figure 6 shows the relationship between the peak strength and the compaction rate (number of blows).

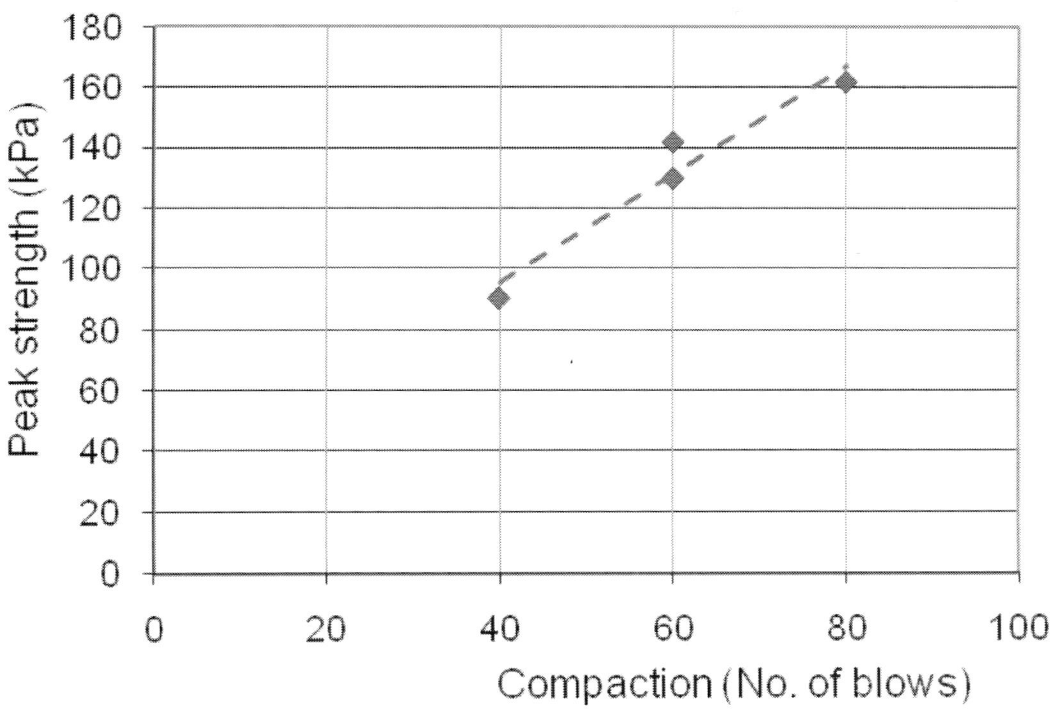

Figure 6: Relationship between the peak strength and compaction rate of the wall/samples.

The results obtained for the maximum unconfined compression strength were smaller than those usually observed on dry rammed earth and cob due to the fact that the wall/samples were tested with moisture content of around 7.8% as well as due to the small scale of the specimens. The authors decided to test the samples with the water content of the molding process because this is the most critical situation and the site that motivated the study (in Perthshire, Scotland), presented a high level of relative humidity.

EXPERIMENTAL SECTION: FLOODING SIMULATION TEST (FST)

Significant differences in the soil specific volume are observed during wetting and drying of soils [13-18]. Cycles of drying and wetting generate progressive shrinkage in clays and result in a change in fabric during drying, which weakens bonding and, as a consequence, degrades the soil structure [19]. Moreover, cycles of wetting and drying can exhibit hydraulic hysteresis [13,16], implying that two samples of the same soil subjected to the same value of suction (s, the difference between pore air pressure and pore water pressure) can be at significantly different values of degree of saturation (Sr) if one is on a drying path and the other is on a wetting path. Figure 7 presents a soil water retention curve (SWRC) of a soil illustrating hydraulic hysteresis [16]. Another important feature observed during cycles of wetting/drying is the occurrence of irreversible volume changes in the soil (e.g., collapse phenomenon) [13,14].

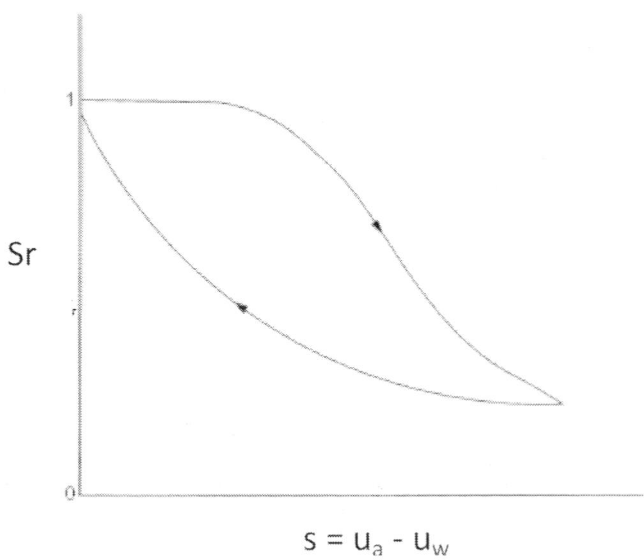

Figure 7: Hydraulic hysteresis in water retention curve of a soil (adapted from [16]): suction (s, difference between pore air pressure and pore water pressure) *versus* degree of saturation (Sr).

A triaxial apparatus was modified at the Soils Laboratory (Heriot-Watt University) to investigate the behavior of the wall/samples under cycles of flooding followed by drying periods, all monitored during long periods of time, around 360 hours. Figure 8 shows the layout of the apparatus. The authors called the proposed test a Flooding Simulation Test (FST). All samples were prepared in small scale (100 mm diameter × 200 mm height) in layers, as the wall/samples molded for the unconfined compression tests following the same procedure as used in the field (procedure used during the repairs in Perthshire, Scotland). The moisture content at preparation was controlled and measured, the number of blows per layer was also controlled and the density of the wall/samples was calculated using the specific gravity obtained at the first stage of the experimental program together with the dimensions and weight of the specimens.

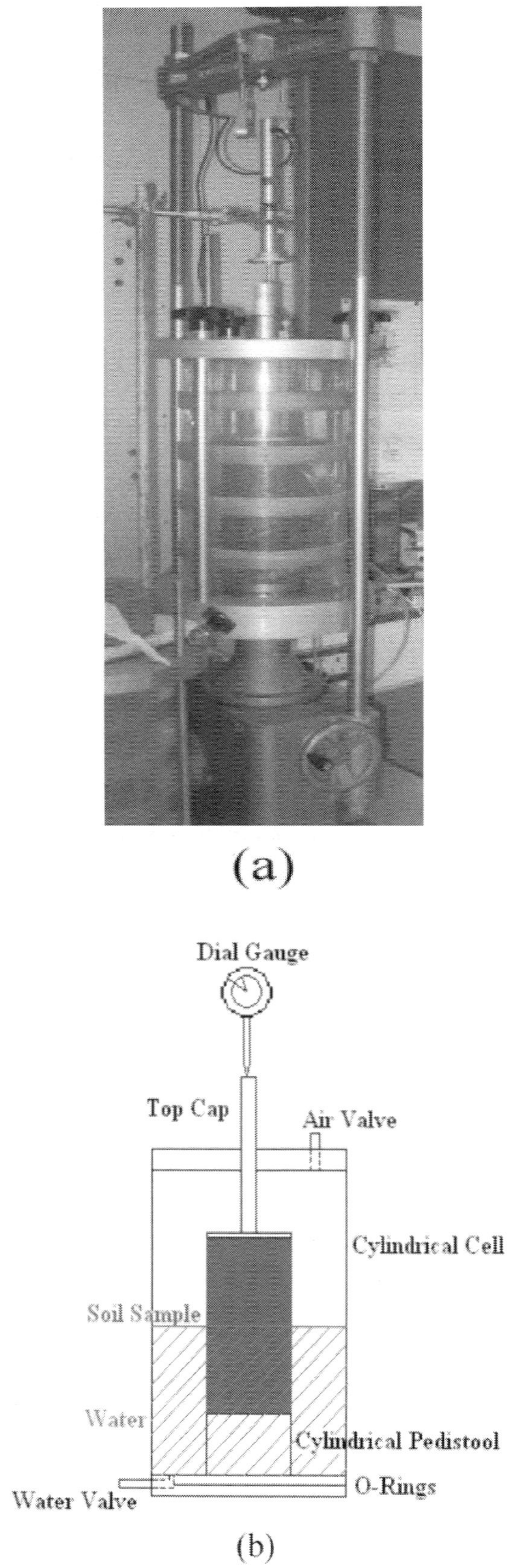

Figure 8: Set up of the flooding simulation test configuration: (a) photograph; (b) diagram (adapted from [8]).

A series of FSTs were carried out for the wall/samples prepared with 60 blows per layer with and without straw as reinforcement material. The wall/samples were exposed to flooding (see Figure 8) for a controlled time of 24, 60 and 120 hours, followed by a drained environment (without submersion in water) also for a controlled time of 72, 60 and 12 hours.

With the aim to investigate the influence of the straw as reinforcement material, one test was carried out with a wall/sample prepared without straw (un-reinforced specimen). The wall/sample was molded with 60 blows per layer, following the same procedures as previously described, and a FST was carried out (Figure 9).

(a)

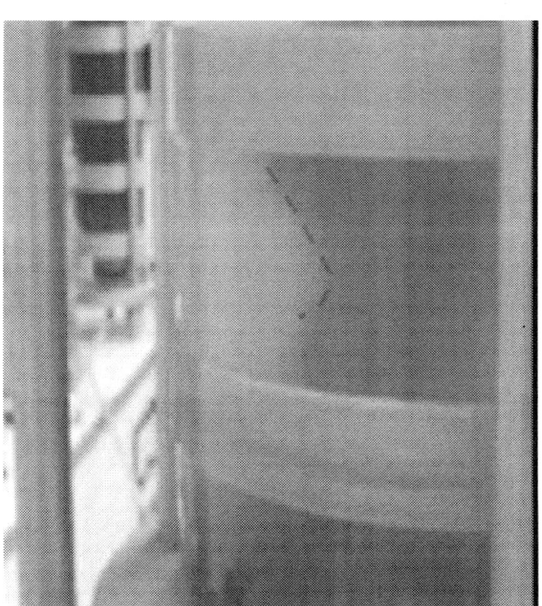

(b)

(c)

Figure 9: Wall/sample prepared without straw. (a) No vertical displacement observed but evident failure; (b) More evidence of failure under the water; (c) Total collapse of the wall/sample.

The tested wall/sample prepared without straw (un-reinforced specimen) collapsed after six days under flooding conditions. There was no change in vertical displacement; however, the external walls of the sample that was under the water failed until a very small diameter of the soil was left. At this point, the weight of the top non-flooded soil was too much for such a small diameter to cope with and failure occurred. The stages of the failure can be observed in Figure 9. The process of failure forming in the flooded part of the wall/sample is evident, and the non-flooded part of the wall/sample can be seen to remain intact until a few moments before total collapse of the structure.

A second wall/sample was molded with 60 blows per layer with straw reinforcing. The FST was executed following cycles of wetting (wall/sample flooded to half of its height) and drying (the water around the wall/sample was drained), as shown in Figure 10. The test duration was 14 days with three cycles of wetting and drying the wall/sample.

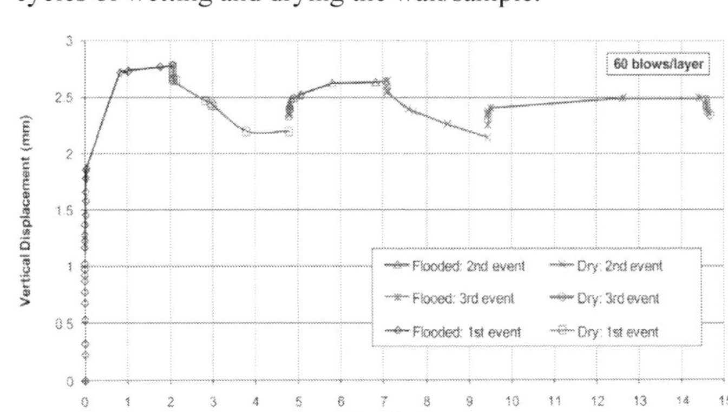

Figure 10: Flooding Simulation Test (FST) results after three cycles of wetting/drying of a wall/sample prepared with 60 blows per layer and straw reinforcing.

Comparing the response after flooding of the wall/sample prepared without straw and the wall/sample prepared with straw (both with 60 blows per layer), it is evident that the straw as reinforcement of the structure has a vital role in the hydro-mechanical behavior of earth structures. It is well known that the fibers (in this case, straw) are added to earth materials mostly to reduce shrinkage and improve tensile strength [20], but this research shows that the straw reinforcement introduces another important benefit to the earth structure. In our research, the fiber (straw) modified the fabric, creating an inter-locking effect that changes the overall behavior of the earth structure when it is exposed to cycles of wetting/drying.

From Figure 10, it is also clear that with the flooding events, the wall/sample presented an increase in volume, followed by a decrease in volume when the water was drained. Keeping in mind that just half of the wall/sample was submerged in water, the first flooding event caused an increase of the vertical height equivalent to 3% of the exposed height under water. The authors believe that this behavior happened because the wall/sample was not loaded (no vertical load was applied to the top of the sample), since it is show in the literature for non-existent or low vertical stresses that the earth/soil can present expansible behavior when wetted [13,17,18,21]. If a load would be introduced to the top of the wall, the authors believe that the wall/sample could present collapsible behavior (sudden reduction of volume by rearrangement of the particles) [14,15,21]. When the water was drained, the wall/samples presented shrinkage; a volumetric reduction due to drying is a typical behavior of materials containing clay [15,16,21]. The wall/sample presented a hydraulic hysteresis [16] due to cycles of wetting/drying (repeated flooding events follow by dry events).

Another problem caused by wetting the earth wall/samples (reinforced by straw) was the growth of rot, as was observed during one of the FSTs after four days under flooding conditions. This shows that the introduction of water generates an environment susceptible to changes in the straw state, seeFigure 11(a).

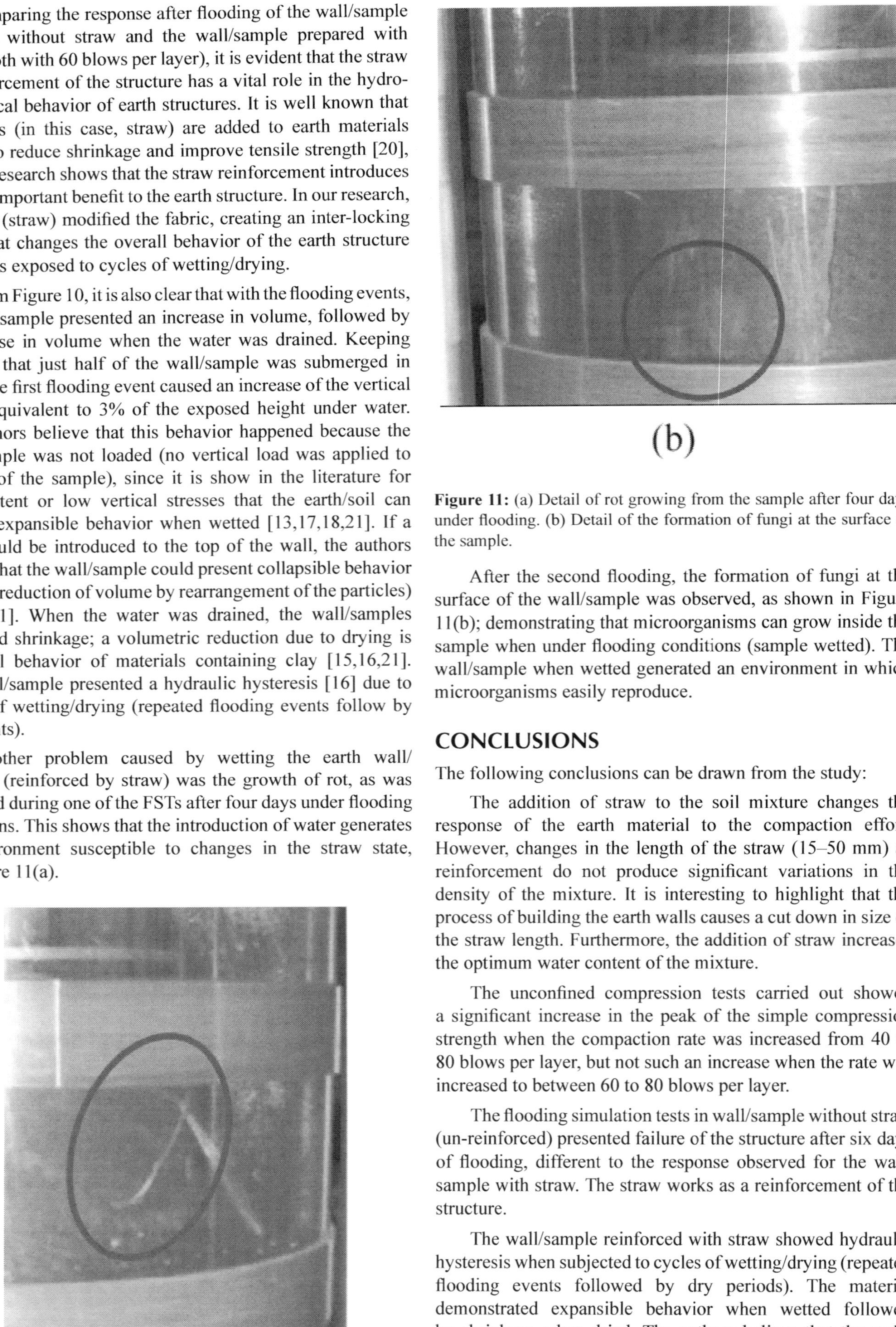

(a)

(b)

Figure 11: (a) Detail of rot growing from the sample after four days under flooding. (b) Detail of the formation of fungi at the surface of the sample.

After the second flooding, the formation of fungi at the surface of the wall/sample was observed, as shown in Figure 11(b); demonstrating that microorganisms can grow inside the sample when under flooding conditions (sample wetted). The wall/sample when wetted generated an environment in which microorganisms easily reproduce.

CONCLUSIONS

The following conclusions can be drawn from the study:

The addition of straw to the soil mixture changes the response of the earth material to the compaction effort. However, changes in the length of the straw (15–50 mm) as reinforcement do not produce significant variations in the density of the mixture. It is interesting to highlight that the process of building the earth walls causes a cut down in size of the straw length. Furthermore, the addition of straw increases the optimum water content of the mixture.

The unconfined compression tests carried out showed a significant increase in the peak of the simple compression strength when the compaction rate was increased from 40 to 80 blows per layer, but not such an increase when the rate was increased to between 60 to 80 blows per layer.

The flooding simulation tests in wall/sample without straw (un-reinforced) presented failure of the structure after six days of flooding, different to the response observed for the wall/sample with straw. The straw works as a reinforcement of the structure.

The wall/sample reinforced with straw showed hydraulic hysteresis when subjected to cycles of wetting/drying (repeated flooding events followed by dry periods). The material demonstrated expansible behavior when wetted followed by shrinkage when dried. The authors believe that the walls could respond differently to flooding if under vertical stress (vertical load on the top of the wall). The authors would like to highlight this behavioral feature has major importance to predict the hydro-mechanical response and long-term life of such structures.

Under flooding conditions, the wall/sample reinforced with straw showed formation of fungi at the surface and some rot growth from the sample was also observed. When wetted, the earth structures produce an environment where microorganisms easily can reproduce.

This work has shown that the structural integrity of earth materials has a certain capacity to resist failure during flooding conditions when a reinforcement material such as straw is used. It must be emphasized that this is influenced by many parameters, including: mixture composition, compaction rates and the nature of the reinforcement utilized.

Future work will be undertaken to investigate the effects of flooding on earth walls under vertical load as well as studying different compaction rates, suction profiles and building procedures.

ACKNOWLEDGMENTS

The authors thank Alastair MacFarlane for general assistance during the experimental program.

REFERENCES

1. Berge, B. *The Ecology of Building Materials*, 2nd ed.; Architectural Press: Oxford UK, 2009.
2. Perroni, C. Prefabricated earth constructions in the UK and Europe. *Struct. Eng.* 2008, *86*, 32–39.
3. Little, B.; Morton, T. *Building with Earth in Scotland: Innovative Design and Sustainability*; Scottish Executive Central Research Unit: Edinburgh, UK, 2001.
4. Weismann, A.; Bryce, K. *Building with Cob: A Step by Step Guide*; Green Books: Dartington, UK, 2006.
5. Jaquin, P.A.; Augarde, C.E.; Gerrard, C.M. *Analysis of Historic Rammed Earth Construction*; Louranco, P.B., Roca, P., Modena, C., Agrawal, S., Eds.; Structural Analysis of Historical Constructions: New Delhi, India, 2006.
6. Medero, G.M.; Forster, A.; Morton, T.; Kennedy, J.; Buckman, J. Cob construction in Scotland: Case study old school house at Cottown, Perthshire. Proceeding of the 1st Mediterranean Conference on Eart Architecture, Cagliari, Italy, 13–16 March 2009; UNESCO World Heritage Centre: Cagliari, Italy, 2009.
7. Hall, M.; Djerbib, Y. Moisture ingress in rammed earth: Part 1—The effect of soil particle-size distribution on the rate of capillary suction. *Construct. Build. Mater.* 2004, *18*, 269–280.
8. Forster, A.; Medero, G.M.; Morton, T.; Buckman, J. Traditional cob wall: Response to flooding.*Struct. Survey* 2008, *26*, 302–321.
9. Williams-Ellis, C.; Eastwick-Field, C.; Eastwick-Field, E. *Building in Cob, Pise, and Stabilised Earth*, reprinted ed.; Donhead Publishing: Shaftesbury, UK, 1999.
10. Reen, K. *Conservation of Earth Buildings in Scotland: The Old Schoolhouse, Cottown, Perthshire*; Kirsty Reen Reports on a Generally Successful Project, Institute of Historic Building Conservation: Tisbury, UK, 1999. Available online: www.ihbc.org.uk/context_archive/63/schoolhouse/cottown.html (accessed on 2 December 2010).
11. Morton, T.; Little, R. *Mixing It with Lime and Clay*; Building Limes Forum: Edinburgh, UK, 2008; Volume 14.
12. *BS 1377-4: Methods of Test for Soils for Civil Engineering Purposes—Part 4: Compaction-Related Tests*; British Standards Institution: London, UK, 1990.
13. Medero, G.M.; Schnaid, F.; Gehling, W.Y.Y. Oedometer behaviour of an artificially cemented highly collapsible soil. *J. Geotech. Geoenviron. Eng.* 2009, *135*, 840–843.
14. Medero, G.M.; Schnaid, F.; Gehling, W.Y.Y.; Gallipoli, D. Analysis of the mechanical response of an artificial collapsible soil. Proceedings of the International Symposium on Experimental Evidence Towards Numerical Modelling of Unsaturated Soils, Weimar, Germany, 18–19 September 2003; Springer: Berlin, Germany, 2004; pp. 135–146.
15. Sivakumar, V.; Tan, W.C.; Murria, E.J.; McKinley, J.D. Wetting, drying and compression characteristics of compacted clay. *Geotechnique* 2006, *56*, 57–62.
16. Wheeler, S.J.; Sharma, R.S.; Buisson, M.S.R. Coupling of hydraulic hysteresis and stress-strain behaviour in unsaturated soils. *Geotechnique* 2003, *53*, 41–54.
17. Sharma, R.S. Mechanical Behaviour of Unsaturated Highly Expansive Clays. Dphil Thesis, University of Oxford, Oxford, UK, 1998.
18. Alonso, E.E.; Lloret, A.; Gens, A.; Yang, D.Q. Experimental behaviour of highly expansible double-structure clay. Proceedings of the 1st International Conference on Unsaturated Soil, Paris, France, 6–8 September 1995; Balkema: Rotterdam, The Netherlands, 1995; pp. 11–16.
19. Cafaro, F.; Cotecchia, F. Structure degradation and changes in the mechanical behaviour of a stiff clay due to weathering. *Geotechnique* 2001, *51*, 441–453.
20. Morton, T. *Earth Masonry*, 1st ed.; BRE Press: Watford, UK, 2008; pp. 43–125.
21. Haghighi, A.; Medero, G.M.; Woodward, P.K.; Laloui, L. Effect of temperature on collapse potential of kaolin clay. Proceedings of the International Conference on Unsaturated Soils, Barcelona, Spain, 6–8 September 2010; Taylor & Francis Group: London, UK, 2010; pp. 543–548.

CHAPTER 6

STRENGTH BEHAVIOR OF MUD BRICK IN BUILDING CONSTRUCTION

Farraj Al-Ajmi, Hany Abdalla, Magdi Abdelghaffar, Jamal Almatawah

Civil Engineering Department, College of Technological Studies, The Public Authority for Applied Education & Training, Kuwait

ABSTRACT

The use of earth construction is well established in energy efficient housing. Mud bricks consist of clay, water, and binding material such as rice husks or straw. The advantages of mud brick include its low-cost and great thermal behavior. Although mud brick is considered one of the oldest construction materials, engineers and builders do not have enough information about its mechanical properties. Also there is no accurate design code to follow before construction. This study is devoted to enhance the low compressive strength of mud brick without sacrificing its low thermal conductivity properties. The experimental program in this research includes the use of different admixtures to increase the compressive strength of the basic mud mix. The experimental results show that the increase of cement ratio, as ingredient to a certain limits, can lead to an optimum compressive strength of the brick.

INTRODUCTION

Earth as mud bricks, has been used in building construction for thousands of years and approximately 30% of the world's present population still live in earthen shelters [1] . Mud brick is an inexpensive, environmentally- friendly and abundantly-available building material. It has been used extensively for building construction around the world, particularly in extreme hot, dry desert climates like that of most Arabian countries [2] . In these countries, mud bricks are made by blending mud and water together into a goopy mixture. Traditionally straw is added to improve tensile strength, and may prevent mud bricks from cracking. The mud brick is then shaped in a mold of almost any size or shape, or by hand. The brick mixtures are then laid to dry in the heat of the sun for about 25 days before use [3] - [6] . Figure 1 shows new unlaid mud bricks in the Jordan River, while Figure 2 shows buildings made from mud bricks in an old town in Niger.

Mud bricks have several advantages over other conventional building materials, e.g., concrete masonry. These advantages include: a very minimal manufacturing process; skilled labor is not necessary; mud is available from natural resources; inexpensive construction materials; and mud structures are able to perform satisfactorily under hot environmental conditions [7] . However, there are many disadvantages in using mud bricks as building materials. These include: mud brick may tend to erode under rain impact; absorption of water causes swelling of mud brick, while evaporation of water from the mud brick gives rise to shrinkage and cracking; and mud brick is a relatively fragile material, which cannot resist earthquake hazards.

Citation: Al-Ajmi, F. , Abdalla, H. , Abdelghaffar, M. and Al-matawah, J. (2016) Strength Behavior of Mud Brick in Building Construction. Open Journal of Civil Engineering, 6, 482-494. doi: 10.4236/ojce.2016.63041.

It is known that mud brick technology has been widely used in desert countries due to the widespread existence of silt and clay deposits in these areas. The importance of using mud brick as a building material, however, has not been seriously investigated through scientific experimentation. Therefore, this study aims to investigate: improvements in mud brick consolidation by increasing its durability; and imparting water penetration to mud brick. In this research, the effect of different mud brick components on strength and thermal conductivity properties are investigated. The basic mixture consists of mud (clay and sand), straw and water. Many additives are provided to the basic mixture in order to improve brick properties.

The overall aim of this study is to determine, through extensive experimental investigation, the effect of different mud brick ingredients on strength and absorption, and utilize these results in a forthcoming study on thermal-conductivity properties. With this in mind, the specific objectives of the study are to improve mud brick consolidation by using materials such as soluble silicate, ethoyl silicate, silanes or siloxanes, isocyanates and various polymers agents to treat mud brick permeability. In addition, the effect of using metallic fibers to improve mud brick durability is investigated.

Figure 1: New unlaid mud bricks in the Jordan river.

Figure 2: Mud brick building in old town in Niger.

MATERIALS AND METHODS

Materials

The materials used in this study for mud brick production were sand and mud as the main matrix, and hay, cement, lime, Bitumen, steel fibers, and steel filing as fibrous materials.

Methods

In this experimental program, physical tests were carried out on natural mud obtained from Saudi Arabia. The objective of these tests was to analyze and investigate the physical properties of the material used in manufacturing the bricks. A sample of the soil material used in this research is shown in Figure 3. A sieve analysis test was performed on the specimen to determine the grading and the ratio of fines. The soil material was washed on a No. 200 sieve. The results of this test showed a relative ratio of sand to fines as follows: % Sand = 8.9% Fines = 91.1%.

Plastic limit and liquid limit tests were carried out on the soil specimen. The plastic limit and liquid limit for the soil specimen were found to be 40.7% and 81.5%, respectively.

Many factors were considered in the experimental program. These included changing the ingredients of the mixture to reach an optimum compressive strength for the produced brick. The variables considered can be divided into the following ten groups, as shown in Tables 1-10, respectively.

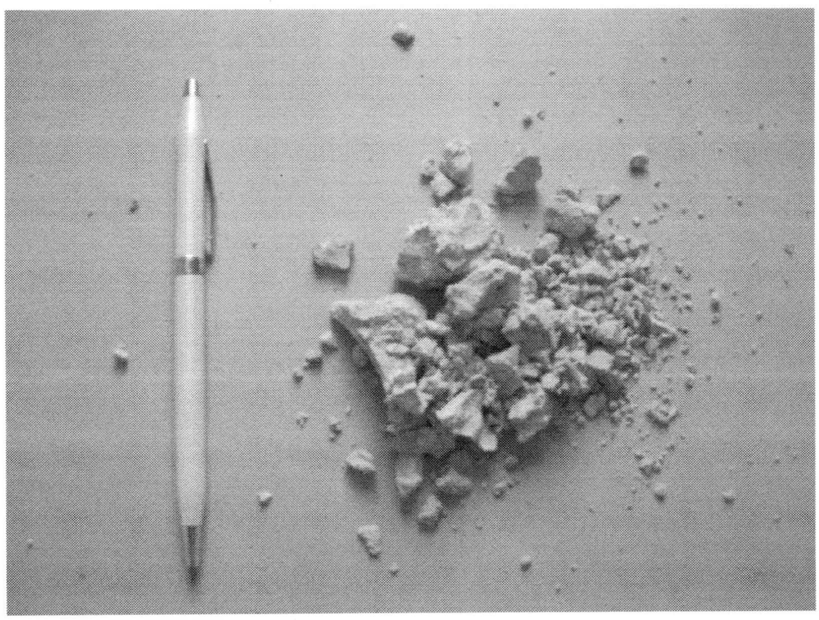

Figure 3: Tested mud.

Table 1: Description of Group 1 (Sand/Mud ratio with constant Hay)

Test No.	Wt. Mud (gm)	Wt. Sand (gm)	Sand/Mud Ratio	Wt. Water (gm)	Wt. Hay (gm)	Av. Wt. Sample (gm)	Av. Load (kN)
1	4000	200	0.05	2000	20	1043	11.3
2	3820	382	0.10	2000	20	1079	12.1
3	3650	548	0.15	2000	20	1080	15.2
4	3500	700	0.20	2000	20	1117	6.7
5	3250	975	0.30	2000	20	1159	4.8
6	3000	1200	0.40	2000	20	1165	1.6
7	2800	1400	0.50	2000	20	1150	0.0
8	2550	1658	0.65	2000	20	1143	4.2
9	2325	1860	0.80	2000	20	1131	0.0
10	2099	2099	1.00	2000	20	1142	0.0

Table 2: Description of Group 2 (Hay ratio with constant Sand/Mud ratio of 0.15)

Test No.	Wt. Mud (gm)	Wt. Sand (gm)	Wt. Water (gm)	Wt. Hay (gm)	Hay/Soil (%)	Av. Wt. Sample (gm)	Av. Load (kN)
11	3650	548	2000	10	0.24	1112	9.0
12	3650	548	2000	30	0.72	1086	1.6
13	3650	548	2000	50	1.20	1076	2.4

Table 3: Description of Group 3 (Cement ratio with constant Sand/Mud ratio of 0.15—No Hay)

Test No.	Wt. Mud (gm)	Wt. Sand (gm)	Wt. Water (gm)	Wt. Cement (gm)	Cement/Soil (%)	Av. Wt. Sample (gm)	Av. Load (kN)
14	3650	548	2000	126	3	1161	12.3
15	3650	548	2000	252	6	1210	16.5
16	3650	548	2000	378	9	1196	21.0
17	3650	548	2000	504	12	1216	21.3

Table 4: Description of Group 4 (Cement ratio with constant Sand/Mud ratio of 0.15 and constant Hay/Soil ratio of 0.48%)

Test No.	Wt. Mud (gm)	Wt. Sand (gm)	Wt. Water (gm)	Wt. Cement (gm)	Cement/Soil (%)	Av. Wt. Sample (gm)	Av. Load (kN)
18	3650	548	2000	126	3	1148	8.9
19	3650	548	2000	252	6	1184	15.7
20	3650	548	2000	378	9	1178	5.1

Table 5: Description of Group 5 (Lime ratio with constant Sand/Mud ratio of 0.15—No Hay)

Test No.	Wt. Mud (gm)	Wt. Sand (gm)	Wt. Water (gm)	Wt. Lime (gm)	Lime/Soil (%)	Av. Wt. Sample (gm)	Av. Load (kN)
21	3650	548	2000	126	3	1137	0.0
22	3650	548	2000	252	6	1169	2.4
23	3650	548	2000	378	9	1165	8.3
24	3650	548	2000	504	12	1200	9.9

Table 6: Description of Group 6 (Lime ratio with constant Sand/Mud ratio of 0.15 and constant Hay/Soil ratio of 0.48%)

Test No.	Wt. Mud (gm)	Wt. Sand (gm)	Wt. Water (gm)	Wt. Lime (gm)	Lime/Soil (%)	Av. Wt. Sample (gm)	Av. Load (kN)
25	3650	548	2000	126	3	1138	6.1
26	3650	548	2000	252	6	1162	8.0
27	3650	548	2000	378	9	1173	3.0
28	3650	548	2000	504	12	1148	2.4

Table 7: Description of Group 7 (Lime ratio with constant Sand/Mud ratio of 0.15 and constant Hay/Soil ratio of 0.48% with Bitumen)

Test No.	Wt. Mud (gm)	Wt. Sand (gm)	Wt. Water (gm)	Wt. Lime (gm)	Lime/Soil (%)	Av. Wt. Sample (gm)	Av. Load (kN)
29	3650	548	2000	252	6	1095	3.1
30	3650	548	2000	378	9	1084	1.0

Table 8: Description of Group 8 (Lime ratio with no Sand constant Hay/Mud ratio of 0.48%)

Test No.	Wt. Mud (gm)	Wt. Sand (gm)	Wt. Water (gm)	Wt. Lime (gm)	Lime/Soil (%)	Av. Wt. Sample (gm)	Av. Load (kN)
31	4200	0	2000	126	3	1114	7.3
32	4200	0	2000	252	6	1110	7.5
33	4200	0	2000	378	9	1136	6.2
34	4200	0	2000	504	12	1152	11.0

Table 9: Description of Group 9 (Adding Steel Fibers to Mud and Sand with no Hay)

Test No.	Wt. Mud (gm)	Wt. Sand (gm)	Wt. Water (gm)	Wt. Fiber (gm)	Fiber/Soil (%)	Av. Wt. Sample (gm)	Av. Load (kN)
35	3650	548	2000	20	0.48	1105	9.1

Table 10: Description of Group 10 (Adding Steel Filing with Lime/Soil ratio of 12% and Hay/Soil ratio of 0.48%)

Test No.	Wt. Mud (gm)	Wt. Sand (gm)	Wt. Water (gm)	Wt. Filing (gm)	Fiber/Soil (%)	Av. Wt. Sample (gm)	Av. Load (kN)
36	3650	548	2000	400	9.5	1175	8.2

1. Group 1: Changing the Sand/Mud ratio with constant Hay ratio (Tests No.1 to No.10).
2. Group 2: Changing the Hay ratio with constant Sand/Mud ratio (Tests No.11 to No.13).
3. Group 3: Changing Cement ratio with constant Sand/Mud ratio (Tests No.14 to No.17).
4. Group 4: Changing Cement ratio with constant Sand/Mud ratio and constant Hay ratio (Tests No.18 to No.20).
5. Group 5: Changing Lime ratio with constant Sand/Mud ratio (Tests No.21 to No.24).
6. Group 6: Changing Lime ratio with constant Sand/Mud ratio and constant Hay ratio (Tests No.25 to No.28).
7. Group 7: Changing Lime ratio with constant Sand/Mud ratio and constant Hay ratio with Bitumen (Tests No.29 to No.30).
8. Group 8: Changing Lime with no Sand and constant Mud and Hay ratios (Tests No.31 to No.34).
9. Group 9: Adding Steel Fibers to Mud and Sand with no Hay (Test No.35).
10. Group 10: Adding Steel filing to mud, sand, Lime and Hay (Test No.36).

MIXING OF RAW MATERIALS

The raw materials used in this study were mixed together to reach optimum brick properties. These properties include compressive strength and water absorption without sacrificing the superior advantage of the mud brick, which is heat performance.

Sand/Mud Ratio

The effect of the sand/mud ratio on the behavior of the produced brick was investigated in tests No.1 to No.10 in Group 1. Each test was carried out on three identical specimens. The average value of these specimens is shown in Table 1. The tests were performed to find the effect of the sand/mud ratio on the compression strength of the bricks. The objective of these tests was to reach an optimum ratio to achieve a high compression strength for the bricks.

In this group, tests were done on thirty mud brick specimens with different sand/mud ratios ranging between 0.05 to 1.0. The total weight of the sand and mud was always kept constant and equal to 4200 gm for each batch used to make three bricks with a constant sand/mud ratio. The hay ratio was kept constant in all the specimens for this group. The hay weight was 20 gm representing 0.48% of the total weight of the mud and sand. Table 1 shows a description for this test group.

The required weights of brick ingredients were prepared for each batch. First, the mud, sand, and hay were mixed in the soil mixer for few minutes, and then water was gradually added. The mud, sand, and hay were mixed with natural tap water having no impurities until reaching a homogeneous paste. The mix was then placed and compacted in steel cubic molds having internal dimensions of 100 × 100 × 100 mm. After casting the cubic samples, all top surfaces were given a smooth final finish by a straight edge. The brick specimens were taken out of the moulds right after casting and left to dry in the air. Figure 4 shows the brick samples after casting.

Hay Ratio

In specimen Group 2, tests were conducted on nine specimens representing tests No.11 to No.13,Table 2. In these tests, the sand/mud ratio was kept constant at 0.15 while changing the hay weight from 10 gm to 30 gm, and 50 gm, representing ratios of hay/soil of 0.24%, 0.72%, and 1.20%, respectively. It should be noted that a hay ratio of 0.48% was examined in test No. 3 of Group 1.

Cement Ratio

Tests No.14 to No.17 of Group 3 were done to examine the effect of adding cement instead of hay as a bonding material to the basic components of the brick (sand and mud). The sand/mud ratio was kept constant at 0.15 in the four tests. The cement ratios considered were 3%, 6%, 9%, and 12%, and are shown in Table 3.

Cement and Hay Ratio

In specimens of Group 4, the effect of adding cement together with hay as bonding materials was considered. In tests No.18 to No.20, a sand/mud ratio of 0.15 and a hay ratio of 0.48% were kept constant in all tests of this group. The cement/soil ratio was varied at 3%, 6%, and 9%, as shown in Table 4.

Lime Ratio

Tests No.21 to No.24 of group 5 examined the effect of adding lime instead of hay as a bonding material to the basic components of the brick (sand and mud). The sand/mud ratio was again kept constant at 0.15 in the four tests. The lime ratios considered were 3%, 6%, 9%, and 12%, as shown in Table 5.

Lime and Hay Ratio

In specimens of group 6, the effect of adding lime and hay as bonding materials was considered. In tests No. 25 to No. 28 H, a sand/mud ratio of 0.15 and a hay ratio of 0.48% were kept constant in all tests of this group. The lime/soil ratio was varied at 3%, 6%, and 9%, as shown in Table 6.

Lime, Hay, and Bitumen

In specimens of Group 7, the effect of adding lime, hay, and bitumen as bonding materials was considered. In tests No.29 and No.30, a sand/mud ratio of 0.15 and a hay ratio of 0.48% were kept constant in all tests of this group. The lime/soil ratio was varied at 3% and 6%. Bitumen was added to the specimen until the whole mixture turned black, as shown in Table 7.

Figure 4: Brick samples after casting.

Lime and Hay (No Sand)

In specimens of Group 8, the effect of removing the sand from the mixture was considered. The specimens of this group consisted of mud, water, lime, and hay only. The weight of the sand was exchanged for the same weight in mud. This resulted in a ratio of hay to mud of 0.48% as in the previous groups. The lime ratios in tests No. 31, No. 32, No. 33, and No. 34 were 3%, 6%, 9%, and 12%, respectively, as shown in Table 8.

Steel Fibers

Test No.35 of Group 9 examined the effect of adding steel fibers instead of hay as a bonding material to the basic components of the brick (sand and mud). The sand/mud ratio used in this test was 0.15. The fibers ratio considered was 0.48%, which is the same ratio of hay used in the other tests, as shown in Table 9.

Lime, Hay, and Steel Filings

Test No.36 of Group 10 examined the effect of adding steel filings to the lime and hay as bonding materials, along with the basic components of the brick (sand and mud). The sand/mud ratio used in this test was 0.15. The lime ratio was 12%, the hay ratio was 0.48%, while the filings ratio was 9.5%, as shown in Table 10.

EXPERIMENT AND TEST PROCEDURE

The testing procedure investigated the strength behavior of the mud brick samples. Experimental measurements were directed to determine the propagation of cracks with an increase of load, the crack pattern before failure, and the ultimate capacity of the brick samples. A California Bearing Ratio (CBR) testing machine was used to determine the compressive strength of the brick samples,Figure 5. Also shown in the figure is the proving ring attached to the vertical jack. This ring, having a capacity of 28 KN was used to monitor the vertically-applied load. All the specimens were loaded until failure.

TEST RESULTS

Tables 1-10 show the failure load for all the specimens. During increases in load, cracks were formed at sides of the tested specimens. Crack width and length increased and eventually lead to final collapse. Figure 6 shows the typical cracking pattern at failure.

Figure 5: Test setup.

Figure 6: Crack pattern at failure.

Figure 7 shows the effect of increasing the sand/mud ratio on the compressive strength of the produced brick. The results of Group 1 show that average failure load increases with an increase of the sand/mud ratio up to a ratio of 0.15. Increasing this ratio above 0.15 led to a decrease in the compressive strength of the brick samples. At this optimum ratio, the average failure load of the mud brick samples was 15.2 KN with a compressive strength of 1.52 MPa. As such, this sand/mud ratio was kept constant in the subsequent testing stages. Various materials were added to the basic sample consisting of sand, mud, and water with the sand/mud ratio kept constant 0.15.

The ratio of hay was the only variable that changed in group 2 of the tests. This group tested hay percentage ratios of 0.24, 0.48, 0.72, and 1.2. Figure 8 shows the effect of increasing the hay ratio in a basic sample. The figure shows that the optimum hay ratio was 0.48% for the basic sample consisting of sand, mud, and water. This ratio was then kept constant in the following test samples consisting of sand, mud, water, and hay.

In group 3, the hay was exchanged for cement as a bonding material. Figure 9(a) shows that increasing the cement ratio has the effect of increasing the compressive strength.

Comparing Figure 8 and Figure 9(a) shows that the basic sample of sand, mud and water needed cement at 6% to reach the same compressive strength of the basic sample with only 0.48% of hay. This indicates that hay is more effective than cement in increasing compressive strength. Moreover, it is well known that increasing the cement in the mixture has a negative effect on the thermal properties of the produced brick [7] .

In group 4, tests for cement and hay were added together to the basic sample. The hay ratio was kept at the optimum ratio of 0.48%, while the cement ratio was varied from 3% to 9%. The results are shown in Figure 9(b). The figure shows that increasing the cement ratio above 6% decreased load capacity. Moreover, at a cement ratio of 6% the load capacity was 15.7 KN with an increase of only 3% in the load capacity when compared to the sample with hay only without cement.

The hay was exchanged by lime as a bonding material in Group 5. The results are shown in Figure 10(a). The results show that increasing the lime ratio increases the load capacity of the brick samples. In Group 5, there was no hay added to the basic mixture.

In Group 6, lime was added to the basic mixture containing sand, mud, and water, in addition to the hay. The hay ratio was kept equal to the optimum ratio of 0.48% in all specimens of this group.Figure 10(b) shows the effect of increasing the lime ratio to the basic components of the brick containing hay. The results show that increasing the lime ratio above 6% had a negative effect on the load capacity of the brick. Comparing Figure 9(b) and Figure 10(b) for specimens with hay in addition to cement or lime, respectively, both figures show a similar trend for compressive strength. This means that in the case of a basic mix with hay, adding an extra bonding material such as cement or lime slightly increases the load capacity up to a limit, after which the strength decreases.

Figure 11 compares the results when adding cement or lime to the basic mixture without hay. The results show that without hay, the cement is more effective than lime in increasing the compressive strength of the produced brick.

The specimens of Group 7 were similar to those of Group 6, except that liquid bitumen was added to the specimens of Group 7. In this case, the brick mix consisted of sand, mud, water, hay, lime, and bitumen. The results in Table 7 show that the addition of bitumen to the mixture with lime and hay had a negative effect on the compressive strength of the brick.

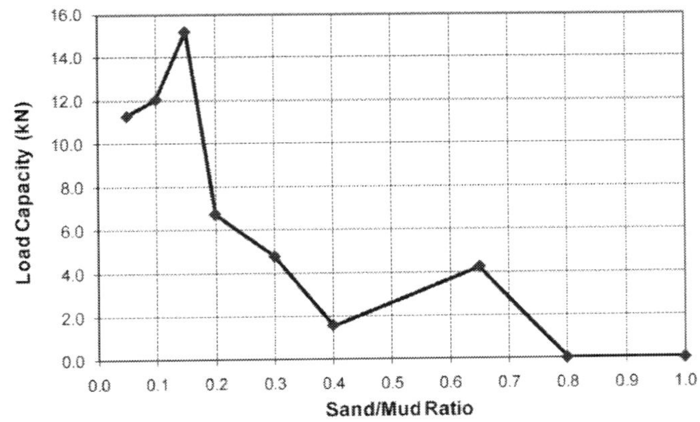

Figure 7: Effect of sand/mud ratio on compressive strength.

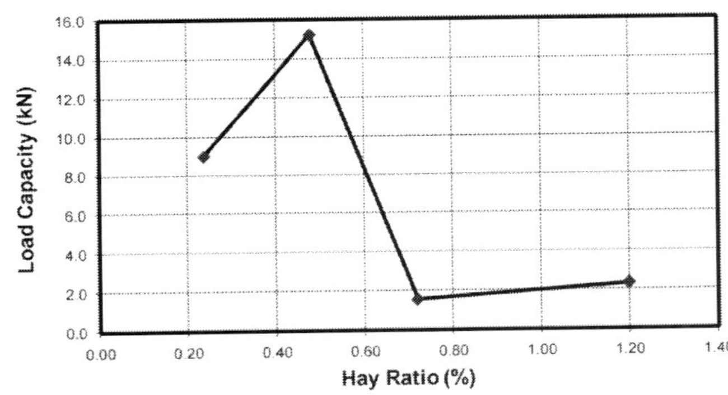

Figure 8: Effect of hay ratio ratio on compressive strength.

The ingredients of the samples of Group 8 are shown in Table 8. The samples of this group have no sand. Lime was added with different ratios to mud, water, and hay. The test results of this group are shown in Figure 12. The results show that increasing the lime ratio leads to an increase in the compressive strength. In Figure 13, a comparison is made between the results of Group 6 containing sand and group 8 with no sand. This figure shows that exchanging the sand with lime had a positive effect on the load capacity of the bricks.

The effect of using different bonding materials in the basic sample consisting of sand, mud, and water is shown in Figure 14. Three types of bonding materials, namely, (hay), (steel fibers with no hay), (lime, hay, and steel filings), were used in the comparison shown in Figure 14. The results show that hay acts well when added to mud and sand to produce bricks.

Table 11 shows an overall analysis for the ten groups of samples considered in this study. The average load of each group was compared to control Group 2 where the samples consisted of a sand/mud ratio of 0.15 with a variable hay ratio. The results show that the major change in compressive strength was achieved in Group 3, where hay was replaced by cement as a binding material.

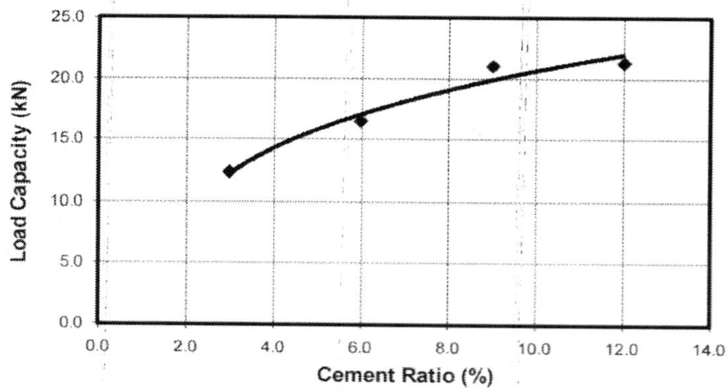

(a)

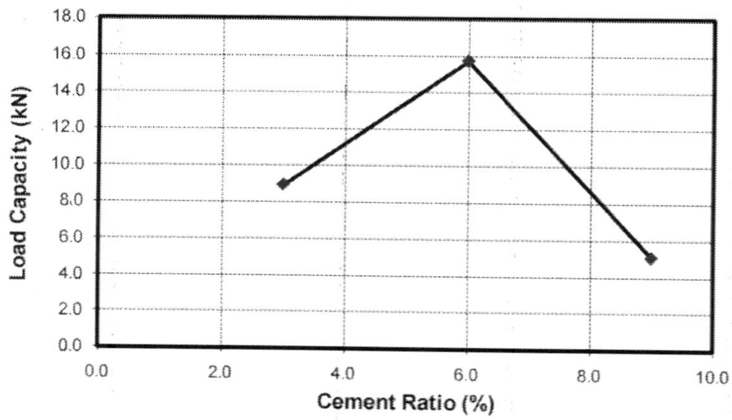

(b)

Figure 9: Effect of cement ratio on compressive strength: (a) without hay, and (b) with hay.

Table 11: Compressive strength analysis

Group No.	Variables Considered	Average Load (KN)	% Change compared to Group 2
1	Sand/Mud ratio	5.90	36%
2	Hay Ratio	4.33	-
3	Cement Ratio	17.78	310%
4	Cement Ratio with Hay	9.90	129%
5	Lime Ratio	5.15	19%
6	Lime Ratio with Hay	4.88	13%
7	Lime Ratio with Bitumen	2.05	-52%
8	Lime Ratio with no Sand	8.00	85%
9	Steel Fibers	9.10	110%
10	Steel Filing	8.20	89%

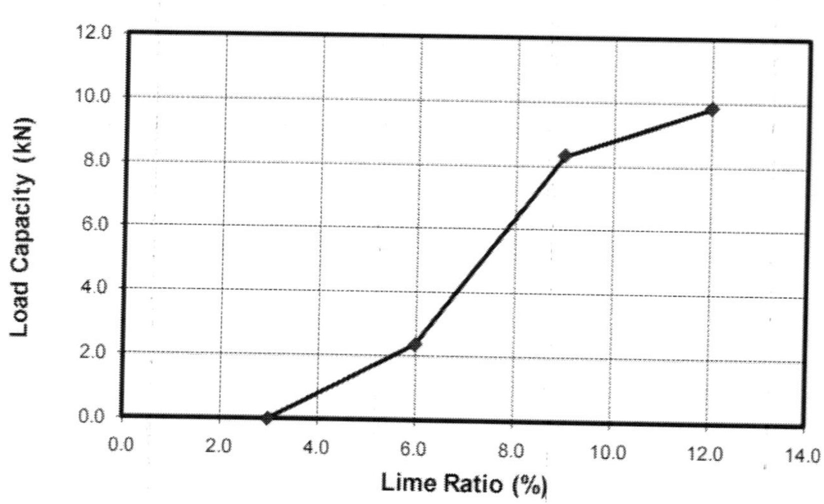

(a)

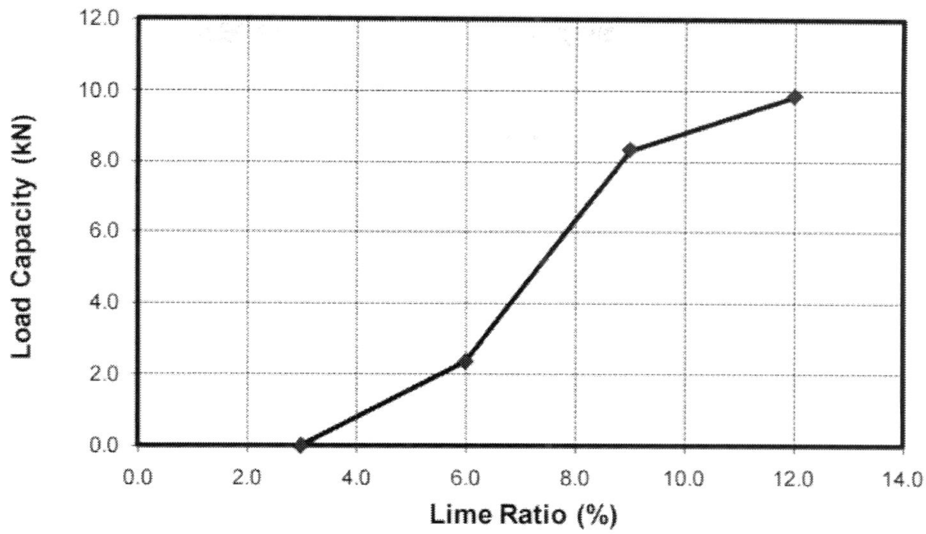

(b)

Figure 10: Effect of lime ratio on compressive strength: (a) without hay, and (b) with hay.

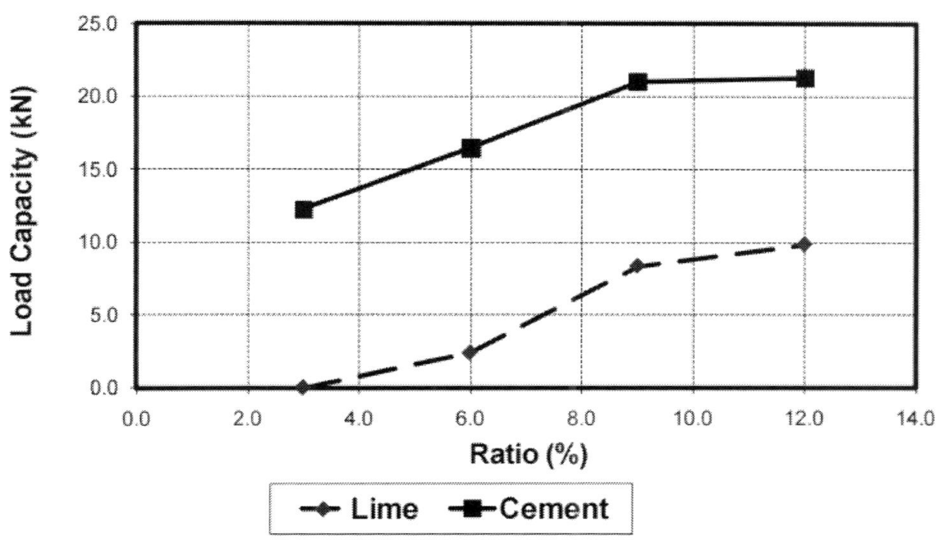

Figure 11: Effect of lime and cement ratios on compressive strength—no hay.

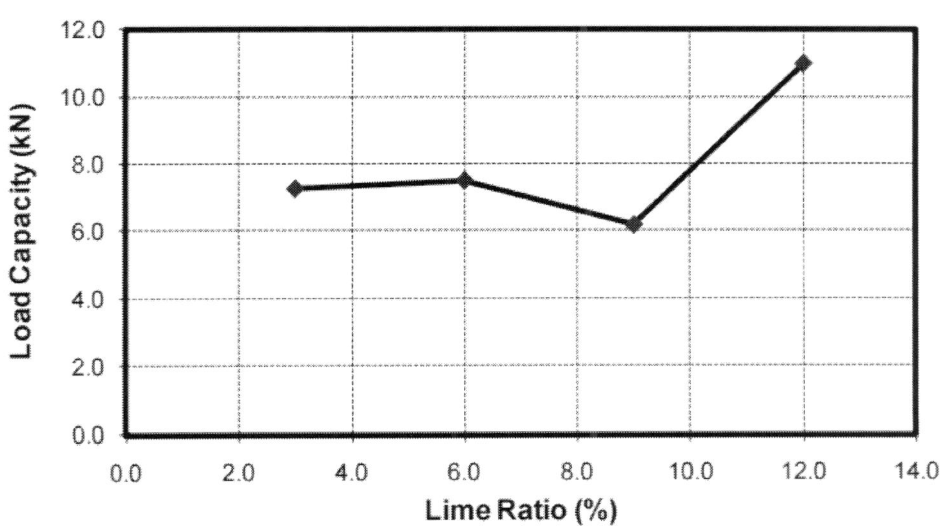

Figure 12: Effect of lime ratio on compressive strength—no sand.

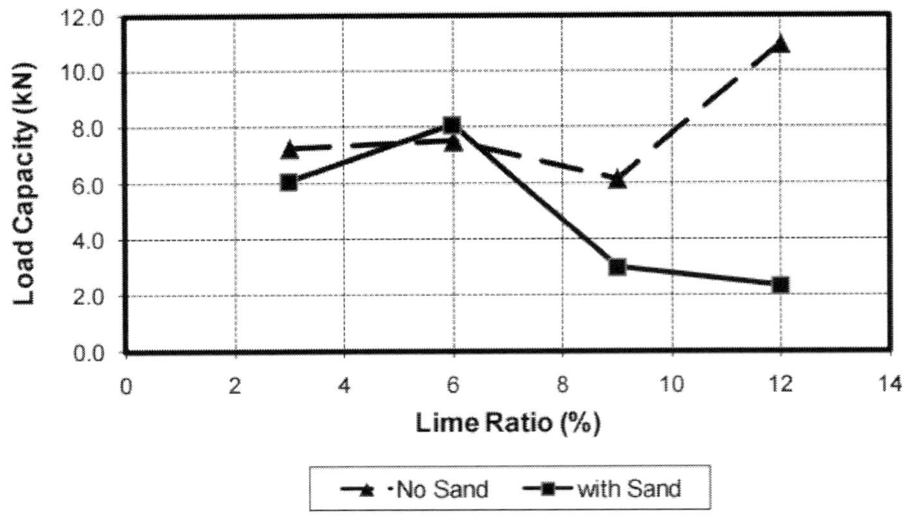

Figure 13: Effect of lime ratio on compressive strength.

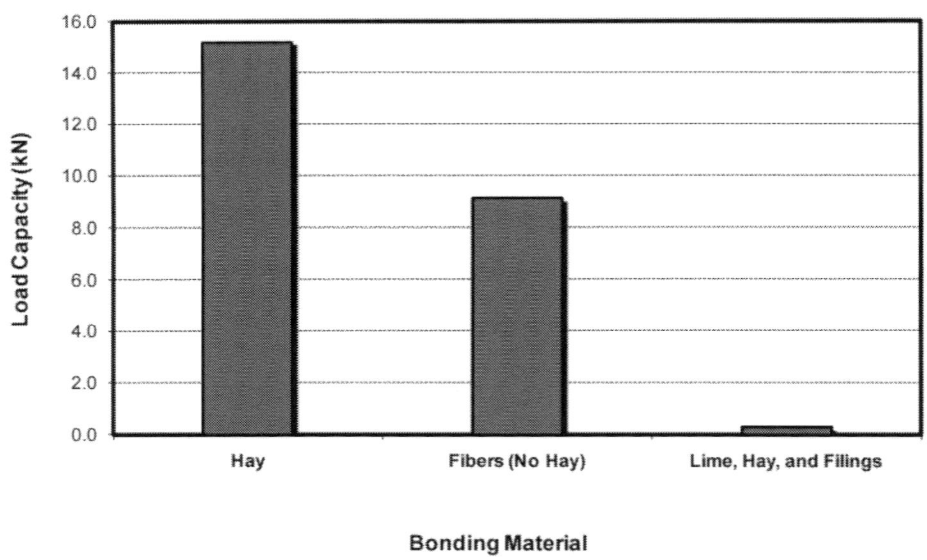

Figure 14: Effect of bonding material on compressive strength.

SUMMARY AND CONCLUSIONS

The compressive strength of mud brick was examined in this study. More than ninety tests were conducted to derive the optimum load capacity of the produced brick. The basic contents of the brick samples were mud, sand, and water. Different bonding materials were added to the basic brick mixture to reach the highest compressive capacity without sacrificing the initiative thermal property. The bonding materials considered in this study were: hay, cement, lime, bitumen, steel fibers, and steel filings. The results from the experimental investigation gave the following conclusions:

1. The optimum sand to mud ratio was 0.15 by weight. Increasing this ratio above 0.15 led to a decrease in the compressive strength of the brick samples.
2. Adding hay to the basic sample of mud, sand, and water had a great effect on the compressive strength of the produced brick. The optimum ratio of hay was 0.48% of the total weight of the sand and mud. Adding cement to the basic contents of the brick sample led to an increase in the compressive capacity of the produced brick. However, increasing the cement ratio might not be preferable due to its negative impact on thermal performance.
3. In case of basic specimens with hay, adding an extra bonding material such as lime or cement slightly increased the load capacity up to a certain limit, then compressive strength decreases.
4. Using steel fibers or steel filings as bonding materials did provide a notable increase in the load capacity of the mud brick.
5. The results showed that the major change in compressive strength was achieved in Group 3, where hay was replaced by cement as a binding material.

REFERENCES

1. Cofirman, R., Agnew, N., Auiston, G. and Doehne, E. (1990) Adobe Mineralogy Characterisation of Adobes from Around the World. In: Cruces, L., Ed., 6th International Conference on the Conservation of Earthen Architecture, Las Cruces, 14-19 October 1990, 424-429.

2. Al-ajmi, F., Hanby, V. and Loveday, D. (2002) Thermal Performance of the Sub-Soil Environment in Dry Desert Climate. ASHRAE Transactions, 108, 395-405.

3. Acostaa, I., Aineto, M., Romerob, M. and Rincon, J.M. (2002) Utilisation of IGCC Slag and Clay Steriles in Soft Mud Bricks (by Pressing) for Use in Building Bricks Manufacturing. Waste Management, 22, 887-891. http://dx.doi.org/10.1016/S0956-053X(02)00075-2

4. Ren, K.B.D. and Kagi, A. (1995) Upgrading the Durability of Mud Bricks by Impregnation. Building and Environment, 30, 433-440.

5. Al-awadhi, S. and Sayigh, A. (1990) The History Buildings Materials in Kuwait and Its Suitability to Local Climate. In: Sayigh, A.A.M., Ed., Energy and Environment into the 1990s, World Renewable Energy Company Limited, 4.

6. Kuwait Institute for Scientific Research (KISR) Lab. (2006) Test for Thermal Conductivity Measurements. Report Submitted to Dr. Farraj Al-ajmi.

7. Abu-Hamdeh, N.H., Khdair, A.I. and Reeder, R.C. (2001) A Comparison of Two Methods Used to Evaluate Thermal Conductivity for Some Soils. International Journal of Heat and Mass Transfer, 44, 1073-1078.

CHAPTER 7

SUSTAINABLE USE OF TEPETATE COMPOSITE IN EARTHEN STRUCTURE

T. López-Lara, Juan Bosco Hernandez-Zaragoza, Jaime Horta, Eduardo Rojas Gonzalez, Carlos Lopez-Cajun, and Gerson Ramirez

Postgraduate Division, Faculty of Engineering, Autonomous University of Queretaro, Cerro de las Campanas S/N, Colonia Niños Héroes, 76010 Querétaro, QRO, Mexico

ABSTRACT

One of the best indicators for construction sustainability is the use of earthy local materials which are completely recyclables and savers of energy during their life cycle. Tepetate is an underestimated earth-natural material, vast and economic, used only in a compacted form in backfills for layers of low resistance in pavements and platforms of buildings. This volcanic soil, named in different ways in several countries, is found in the central region of Mexico. Its resistance as compacted material is very low, of the order of 0.08 MPa. In this work, an improved sustainable-tepetate composite, using CaOH, is presented. This research includes the determination of mechanical properties as well as the physicochemical characterization of the sustainable-tepetate composite behavior. It can be concluded that the strength of the proposed composite increases significantly, immediately after treatment and with time. X-Ray Diffraction shows that all the mineralogical phases prevail in the natural tepetate and only a new phase appeared (calcite), which increases with time. This and the reaction of CaOH with clay content are very likely associated with the continuous strength increase of the composite.

INTRODUCTION

Construction based on earthy materials is the most important, since these are natural, sustainable, and plenty in any region of the world. Unfortunately, the applications of an earthy material, in general, depend on their mechanical properties in their pure natural form. This fact makes them of limited applications, which leads to be used only as fills in platforms. One example of this kind of materials is the tepetate. However, there is a potential of being used as a composite via the use of other economical and sustainable materials for constructions such as the CaOH. Indeed, CaOH is a material being used from ancient times. There are several antique constructions as bridges, castles, and aqueducts built with stones and CaOH as joining material. These constructions have lasted for centuries and they still can be seen nowadays. The previous shows their long useful life.

Moreover, CaOH has a wide field of applications, both, in industry (paper, paints, etc.) and construction (it decreases soils expansion and increases strength on pavements). It is also used as fertilizer, metallurgy, water treatment, glass manufacture, and so forth.

In this work, CaOH was used to improve the mechanical properties of natural tepetate. This is an economic and vast volcanic soil used in construction only as backfill material.

Citation: T. López-Lara, Juan Bosco Hernandez-Zaragoza, Jaime Horta, Eduardo Rojas Gonzalez, Carlos Lopez-Cajun, and Gerson Ramirez, "Sustainable Use of Tepetate Composite in Earthen Structure," Advances in Materials Science and Engineering, vol. 2013, Article ID 806387, 6 pages, 2013. doi:10.1155/2013/806387.

BACKGROUND

The use of earth on site as a building material saves manufacturing cost, time, energy, environmental pollution, and transportation cost. Earth can also be used in the construction of low cost sustainable houses which is significantly cheaper than using conventional bricks. Selected soil (sand and clay) is mixed with water to the correct proportions and then placed in a hydraulic or hand operated compressing machine to produce compressed blocks, which, after curing, is used for building walls [1, 2].

Tepetates have been reported in several countries, mainly in Latin America. They have been denoted with different names such as silcrete in the USA; talpetate in Nicaragua; hardpan, duripan, and cangahua in Colombia and Ecuador; cancagua, moromoro, tosca, and ñadis in Chile; hardpan in Perú; and kora and masa in Japan [3].

Tepetate is a word derived from the Nahuatl tepétlatl, whose root tetl means stone and pétlatl, petate. Literally it has been translated as "stone-petate," "stone-like," or "soft-rock." For the aztecs, this word was included within their classification of materials of a type of soil for agriculture, hard to work [4]. However, when the Spaniards arrived to Mexico, the word tepetate was synonymous of agriculture-useless because of its low quality [5].

In Mexico there exists a great amount of materials denoted as tepetates. These occupy approximately 30% of the surface of the whole country [6, 7], either on the surface or in the first meter deep [8]. Tepetates are interstratified volcanic banks, including paleosoils of different nature characterized by horizontal boundaries and well defined between the layers. Tepetates of the Mexican central plateau have been described as massive, compacted and hard, natural formations, and cemented for several chemical agents, including clays and silicates [3]. Tepetates classification has been based on the type of cementing substance. When the cementing substance is mainly silica (SiO_2), the tepetates are denoted "duripan"; when is calcium carbonate ($CaCO_3$), the tepetates are denoted "petrocalcic" type; when is calcium sulfate ($CaSO_4$), they are denoted "petrogypsic" type; when are salts, they are denoted "petrosalt" type; when is iron, they are denoted "petroferric" type and when is clay, they are denoted "fragipan" type. Tepetates are originated basically from volcanic materials that have been cemented or compacted, as consequence of three main processes: (1) consolidation of mineral particles provoking compaction; (2) nature of the pyroclastic materials consolidated at the instant of being deposited; (3) hardening by pedologic (sedimentary) processes which produce cementing substances in solution [6, 7].

Several works on Mexican tepetates have been reported in the literature. Most of them have been based in their physical and chemical characterization [9–11]. In most of those it reports is assessed that tepetate is a matrix composed of sand, lime, and small percentages of clays. However, once in a while they can exhibit high content of clays. This variability leads to a problem for work-rehabilitation of tepetates for each textural type generates an independent physical and mechanical behavior. Clearly, this fact requires a specific investigation. Some of their reports were based on micromorphological characteristics [12] which have a distribution related to gross and fine particles, simple porphyritic, and morphology of peds in subangular blocks suggesting a reorganization of the basal mass. There are also reports focused on the cementing substance identification, the use of soil, and its ecological significance [6, 7].

Some authors have made reference to the relief, climate, and pedologic origin of tepetates [13, 14]. As a matter of fact, it has been determined that the secondary hardening happens under template or semiarid conditions, This climate condition allows the liberation of compounds. The liberated compounds are lixiviated and deposited and then, they act as cementing substance (free silica, calcium carbonate) in the soil. These are particularly efficient if in the matrix of the horizon where they are deposited, are absorbed by clay. Tepetates are formed, preferably, in subhumid climates (annual precipitation less than 800 mm), characterized for a dry season lasting from four to six months [14], where, in general, evapotranspiration is greater than the precipitation.

There are also a great amount of studies about its agrological properties (important for agronomy). Indeed, since the pre-Hispanic age there have been some attempts for using them in agriculture, by breaking them up and fertilizing them [15]. This is so because they have a very low content of organic matter, nitrogen, and phosphorous, which makes it difficult for using them in agricultural production [16]. Some studies have been on tepetate taxonomy; for example, [17] reported that in the lower level, known as generic, there were kinds of soils that can be located in the group of work soils as well as in the nonwork soils. This double inclusion is exemplified with the materials named as tepetates. About its classification and geological origin [6, 18], tepetates are originated from old deposits of volcanic ashes, deposited in situ or reworked, which were subjected to both processes, namely, diagenesis (hydroconsolidation) and pedogenesis that contributed to its compaction and/or cementation.

On the other hand, tepetates below soils produce lithological discontinuity that block the water infiltration and favor the lateral leakage, marking a surface where slides are prone. As well, tepetate could favor erosion and prevent aquifers reloading [19].

In 1996, while the III International Symposium of Hardened Volcanic Soils was celebrated at Quito, Ecuador, it was proposed to characterize tepetates as a hardened horizon, of volcanic origin, whose material is basically composed of pyroclastic materials, or fluids, or else as degraded volcanic soils.

Based on research done for several authors [9–11], it is possible to infer that, independently of its origin, tepetates always show common physical, mechanical, and chemical properties among them. Its compaction or cementation should be highlighted, which are reflected in high apparent densities (16.67–18.63 KN/m^3), low porosity (from 13% to 24%), and its hydraulic conductivities and holding low humidity. These characteristics limit significantly the fast incorporation of primary plants, and generate a constant soil erosion.

Llerena [20] and García [21] considered tepetates of the Mexican Valley as pumite fragments of the Terciary or Cenozoic in process of weathering. Valdés [22] mentioned that tepetates of the Mexico were formed due to the alluvial sediments that lately were consolidated. Rodríguez et al. [23] noticed that the results reported by the aforementioned author lack sufficient precision and left questions about the hardened layers, concluding that tepetates may have diverse origins.

Zebrowski [3] states two geological processes for explaining the horizons hardening, as follows.

a. Simple consolidation-compaction or by the hydroconsolidation of volcanic materials transported by the water. In both cases, there always exists an increase of the apparent density material, a greater hardness, and consequently, a decrease of porosity.
b. Hardening of volcanic materials at the instant of their deposit and posterior cooling (pyroclastic flows) is another.

These natural deposits, called tepetate in Mexico, have very low permeability and the overlying soil erodes rapidly when cultivated. Unconfined compression varied from 0.29 to 15.7 MPa, and the mean was 2.42 MPa. The strongest, most indurated tepetate appears to occur when disseminated carbonates combine with silica as cementing substances [24]. The reported properties are circumstantial and depend on the geological consolidation and type of cementing substance.

As construction material, the properties of tepetates show a wide range of values, thus making it less reliable as homogeneous material for construction. Adding, it has a very low strength as construction material [25]. The properties of Tepetate depend on the volcanic origin of this material, mainly through phreatomagmatic eruptions. As construction material, 80% of the tepetates are classified as silty sand according to the Unified System of Soil Classification with no plasticity [26].

About problematic soils used in embankments for sustainable pavements, there is a study for bituminous sand materials where their high bitumen content makes oil sand materials problematic for field operations [27]. Then, about expansive soils there is a study on their engineering properties as subgrade when stabilized by using different percentage of hydrated lime on thickness of pavement structural system. The results suggest that the highly strength, lowest swelling and small thickness of pavement were determined with an optimum percentage of lime content of 6% [28].

CaOH is a natural linking material in the preparation of mortars for construction. Indeed their usage goes back to ancient times. Developed countries specify for construction in seismic zones the compulsory use of CaOH in mortars because of their unique features of adherence and strength for sustaining diagonal tensions.

Up to the industrial revolution and the cement discovery in 1824 at Portland, UK, CaOH had been the main link for construction in mortars, stuccos, and paints. Due to the limited facility of transportation, constructors applied the local material but they knew a wide spectrum of tricks for correcting the effects of each of the found CaOH and producing the mortars with the required quality in each application case, such as the control of speed for solidification, hardness, and the degree of waterproofing [29].

Lime treatment in clay soils can be explained by two types of chemical reactions that occur when lime is added to a wet soil: short-term reactions known as soil improvement or modification consisting of cation exchange and flocculation and the long-term reaction known as stabilization/solidification: the pozzolanic activity. During the cation exchange, the highly alkaline environment produced by the addition of lime causes silica and alumina to be dissolved out of the structure of the clay minerals and to combine with the calcium to produce new cementitious compounds: calcium silicate hydrates, calcium aluminate hydrates, and calcium aluminosilicate hydrates. The calcium hydrosilicates contribute substantially to the strength of the stabilized soil material and are of varying composition. These reactions contribute to flocculation by bonding adjacent soil particles together and strengthen the soil with curing time [30, 31].

EXPERIMENTAL METHODOLOGY

Soils Samplings

Tepetate was taken from one of the large active banks of Queretaro, QRO, Mexico. It should be pointed out that this location was selected because tepetate was being used for backfills in residential areas and also as layers for base or subbase of pavements. The place is known as "Mompani," having geographical coordinates: latitude 20°39′6.90″N, longitude 100°28′29.61″W, and a height of 1905 m above sea level. The CaOH used was of the commercial type.

Procedure

After extracting the material from "Mompani," the index and mechanical properties were determined, following the procedure set by the American Society for Testing and Materials (ASTM):(i)gradation (size of aggregates) [32];(ii)liquid limit, plastic limit, and index limit (material plasticity) [33];(iii) classification by the Unified System for Soil Classification (it defines soil type) [34];(iv)proctor Standard Compaction (it determines the maximum dry density and optimum humidity of the material for obtaining the greater strength) [35];(v) unconfined compressive strength (material strength) [36].

Then, the physicomechanical behavior of the sustainable-tepetate composite using CaOH was studied, by performing the tests listed as follows:(i)proctor Standard Compaction [35];(ii)unconfined compressive strength at different ages [36].

Finally the physicochemical characterization of natural tepetate and the sustainable-tepetate composite was done via X-Ray Diffraction.

RESULTS

Index and Mechanical Properties of the Natural Tepetate under Study

Gradation was made by the dry mechanical method. It was determined using sieve analysis. With this method, the particles distribution can be quantified starting from the larger ones (75 microns, retained in sieve N°200) [32]. The results of soil gradation were graves (23.22%), sands (42.55%), and fines (34.23%).

The plasticity properties (Liquid Limit, LL, Plastic Limit, PL, and Plastic Index, PI) were determined for the bank of selected material [33]. The values obtained were LL = 32.45%, PL = 31.38%, and PI = 1.07%. The soil classification was obtained by the plasticity chart. From this, one had that the fines are limes of low plasticity (ML). Then, with the plasticity properties and the gradation (45.5% sands and 32.36% of fines) the soil can be classified as a silty sand (SM) [34]. Since tepetate is a material placed and compacted in the construction site, its ideal compaction was determined via Proctor Standard test [35]. The maximum specific dry weight was 12.67 KN/m^3 with an optimum humidity of 28.5%. The strength of natural tepetate was measured to 0 days and was 0.082 MPa [36].

Properties of the Sustainable-Tepetate Composite

Proctor Standard Compaction

The Proctor Standard test [35] was made for the sustainable-tepetate composite with different percentages of CaOH (2%, 4%, 6%, 8%, 10%, 12%, 15%, and 20%, with respect to the soil dry weight). These are shown in Table 1. From that table, it can be seen that, in general with the increase of CaOH, the optimum humidity tends to decrease and its maximum specific-dry weight increases slowly. Moreover, it can be seen that the increment of CaOH disintegrates tepetate, easing and increasing the compaction. The optimum moisture of most of the composites is between 26–28.5%, and from 12.47 to 13.18 KN/m^3 of maximum dry unit weight.

Table 1: Materials standard compaction

Table 1: Materials standard compaction.

Material	Maximum dry unit weight (KN/m^3)	Optimum moisture (%)
Natural tepetate	12.67	28.5
Tepetate-CaOH 2%	12.47	27.5
Tepetate-CaOH 4%	12.55	28
Tepetate-CaOH 6%	12.73	27.8
Tepetate-CaOH 8%	12.94	26.5
Tepetate-CaOH 10%	12.79	27.5
Tepetate-CaOH 12%	12.81	28.5
Tepetate-CaOH 15%	13.09	26.5
Tepetate-CaOH 20%	13.19	25.5

Unconfined Compression Strength of Sustainable-Tepetate Composite

The Unconfined Compression Strength Standard test [36] was made for the sustainable-tepetate composite with different percentages of CaOH. Probes of sustainable-tepetate composites were compacted for different CaOH percentages. For this, their dry unit weight and optimum humidity, obtained by the Proctor standard, were applied. Probes were prepared for several periods of time, namely, for 0, 7, 14, 28, 42, 56, 70, 112, and 126 days, as shown in Table 2 and Figure 1. Tepetate strength test was made only to 0 days and the result was 0.08 MPa.

Table 2: Resistance of sustainable-tepetate composite in time

Table 2: Resistance of sustainable-tepetate composite in time.

Resistance material (MPa)	Time (days)								
	0	7	14	28	42	56	70	112	126
Tepetate-CaOH 2%	0.09	0.46	0.53	0.27	0.30	0.34	0.31	0.33	0.45
Tepetate-CaOH 4%	0.50	1.18	1.21	1.50	1.66	1.78	1.72	1.92	1.98
Tepetate-CaOH 6%	0.33	0.95	1.29	1.85	2.35	2.83	2.64	2.66	2.73
Tepetate-CaOH 8%	0.38	1.22	1.52	2.17	3.02	3.86	3.58	3.41	4.17
Tepetate-CaOH 10%	0.44	1.36	1.82	2.97	3.54	4.14	4.12	4.83	4.04
Tepetate-CaOH 12%	0.47	1.24	1.78	2.88	3.51	4.03	4.00	4.44	3.97
Tepetate-CaOH 15%	0.36	1.09	1.46	2.79	3.44	3.93	3.91	4.10	3.96
Tepetate-CaOH 20%	0.32	0.93	1.22	2.30	3.45	4.05	3.80	3.91	3.85

Figure 1: Strength time variation of the sustainable-tepetate composite.

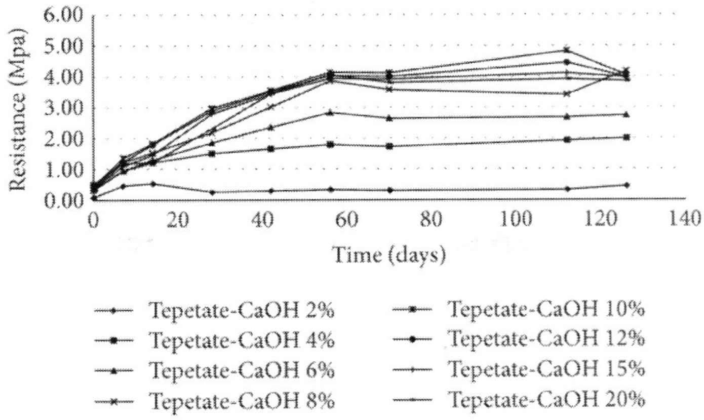

Figure 1: Strength time variation of the sustainable-tepetate composite.

From Figure 1, it is apparent that the strength increases when the amount of CaOH in the composite also increases. The increase of strength with time is also noticeable, and the same happens with clays [37]. However, there is an optimum amount (10%) of CaOH leading to the maximum strength. As time goes on, the composite strength increases significantly, reaching to 2.94 MPa at 28 days and up to 4.21 MPa at 56 days, stabilization time. The increase of resistance of the tepetate is very likely for its clay content. The treatment of clay with lime increases its resistance with the time due to the formation of calcium silicate hydrates, calcium aluminate hydrates, and calcium aluminosilicate hydrates. These reactions contribute to flocculation by bonding adjacent soil particles together and strengthen the soil with curing time.

X-Ray Diffraction (XRD)

X-Ray Diffraction was done with a diffractometer Bruker D8-Advance at 30 KV and 30 mA. Samples of natural tepetate and the sustainable-tepetate composite at 15 and 30 days were analyzed for identifying minerals at the outset and those formed after some time.

Natural Tepetate. X-Ray Diffraction to natural tepetate was done. The study shows that several minerals as the halloysite, hematite, and quartz can be identified. The halloysite is a type of clay with plasticity much greater than the kaolinite. Hematite is a product of contact metamorphism and iron formations and is a common cementing substance in sedimentary rocks, frequently abundant in weathered iron minerals. This is an outcome of the goethite decomposition, which was found in the composite.

Quartz was another mineral found. Actually, this is the main component of sands [6] and was also identified at the tepetate XRD.

Sustainable-Tepetate Composite at 15 Days. The diffractogram of better sustainable-tepetate composite (10% CaOH) at 15 days shows that all the phases of the natural tepetate are present, among them, quartz, hematite, and halloysite (from white to light gray), but a new phase (calcite) appeared as result of the reaction of the CaOH with water.

The use of CaOH with expansive-clay soils, a treatment applied generally to inhibit volumetric instability, showed an increase of strength. Additionally, it is known that such a mixture generates calcite formation, which increases with time [37]. Therefore, it is very likely that the calcite is responsible of the strength composite.

The light gray and white colors of the tepetate seem to be originated by the absence of weathering of the original rock of gray and white colors, deposits of $CaCO_3$, arising from salts or as result of iron removal, leaving great amount of minerals rich in SiO_2 as quartz, feldspar, and kaolinite [6, 7].

Sustainable-Tepetate Composite at 30 Days. The diffractogram of the sustainable-tepetate composite (10% CaOH) at 30 days shows that all the phases of the natural tepetate are present, but the calcite shows an increment with respect to 15 days, because the deflection or peak of the calcite grows in the diffractogram [37].

CONCLUSIONS

The strength of the natural tepetate was 0.08 MPa. It was constituted by graves (23.22%), sands (42.55%), and fines (34.23%).

The mechanical behavior of the sustainable-tepetate composite was improved with a maximum 10% of CaOH. Indeed with this percentage of CaOH, the maximum strength was obtained. After this percentage the strength showed no improvement. Additionally, a significant increase of strength was recorded as time went on. In fact, at 28 days the strength was 2.94 MPa whereas at 56 days, time for stabilization, the strength was 4.21 MPa. Thus, it is apparent that the inclusion of CaOH produced an optimum strength of the tepetate, by improving their mechanical properties in structures of compacted soils. This methodology creates a new sustainable composite using tepetate.

From the X-Ray Diffraction it was inferred that the natural tepetate was constituted mainly by halloysite, quartz, and hematite. These correspond to the gradation of soil with a better definition (weathered sands and halloysite clay). Since there was presence of iron minerals (hematite) and clay (halloysite), the tepetate could be classified as of the petroferric type with some halloysite clay. In the sustainable-tepetate composite with CaOH, all the phases of natural tepetate appeared. However, a new phase formed (calcite), which increased with time, could be associated very likely with the increase of strength of the composite with time. At the same time, the reactions of CaOH with clay content contribute to flocculation by bonding adjacent soil particles together and strengthen the soil with curing time.

The sustainable-composite tepetate offers an alternative construction material because of local construction material, being lasting, resistant, and economical, of low consumption energy in its preparation. Moreover, it is one hundred percent recyclable and thus can be used for reinforced platforms or for block elements.

REFERENCES

1. M. S. Zami and A. Lee, "Using earth as a building material for sustainable low cost housing in Zimbabwe," The Built & Human Environment Review, vol. 1, pp. 40–55, 2008.
2. R. Sangeeta and C. Swaptik, "Earth as an energy efficient and sustainable building material," International Journal of Chemical, Environmental & Biological Sciences, vol. 1, no. 2, pp. 257–261, 2013.
3. C. Zebrowski, "Los suelos volcánicos endurecidos de América Latina," Terra Latinoamérica, vol. 10, pp. 15–23, 1992.
4. Ch. Gibson, Los Aztecas Bajo El Dominio Español, 1519–1810, Siglo XXI, Coyoacán, México, 13 edition, 1996.
5. C. A. Ortiz-Solorio, Los levantamientos etnoedafológicos [Ph.D. thesis], Colegio de Postgraduados, Instituto de Recursos Naturales, Montecillo, México, 1999.
6. D. Flores-Román, A. González-Velázquez, J. R. Alcalá-Martínez, and J. E. Gama-Castro, "Los tepetates,"Revista De Geografía, vol. 3, no. 4, pp. 37–42, 1991.
7. D. Flores-Román, A. González-Velázquez, J. R. Alcalá-Martínez, and J. E. Gama-Castro, "Duripans in the subtropical and temperate subhumid climate of the Trans-Mexico Volcanic Belt," Revista Mexicana De Ciencias Geológicas, vol. 13, no. 2, pp. 228–239, 1996.
8. E. G. Guerrero, M. J. L. Luna, and O. E. Caballero, "Distribución de los tepetates de la República Mexicana escala 1:4 000 000," Terra, vol. 10, pp. 131–136, 1992.
9. G. Miehlich, "Formation and properties of tepetate in the central highlands of Mexico," Terra, vol. 10, pp. 137–144, 1992.
10. H. D. Peña, M. E. Miranda, C. Zebrowski, and M. H. Arias, "Resistencia de los tepetates de la vertiente occidental de la Sierra Nevada," Terra, vol. 10, pp. 164–177, 1992.
11. D. J. Etchevers, M. Rosa, López-Claude, C. Zebrowski, and D. Peña, "Características químicas de tepetates de referencia de los estados de México y Tlaxcala, México," Terra, vol. 10, pp. 171–182, 1992.
12. R. Hessmann, "Micromorphological investigations on "tepetate" formation in the "toba" sediments of the state of Tlaxcala (México)," in Suelos Volcánicos Endurecidos, vol. 10 of Terra, pp. 145–150, 1992.
13. Geissert, "Los tepetates de Xalapa, Veracruz México: relación con el relieve modelado actual y esquema cronológico," Terra, vol. 10, pp. 221–225, 1992.
14. D. Dubroeucq, P. Quantin, and C. Zebrowski, "Los tepetates de origen volcánico en Mexico. Esquema Preliminar de clasificación," Terra, vol. 7, pp. 1–13, 1989.
15. S. Rodriguez-Tapia, C. A. Ortíz-Solorio, C. Hidalgo-Moreno, and M. C. Gutiérrez-Castorena, "Los tepetates de la ladera oeste del cerro Tláloc: saprolita, sin endurecimiento pedológico," Terra. Latinoamérica, vol. 22, no. 1, pp. 11–21, 2004.
16. J. D. Álvarez-Solís, R. Terrera-Cerrato, and J. D. Etchevers-Barra, "Actividad microbiana en tepetate con incorporación de residuos orgánicos," Agrociencia, vol. 34, no. 5, pp. 523–532, 2000.
17. B. J. Williams and C. A. Ortiz-Solorio, "Middle American folk soil taxonomy," Annals, Association of American Geographers, vol. 71, no. 3, pp. 335–358, 1981.
18. T. J. Nimlos and C. Ortiz-Solorio, "Tepetate: the rock mat," Journal of Soil & Water Conservation, vol. 42, no. 2, pp. 83–86, 1987.
19. J. E. Gama-Castro, E. McClung de Tapia, E. Solleiro-Rebolledo et al., "Incorporation of ethnopedological knowledge in the study of soils in the Teotihuacan Valley, Mexico," Eurasian Soil Science, vol. 38, no. 1, pp. S95–S98, 2005.
20. L. D. Llerena, El distrito de conservación del suelo y agua de Chapingo, México [Ph.D. thesis], Escuela Nacional de Agricultura, Chapingo, México, 1947.
21. E. A. García, Estudio de los suelos tepetatosos y las posibilidades de recuperación agrícola [Ph.D. thesis], Escuela Nacional de Agricultura, Chapingo, México, 1961.
22. L. A. Valdés, Características morfológicas y mineralógicas de los suelos de tepetate de la Cuenca de México [M.S. thesis], Colegio de Postgraduados, Escuela Nacional de Agricultura, Chapingo, México, 1970.
23. S. Rodríguez, M. C. Gutiérrez-Castorena, C. Hidalgo, and C. A. Ortiz-Solorio, "CIntemperismo en Tepetates y en cenizas volcánicas y su influencia en la formación de Andisoles," Terra, vol. 17, pp. 97–108, 1999.
24. T. J. Nimlos, "The density and strength of Mexican tepetate (duric materials)," Soil Science, vol. 147, no. 1, pp. 23–27, 1989.
25. T. López-Lara, J. B. Hernández-Zaragoza, J. Horta-Rangel, E. Rojas, C. López-Cajún, and D. Rosales-Hurtado, "Tepetate as construction material," Journal of Materials in Civil Engineering, 2012.
26. T. López-Lara, J. B. Hernández-Zaragoza, J. Horta-Rangel, E. Rojas, and D. Rosales-Hurtado, "Geocharacterisation of "Tepetates"," European Journal of Environmental and Civil Engineering, 2013. ·
27. J. Anochie-Boateng, E. Tutumluer, and S. H. Carpenter, "Case study: dynamic modulus characterization of naturally occurring bituminous sands for sustainable

pavement applications," International Journal of Pavement Research and Technology, vol. 3, no. 6, pp. 286–294, 2010.

28. A. M. Adam, I. A. Ibrahim, A. J. Alhardllo, A. M. Hadi, and M. Y. Ibrahim, "Effect of hydrated lime on behavior of expansive soil as subgrade of flexible pavement structural system," in Sustainable Construction Materials, S. Wu, L. Mo, B. Huang, and B. F. Bowers, Eds., pp. 64–76, American Society of Civil Engineers, 2012.

29. O. F. Arredondo, Estudio De Materiales: VI. Cerámica Y Vidrio, Instituto Eduardo Torroja de la Construcción y el Cemento, Madrid, Spain, 9nd edition, 1980.

30. J. A. Lasledj and M. Al-Mukhtar, "Effect of hydrated lime on the engineering behavior and the microstructure of highly expansive clay," in Proceedings of the 12th International conference of International Association for Computer Methods and Advances in Geomechanics, pp. 3590–3598, Goa, India, 2008.

31. J. Ninov, I. Donchev, A. Lenchev, and I. Grancharov, "Chemical stabilization of sandy-silty illite clay,"Journal of the University of Chemical Technology and Metallurgy, vol. 42, no. 1, pp. 67–72, 2007.

32. Standard Test Method for Particle-Size Analysis of Soils, ASTM D422-63, Annual Book of ASTM Standards, 2007.

33. Standard Test Methods for Liquid Limit, Plastic Limit and Plasticity Index of soils, ASTM D4318-10,Annual Book of ASTM Standards, 2010.

34. Standard Test Method for Classification of Soils for Engineering Purposes, ASTM D 2487-93, Annual Book of ASTM Standards, 1993.

35. Standard Test Methods for Laboratory Compaction Characteristics of Soil Using Standard Effort (12400ft-lbf/ft3(600kN-m/m^3)), ASTM D698-00a, Annual Book of ASTM Standards, 2000.

36. Standard Test Method for Unconfined Compressive Strength of Cohesive Soil, ASTM D2166-00, Annual Book of ASTM Standards, 2000.

37. T. López-Lara, J. Horta-Rangel, J. B. Hernández-Zaragoza, and V. M. Castaño, "Properties of waste soil-hydrated lime composite," Mechanics of Time-Dependent Materials, vol. 10, no. 2, pp. 155–163, 2006.

CHAPTER 8

ENHANCING SUSTAINABILITY OF RURAL ADOBE HOUSES OF HILLS BY ADDITION OF VERNACULAR FIBER REINFORCEMENT

Vandna Sharma[a], Hemant K. Vinayak[b], Bhanu M. Marwaha[a]

[a] Department of Architecture, National Institute of Technology, Hamirpur 177005, Himachal Pradesh, India

[b] Department of Civil Engineering, National Institute of Technology, Hamirpur 177005, Himachal Pradesh, India

ABSTRACT

Adobe is a commonly used building material in rural houses of district Hamirpur of the North Indian state of Himachal Pradesh. Adobe is a sustainable material but has limitations of building smaller room sizes and requires frequent maintenance which is not suitable for modern lifestyle. These become main reasons for rejection of adobe as a building material. Initial investigation comprising of water content analysis, specific gravity analysis, grain size analysis, plastic limit and liquid limit analysis, maximum dry density check reveals that soil is sand clay and its low compressive strength shall be increased for enhancing its sustainability. For this purpose, stabilization with natural fibers of *Pinus roxburghii* and *Grewia optiva* in 0.5%, 1%, 1.5% and 2% proportions is proposed. Total 360 cubical and cylindrical shaped samples of both stabilized and unstabilized compositions were prepared and tested in a laboratory according to Indian standards. Unconfined compressive strength tests and maximum Stress Carrying Capacity tests were conducted after 07 days, 14 days, 28 days, 56 days and 90 days of casting. Results reveal that compressive strength of soil increases by 131–145% with addition of fiber *P. roxburghii* and 225–235% with addition of fiber *G. optiva* for cubical and cylindrical samples respectively. Increased compressive strength also results in a reduced thickness of traditional mud walls thereby increasing internal room size which would suit to changed modern lifestyle requirements. Enhanced properties of adobe will result in wider acceptance of adobe as a building material that will make development of rural housing more sustainable on a wider scale.

INTRODUCTION

History of human settlements shows varied forms of human habitat developed as per the needs of people in consideration with climatic factors (Houben and Guillaud, 1994). Different topographical and geographic conditions lead to the development of habitat in many different forms involving use of local materials and techniques however among all, earth has remained a prominently used traditional building construction material (Bui et al., 2009). Earth has several advantages such as easy-to-construct with material, easy availability, thermal conduciveness and low cost maintenance features. Given these advantages, earth has preferentially been used for construction purposes especially in rural areas (Taylor and Luther, 2004). Earth is used in many forms as per the geographical and climatic factors of the area such as rammed earth-walls, cob walls, adobe bricks (Bansal and Minke, 1988 and Niroumand et al., 2013). Vernacular architecture of the area specifies use

Citation: Vandna Sharma, Hemant K. Vinayak, Bhanu M. Marwaha, Enhancing sustainability of rural adobe houses of hills by addition of vernacular fiber reinforcement, International Journal of Sustainable Built Environment, Volume 4, Issue 2, December 2015, Pages 348-358, ISSN 2212-6090, http://dx.doi.org/10.1016/j.ijsbe.2015.07.002.

of earth in different ways for building purposes. Hilly areas of North Indian state Himachal Pradesh involve the use of earth in the construction of Dhajji wall, Kath-Kuni wall, Dhol-Maide, Doriya and Farque style walls as discussed byHanda (2002) and Kumar and Pushplata, 2013a and Kumar and Pushplata, 2013b which is different from rammed earth and bamboo walls of houses in parts of North-east India as investigated by Singh et al. (2011).

However use of earth has specific constraints in terms of strength and durability which restrict its use on a wider scale. This traditional material is being widely replaced by stronger and more durable construction materials globally as discussed by Foruzanmehr and Vellinga (2011) and Kairamo (1975).

Earth is used in the form of adobe bricks in construction of rural houses in district Hamirpur of the North Indian state of Himachal Pradesh. However use of adobe has reduced significantly in house construction over the years due to limitation of small sized rooms buildable with adobe and frequent maintenance requirement of adobe walls which is not suitable for modern lifestyle. The present study is focused on addressing the problem of strength aspects of earth thereby improving its compressive strength property by addition of natural and vernacular fibers. In practical sense the increase in compressive strength of wall material results in reducing the thickness of traditional mud walls thereby resulting in an increase in the internal room size. Due to increased compressive strength, improved adobe material can sustain a load of traditional roof truss made up of bamboo, wood or steel of longer spans easily. Both reinforcing fibrous materials being sustainable would produce such construction materials that would also be sustainable and would lead to the development of sustainable rural housing. The research was conducted with objectives to: (1) find the maximum stress carrying capacity of fiber reinforced soil samples (180 cubical samples and 180 cylindrical samples) and (2) investigate unconfined compressive strength of cylindrical soil samples.

Previous studies on enhancement of properties of soil using natural fibers includes use of fibers: jute, sisal, straw, rice-husk, sugarcane bagasse (Ramírez et al., 2012 and Khosrow et al., 1999), chopped barley straw (Parisi et al., 2015), processed waste tea (Demir, 2006), vegetal (Achenza and Fenu, 2006), oil palm empty fruit bunches (Kolop et al., 2010), lechuguilla (Juárez et al., 2010), pineapple leaves (Chan, 2011), cassava peel (Villamizar et al., 2012) and *Hibiscus cannabinus* (Millogo et al., 2014). However no study regarding use of fibers of *Pinus roxburghii* and *Grewia optiva* as reinforcement in low compressive soil for enhancement of its mechanical properties has been conducted so far which involves checking of maximum stress carrying capacity and comparative strength assessment of cylindrical and cubical fiber reinforced soil samples.

The paper has been organized into two sections: Sections 3 and 4 and Section 6 and 7. In the first half section of the paper, important aspects of traditional material earth relative to its usage in the present context has been discussed which includes building regulation and codes of practice regarding use of earth, benefits of earth in terms of thermal comfort, constraints in terms of mechanical properties of compressive strength and durability and measures to improve the strength and durability properties. The second half section of the paper reports the experimental investigation for improving strength properties of soil of district Hamirpur of state Himachal Pradesh of North India. The investigation results show that compressive strength of soil can be considerably increased by inclusion of natural fibers. This result would be useful to produce fiber mix adobe bricks with improved compressive strength properties. Experimental investigations for checking the durability of same mix specimens have been identified as follow up research.

BUILDING REGULATIONS FOR SUSTAINABLE BUILDING MATERIAL: EARTH

For propagating use of earth on a large scale for wider acceptability by people; planning, designing and technical guidelines regarding its use are framed in the form of comprehensive codal provisions and legal regulations. In some countries they are mandatory while in others they are advisory guidelines for all construction works with earth (Middleton, 1992 and Torgal and Jalali, 2012) as given in Table 1. These regulations give details of usage process of earth in various forms for general construction purposes and for construction in seismic activity areas.

Table 1: Description of country specific guidelines/codes for earth construction (Middleton, 1992 and Torgal and Jalali, 2012)

S. N.	Country	Relevant guideline/code
1	United States	ASTM International: 2010 – Standard Guide for Design of Earthen Wall Building Systems (ASTMD 559-57)
2	New Zealand	SNZ: Standards New Zealand (NZS 4297:1998, NZS 4298:1998, NZS 4299:1998)
3	New Mexico	State Regulations "Rammed Earth And Adobe Based Construction" – 1999 (CID: NMAC 1474)
4	Spain	AENOR: Spanish Association For Standardization And Certification 2008 (UNE 41410:2008-12-10)
5	Colombia	ICONTEC, 2004 (NTC 5324)
6	Australia	Australian Regulations "Bulletin 5" – 1952 by CSIRO Replaced by "Australian Earth Building Handbook" 2002

S. N.	Country	Relevant guideline/code
7	South Africa	ARSO: African Regional Standards: 1996 (ARS 670, ARS 683)
8	Kenya	KEBS: Kenya Bureau of Standards (KS 02-1070)
9	Tunisia	INNORPI: Tunisian Standards: 1996 (INNORPI: NT 2133, INNORPI: NT 2135)
10	France	AFNOR: 2001 (AFNOR:XP P13-901)
11	India	IS 13827:1993, IS 1725 1982 (reaffirmed 2002), General Building Construction, NBC Handbook (IS 1725 1982)

EARTH: ADVANTAGES AND CONSTRAINTS

Earth walls have proven to be better thermal insulators as compared to brick or cement concrete block walls (Goodhewa and Griffiths, 2005 and Binici et al., 2007). Study performed by Shukla et al. (2009) shows that earth has low embodied energy values along with low operational energy and transportation energy values as compared to modern construction materials. Studies by Mohammed (2004) and (Tiwari et al., 1996) further show the effectiveness of earth in maintaining comfortable indoor thermal environment of houses in relation to outdoor environment. However, earth as a building material suffers from some drawbacks in mechanical properties. Limitation in the form of low compressive strength and durability against weathering agents restricts its use on large scale (Alfred and Ngowi, 1997). This limitation coupled with problems of frequent and tedious maintenance of earth structures forms the reasons for replacement of use of this material with new and modern materials like cement, concrete, steel etc. (Quiteria et al., 2012).

IMPROVING MECHANICAL PROPERTIES OF EARTH

Research studies (Binici et al., 2005, Morel et al., 2007, Aubert et al., 2013, Jaquin et al., 2009 and Illampas et al., 2014) indicate that both compressive strength and durability of earth can be increased by addition of either stabilizers or reinforcement in the form of natural/artificial fibers or by adding both stabilizer and fiber reinforcement. Natural fibers and stabilizers generally used for soil reinforcement are fibers of sisal, jute, cow-dung, rice husk, sugarcane bagasse etc. (Khosrow et al., 1999, Ashour et al., 2011, Marín et al., 2010 and Ramírez et al., 2012) while man-made stabilizers and reinforcement include use of cement, lime, rubber fibers, plastic waste etc. (Consoli et al., 2002).

Use of a particular type of stabilizer and its consequent behavior in the improvement of compressive strength of soil is different from the other owing to the difference in its chemical composition and properties. However addition of natural fibers is more beneficial than addition of artificial fibers or stabilizers since natural materials are lower in embodied energy and toxicity than man-made materials as shown by study by John et.alJohn et al. (2005). The study discusses that natural fibers require no or very less processing and can be used in crude form also. Therefore they are more environment friendly than artificial fibers. Moreover use of these natural stabilizing materials in production of construction materials results in the manufacture of materials which themselves are sustainable. Choice of use of natural fibers depends upon their availability and rate of acceptance of resulting product in the area. In this regard, vernacular natural fibers always have added advantage of being easily available and known to common masses who would be their potential users. Regarding economy, transportation and handling costs of vernacular materials are negligible which makes this technology easy-to-work with type (Morela et al., 2001).

Various compressive strength tests of earth include triaxial tests, unconfined compressive strength tests, CBR (California bearing ratio) tests, direct shear tests, tensile and flexural test (Lima et al., 2012, Nowamooz and Chazallon, 2011, Morel and Pkla, 2002 and Miccoli et al., 2014) and durability tests which include wire brush test, spray test, drip test, capillary absorption test, wetting and drying test, freezing and thawing test (Heathcote, 1995, Ogunye and Boussabaine, 2002 and Guettala et al., 2006), condensation and weathering tests (Ren and Kagi, 1995). Studies (Quagliarini and Lenci, 2010, Ghavami et al., 1999, Consoli et al., 1998 and Ranjan et al., 1999) indicate the relation between fiber length and strength properties that increase in fiber length increases shear strength properties up to a certain limit. The reason may be the increase in contact area with the soil which results in improvement in strength and stiffness of soil.

EXPERIMENTAL INVESTIGATIONS

Presently the district comprises 60% rural houses (Revised Development Plan, 2012), out of which nearly 70% are made of adobe, bamboo and wood as shown in Fig. 1. However, now people are shifting toward use of modern materials like cement concrete owing to its better compressive strength and durability properties. This has produced serious effects on usage of earth in construction of rural houses which is increasingly being abandoned due to problems in its use in terms of tedious repair and maintenance, low flexibility in design and low durability. In order to solve the problem of low strength of soil which results in low compressive strength property of resultant adobe, the present study is focused on improving the compressive strength of sand clay soil of district Hamirpur of the state Himachal Pradesh of North India using natural vernacular fibers of*G. optiva* popularly known as 'Beul' and *P. roxburghii* popularly known as 'Chir Pine'.

A series of unconfined compressive strength tests and flexure tests were conducted for different specimens comprising 2.5% cement and randomly mixed varied proportions of both the fibers after 7, 14, 28, 56 and 90 days of casting. The test results were calculated as an average of strength results of four specimens as specified by Indian standards (IS: 2720-10, 1991).While Unconfined Compression Strength tests were conducted in Load Frame testing machine with capacity up to 50 kN using Indian Standard codal provisions (IS: 2720-10, 1991), the maximum stress carrying capacity was tested in compression and flexure testing machine after preliminary checks of physical properties of the soil.

Figure 1: Rural adobe houses of district Hamirpur, Himachal Pradesh, India.

Materials

Soil

Materials used in this investigation are soil, cement (grade M43), and indigenous natural fibers of *P. roxburghii* and *G. optiva*. Soil was brought from NIT Hamirpur (near gate 1) as shown in Fig. 2. Initial investigation comprising water content analysis (IS: 2720-2, 1973), specific gravity analysis (IS: 2720-3, 1980), grain size analysis (IS: 2720-4, 1985), plastic limit and liquid limit analysis (IS: 2720-5, 1985) and maximum dry density and optimum water content tests (IS: 2720-7, 1985) for checking physical and mechanical properties of soil were conducted as shown in Fig. 3. Results showed that soil is sand clay designated by SC as per Indian Standard classification (IS 1498-1970) given in codal provision IS: 2720-4 (1985) and of low compressive strength. Tests revealed that soil has a water content of 11.48%, specific gravity 2.67, plastic limit 17.78%, liquid limit 23.29%, plasticity index (PI) 5.512, optimum moisture content 12.5% and maximum dry density 18.8 kN/m^3 as reported in Table 2. Specific gravity of M43 grade cement used is 3.10. Soil used in manufacture of adobe is taken from either fields or grazing/pasture areas. Since the study area is located in hilly terrain, therefore all areas are not suitable for cultivation purposes for the reasons of high slopes leading to uneconomical cultivation. Moreover such areas are not fragile hills or large mountain areas rather they are flat/gently sloping areas located in pastures or near farm areas. Therefore preferably those hillside areas where topography does not allow both cultivation and construction activities and also are not ecologically fragile, are favored for excavation and procurement of soil. Materials used in investigation are shown in Fig. 2.

Figure 2: Materials used in investigations: sand clay soil and fibers of *Pinus roxburghii* and *Grewia optiva*.

(a) Sieve analysis (b) Plastic limit test (c) Standard Proctor Test

Figure 3: Different tests conducted for physical properties of earth.

Table 2: Physical properties of selected soil

Property name	Value
Water content (%)	11.48
Liquid limit (%)	23.29
Plastic limit (%)	17.78
Plasticity index (PI)	5.512
Optimum moisture content (OMC) (%)	12.5
Maximum dry density (MDD) (kN/m^3)	18.8
Specific gravity of soil solids (G)	2.67
Classification as per Indian Standard	SC

Fibers

Fibers of *P. roxburghii* and *G. optiva* are native to the region. Moreover they both are treated as waste materials which require frequent disposal. Large *P. roxburghii* forest area requires prescribed fires in a controlled manner by local forest department every year to lower the danger of spread of natural forest fire during every summer season. At global scale also, prescribed fires are the most common practice to counteract the spread of natural forest fires, however it results in harmful emissions. Therefore with the dual purpose of minimizing negative impacts associated and taking advantage of its tensile strength, fibers of *P. roxburghii* have been proposed for manufacturing of improved soil blocks to advantageously use this forest waste product. Similarly fibers of*G. optiva* are used only for rope making that has also been limited nowadays due to use of plastic rope. After use of leaves of *G. optiva* as fodder for animals, its fibrous stem and branches are thrown away as waste materials which are the main source of fibers from this plant. In this way, fibers of *G. optiva* can be a lucrative option to be used for producing improved soil blocks. Therefore utilization of both waste fibrous materials would result in enhancing adobe block properties leading to the production of more sustainable materials.

Studies performed by Kumar et al. (2006) and Casagrande et al. (2006) showed an increase in peak strength value of soil with addition of fiber reinforcement. Compressive strength of fiber reinforced mud bricks is more than that of customary unreinforced mud bricks as shown by Binici et al. (2005). A similar study by Babu et al. Babu and Vasudevan (2008) showed that addition of discrete coir fibers by about 1–2% by weight in the soil increases the strength and stiffness of tropical soil. A research by Dasaka et al. (2011) on the relation between soil strength and length of fiber showed that variation in the length of fibers and content in the soil results in improvement in strength characteristics of soil. Length of fibers therefore plays a significant role in compressive strength improvement of soil. In this investigation, fibers of size 30 mm for both *P. roxburghii* and *G. optiva* were used for random mixing. Physical and engineering properties of both the fibers *P. roxburghii* (Chir Pine) and *G. optiva* (Beul) were checked and reported as given in Table 3.

Table 3: Physical and engineering properties of fibers: *Pinus roxburghii* (Chir Pine) and *Grewia optiva* (Beul)

Property	Value (*Pinus roxburghii*)	Value (*Grewia optiva*)
Tensile strength (N)	11.15	15.35
Shape	Triangular cross-section	Circular
Diameter (mm)	.48	.03
Length (mm)	324	730
Color	Dark brown	Light brown

Standard proctor tests and flexural strength tests were conducted (IS: 2720-7, 1985 and Kumar et al., 2006) and compaction curves were developed. The soil used in this investigation has a composition of coarse grain particles with values as .0756 for sand and .0005 for clay as per Indian Standard classification of soil (IS 1498-1970). Study by Ranjan et al. (1996) discusses the relation between grain size of given soils and the fiber-bond strength. The study showed that finer sand particles have more fiber bond strength than coarse grained soils and silts have even better performance.

Specimen Preparation for Testing

Specimens with two types of geometrical shapes were prepared namely: cylindrical and cubical. Cylindrical specimens for this investigation were prepared with dimensions 38 mm diameter and 76 mm height and cubical samples were prepared with dimensions 190 mm × 90 mm × 90 mm. All 360 samples (180: cylindrical and 180: cubical) were prepared using a mixture of moist soil, moist soil + cement and moist soil + cement + fibers. Specimens were prepared corresponding to optimum moisture content and maximum dry unit weight values. Samples were prepared with and without stabilizers and reinforcement. For each mixture composition, 40 samples (20: cylindrical and 20: cubical) were made as per Indian standards. Specimens were classified according to composition and respective designation as given in Table 4. After preparation all samples were covered with gunny bags and allowed to cure in shade for a period of one week. Before starting with compaction procedure, mold (in both cases) was lubricated from inside to facilitate easy removal of samples from the mold and minimize the chances of sample fracturing during removal. Compaction of mixtures was done in the form of three equal layers. Each layer was filled up to the required fraction of height and compacted. Each layer was scoured before filling the mold with a mixture for subsequent fraction of layer. This was done to ensure that a good bond is established between the layers in each effort. After removal from mold each sample was trimmed to the desired height and fiber sticking out from the sample at the top, bottom and sides was trimmed with scissors.

Table 4: Designation of mixtures

Mix designation	Mix specification	Specimen number
M1	Soil(0% Cement, 0% Fiber)	40
M2	Soil(2.5% Cement, 0% Fiber)	40
M3	Soil(2.5% Cement, 0.5% Fiber: *Pinus roxburghii*)	40
M4	Soil(2.5% Cement, 1.0% Fiber: *Pinus roxburghii*)	40
M5	Soil(2.5% Cement, 1.5% Fiber: *Pinus roxburghii*)	40
M6	Soil(2.5% Cement, 2.0% Fiber: *Pinus roxburghii*)	40
M7	Soil(2.5% Cement, 0.5% Fiber: *Grewia optiva*)	40
M8	Soil(2.5% Cement, 1.0% Fiber: *Grewia optiva*)	40
M9	Soil(2.5% Cement, 2.0% Fiber: *Grewia optiva*)	40
	Total number of specimens	360

Testing of Specimens

Specimens were tested after curing was complete. Tests were conducted for compressive strength and stress carrying capacity after 07, 14, 28, 56 and 90 days of casting. The test results were taken as average of four specimens as per Indian standards (IS: 2720-3, 1980). If the individual variation was found to be more than ±5% of the average, the value was not considered in calculation of the average value. Increase in strength was determined by performing a series of unconfined compressive tests (IS: 2720-3, 1980) with a constant strain rate of 0.04 mm/min. throughout the testing investigation on cylindrical samples and stress carrying capacity was noted by testing cubical samples in compression and flexural load testing machine. The maximum stress carrying capacity of both types of samples was compared to evaluate the difference in strength increase with different shapes. Tests were performed on soil specimens prepared at the maximum dry unit weight determined using the standard Proctor test and optimum moisture content (IS: 2720-7, 1985). Investigations included testing of unreinforced samples (0% fibers), stabilized unreinforced samples (2.5% cement) and stabilized reinforced samples prepared with 2.5% cement and 0.5%, 1%, 1.5% and 2% (by weight of dry soil) of fibers of both *P. roxburghii* and *G. optiva*.

RESULTS AND DISCUSSION

This section discusses the results of Standard Proctor Compaction Test, Stress Carrying Capacity test and Unconfined Compression Strength test results. The resultant stress carrying capacity curves and stress–strain curves were plotted for various cases and compared.

Standard Proctor Compaction Test results

The results of Standard Proctor Compaction Test are given in Table 5. The table shows the results of Proctor Compaction Tests conducted on low compressive sand clay soil with varying contents of cement as 2.5% and 3.5% as per codal provisions (IS: 2720-7, 1985). The table clearly shows that the maximum dry density of soil increases with addition of cement up to 2.5%. However, further addition of 3.5% decreases this value. Therefore for addition of 2.5% cement, the value of maximum dry density (MDD) for given OMC value is more than corresponding values of MDD for addition of 3.5% cement in soil. This resulted in the selection of 2.5% of cement to be used as stabilizer for soil along with fibers for subsequent UCS and stress carrying capacity experimental program.

Table 5: Standard Proctor Compaction Test results.

Mix designation	MDD (kN/m^3)	OMC (%)
M1: Soil	18.8	12.5
M2:Soil + 2.5% cement	19.31	8.2
Soil + 3.5% cement	17.36	12.9

Stress carrying capacity

Testing of Cube Samples

Compression and flexure testing machine were used to investigate maximum stress carrying capacity of 180 cubical samples of composition: soil + 0% stabilizer + 0% reinforcement and soil + 2.5% cement + fibers. The samples were tested after periods of 07, 14, 28, 56 and 90 days to understand the gain in strength of the samples as shown inFig. 4. The figure shows that stress carrying capacity of samples increases with time which also implies gain in strength. It was seen that the maximum gain in strength is shown by samples M9 followed by M4 corresponding to a fiber mix of 2% *G. optiva* and 1% *P. roxburghii*. The results indicate that samples of fiber mix *G. optiva* show maximum gain in strength.

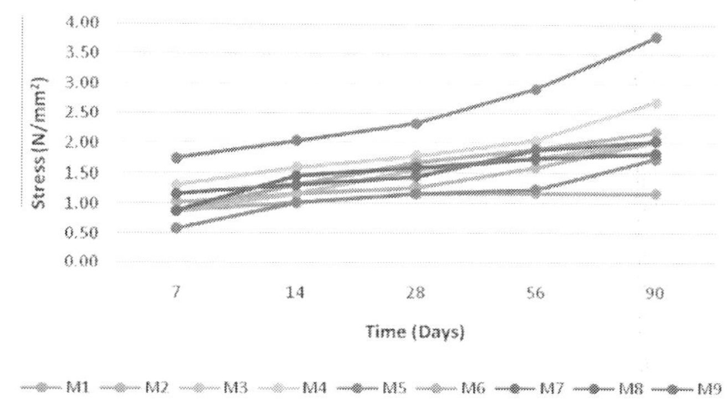

Figure 4: Maximum stress carrying capacity and gain in strength for cubical soil samples.

It can be interpreted from the curves that: (1) samples M1and M2 have brittle behavior and they break immediately into pieces on the appearance of cracks after reaching maximum load carrying capacity, (2) samples M3, M4, M5 and M6 show ductile behavior. The samples did not disintegrate into pieces immediately after appearance of cracks when the maximum load carrying capacity of samples was reached. Instead disintegration was delayed with time and samples with more fiber content showed maximum delay in disintegration, (3) samples M7, M8 and M9 showed similar ductile behavior with delay in disintegration of samples even after appearance of cracks, (4) samples M7, M8 and M9 (of *G. optiva*) showed more ductile behavior than samples M3, M4, M5 and M6 (of *P. roxburghii*).

Comparing stress carrying capacity for cubical and cylindrical samples

Comparing stress carrying capacity (and corresponding compressive strength values) of samples of both geometrical shapes, it was seen that the values are comparable for both types as shown by Figure 4 and Figure 5. The curves clearly show that while cubical samples showed a maximum stress carrying capacity value of 3.8 N/mm², cylindrical samples showed it 2.85 N/mm². It can be interpreted from the figures that stress carrying capacity of soil (which also corresponds to its compressive strength) increases with addition of fiber reinforcement of *P. roxburghii* and *G. optiva*. Furthermore it also indicates that soil samples of fiber *G. optiva* have more strength than soil samples of fiber*P. roxburghii*.

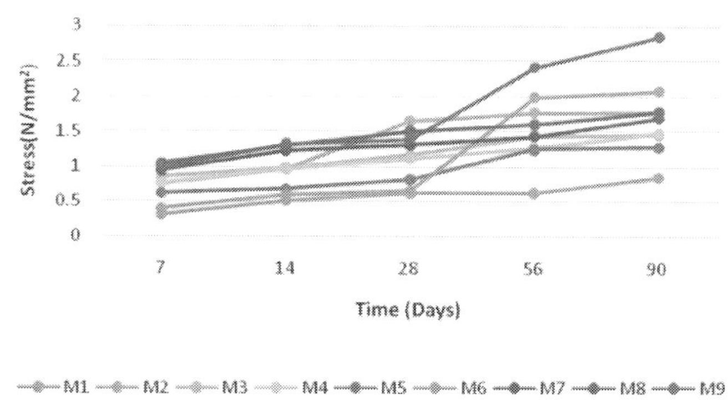

Figure 5: Maximum stress carrying capacity and gain in strength for cylindrical soil samples.

Compressive Strength Test Results

The compressive strength properties of different mix samples of cylindrical shape were recorded and analyzed as shown in Fig. 6. The compressive strength values required by Indian standards for traditional mud brick is 1.0 N/mm². However, the values in the case of the fiber reinforced mud bricks tested in the present study are higher with maximum values of 2.08 N/mm² for *P. roxburghii* and 2.85 N/mm² for *G. optiva*. This also implies that with increased compressive strength, now wall thickness can be reduced for practical purposes, leading to an increased internal room size.

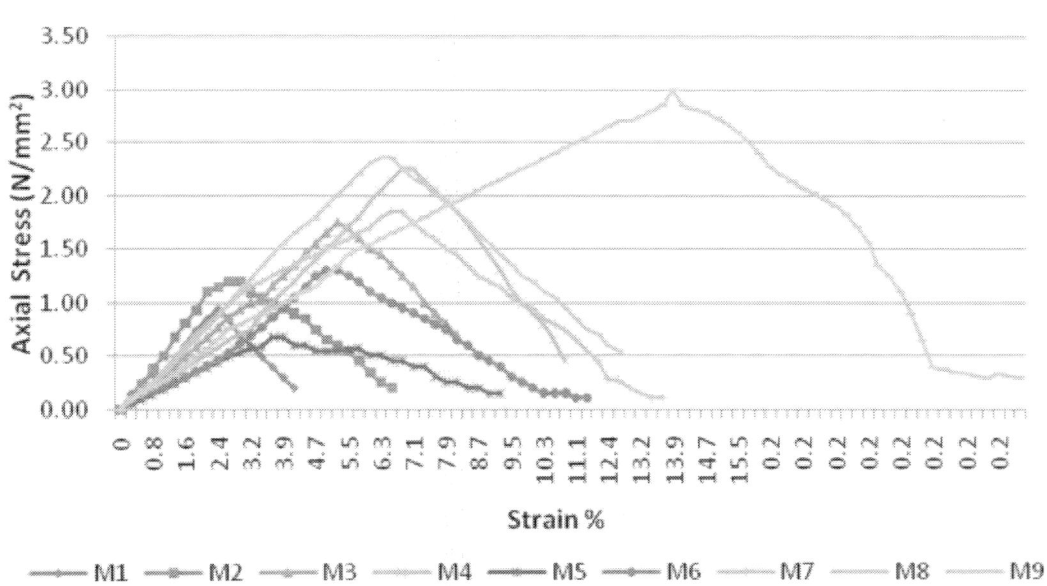

Figure 6: Compressive strength of cylindrical soil samples.

Analysis of Figure 4, Figure 5 and Figure 6 shows that in the case of cubical samples for tests conducted after a period of 07 and 14 days maximum early gain in strength is shown by mix M9 followed by mixes M4, M8, M2, and M7. However for tests conducted after a period of 28 days, 56 days and 90 days the maximum final gain in strength is shown by mix M9 followed by mixes M4, M6, M8, M2. Similarly for cylindrical samples, for tests conducted after a period of 07 and 14 days the maximum early gain in strength is shown by mix M9 followed by mixes M8, M7. For tests conducted after a period of 28 days, 56 days and 90 days the maximum final gain in strength is shown by mix M9 followed by mixes M9, M6, M8, M2 and M7.

It can be interpreted from the curves that inclusion of 0.5–2% of fibers in the soil increases peak compressive strength of the low compressive soil by about 50–225% for cubical samples (base strength of 1.17 N/mm^2 of M1) and 50–235% (base strength of .85 N/mm^2 of M1) for cylindrical shaped samples. The maximum increase in compressive strength is shown by specimen M9 reinforced with fiber 2% *G. optiva* and is 225–235% for cubical and cylindrical samples (with base strength of 1.17 and .85 N/mm^2respectively of soil) followed by specimen M6 reinforced with 2% *P. roxburghii* showing an increase in compressive strength by 87–145% for cubical and cylindrical samples (base strength of 1.17 N/mm^2 and .85 N/mm^2 respectively of soil) and specimen M4 reinforced with 1% *P. roxburghii* showing an increase in compressive strength by 131% for cubical samples (base strength of 1.17 $N/mm^2$of soil). These curves therefore imply that compressive strength of low compressive sand clay soil of district Hamirpur of the North Indian state Himachal Pradesh, can be improved considerably by inclusion of fiber reinforcement in the form of natural vernacular fibers of *G. optiva* and *P. roxburghii* found in the area.

In practical sense this implies vernacular natural materials can be advantageously added in the soil of district Hamirpur (used for manufacture of adobe bricks for construction of houses) in order to increase its compressive strength. The increase in compressive strength of wall material results in a lesser width of walls thereby resulting in an increase in room size. The increased compressive strength further results in supporting traditional roof truss made up of bamboo, wood or steel of longer spans. This would address the problem of narrow room sizes which earlier existed in old houses due to material strength constraint of adobe. This can lead to increased social acceptance mitigating the problems of rejection by common masses due to design of narrow room sizes.

Testing investigation was also carried out to check tensile strength properties of both the fibers. It was found that tensile strength properties of fiber *G. optiva* are better than those of *P. roxburghii*. Better fiber-soil bonding in the case of *G. optiva* specimens gives them more compressive strength than specimens of *P. roxburghii*. Factors that influence fiber-soil bond will be examined at a microscopic level for both types of fibers especially over the period of time.

CONCLUSION

In this study, natural vernacular fibers of *P. roxburghii* and *G. optiva* in proportions of 0.5%, 1%, 2% are added to sand clay soil to enhance its low compressive strength and a total of 360 samples of various compositions are prepared (180 cubical samples and 180 cylindrical samples) and tested in laboratory as per IS (Indian Standard) codes for which the following results are drawn:

1. Unreinforced soil without addition of stabilizer shows brittle behavior in UCS tests.
2. Unreinforced cement stabilized soil specimens also show brittle failure behavior.
3. The maximum stress carrying capacity is shown by samples of mix M9 reinforced with fiber *G. optiva* for both cubical and cylindrical types.
4. Addition of fibers leads to an increase in compressive strength of soil by 50–225% for cubical samples (base strength of 1.17 N/mm^2 of M1) and 50–235% (base strength of .85 N/mm^2 of M1) for cylindrical shaped samples.
5. Increase in compressive strength value of soil is more in the case of specimens reinforced with fiber *G. optiva* which is 225–235% for cubical and cylindrical specimens (with base strength of 1.17 and .85 N/mm^2 respectively of soil) than in the case of reinforcement with fiber *P. roxburghii* which is 131–145% for cubical and cylindrical specimens (with base strength of 1.17 and .85 N/mm^2 respectively of soil).

Considering the above conclusion, fibers of both the vernacular natural materials can be advantageously added in the soil of study area. In practical sense, the increase in compressive strength will reduce the thickness of traditional mud walls thereby increasing internal room size which would suit to changed modern lifestyle requirements. This would help in overcoming the reasons of rejection of this material in terms of constraints of narrow achievable room size by using customary adobe bricks and frequent maintenance of traditional adobe walls.

Since both the materials are indigenous to the area and are sustainable therefore their beneficial additions for increasing compressive strength of soil will produce superior sustainable construction materials that would help to achieve sustainability of rural houses on a wider scale. Experimental investigation for durability of specimens of same composition has been identified as follow up research area which would also address the problems of erosion and frequent repair and maintenance of rural houses owing to poor weatherability.

ACKNOWLEDGMENTS

The authors are thankful to Dr. R.K. Dutta and Er. Prakesh of Civil Engineering Department, and Ar. Aniket Sharma of Architecture Department, NIT Hamirpur for providing necessary help in performance of various experiments.

REFERENCES

1. Achenza, M., Fenu, L., 2006. On earth stabilization with natural polymers for earth masonry construction. Mater. Struct. 39, 21–27.
2. Alfred, B., Ngowi, 1997. Improving the traditional earth construction: a case study of Botswana. Constr. Build. Mater. 11 (1), 1–7.

3. Ashour, T., Georg, H., Wu, W., 2011. Performance of straw bale wall: a case of study. Energy Build. 43, 1960–1967.
4. Aubert, J.E., Fabbri, A., Morel, J.C., Maillard, P., 2013. An earth block with a compressive strength higher than 45 MPa! Constr. Build. Mater.47, 366–369.
5. Babu, G.L.S., Vasudevan, A.K., 2008. Strength, and stiffness response of coir fiber-reinforced tropical soil. J. Mater. Civ. Eng. 20 (9), 571–577.
6. Bansal, N.K., Minke, G., 1988. Climatic Zones and Rural Housing in India. Kern for schungsanlage, Julich, Germany.Binici, H., Aksogan, O., Bodur, N.M., Akca, E., Kapur, S., 2005.
7. Investigation of fiber reinforced mud brick as a building material. Constr. Build. Mater. 19, 313–318.
8. Binici, H., Aksogan, O., Bodur, N.M., Akca, E., Kapur, S., 2007. Thermal isolation and mechanical properties of fiber reinforced mud bricks as wall materials. Constr. Build. Mater. 21, 901–906.
9. Bui, Q.B., Morel, J.C., Reddy, B.V., Ghayad, W., 2009. Durability of rammed earth walls exposed for 20 years to natural weathering. Build. Environ. 44, 912–919.
10. Casagrande, M.D.T., Coop, M.R., Consoli, N.C., 2006. Behaviour of a fiber reinforced bentonite at large shear displacements. J. Geotech. Geoenviron. Eng. 132 (11), 1505–1508.
11. Chan, C.M., 2011. Effect of natural fibers inclusion in clay bricks: physico-mechanical properties. Int. J. Civil Environ. Eng. 3 (1), 51–57.
12. Consoli, N.C., Montardo, J.P., Prietto, P.D.M., Ulbrich, L.A., 1998. Influence of fiber and cement addition on behavior of sandy soil. J. Geotech. Geoenviron. Eng-ASCE 124, 1211–1214.
13. Consoli, N.C., Montardo, J.P., Prietto, P.D.M., 2002. Engineering behavior of sand reinforced with plastic waste. J. Geotech. Geoenviron. Eng-ASCE 128, 462–472.
14. Dasaka, S.M., Sumesh, K.S., 2011. Effect of coir fiber on the stress–strain behavior of a reconstituted fine-grained soil. J. Nat. Fibers 8 (3), 189–204.
15. Demir, I., 2006. An investigation on the production of construction brick with processed waste tea. Build. Environ. 49 (1), 1274–1278.
16. Foruzanmehr, A., Vellinga, M., 2011. Vernacular architecture: questions of comfort and practicability. Build. Res. Inf. 39 (3), 274–284.
17. Ghavami, K., Toledo, F.R.D., Barbosa, N.P., 1999. Behaviour of composite soil reinforced with natural fibers. Cem. Concr. Compos. 21, 39–48.
18. Goodhewa, S., Griffiths, R., 2005. Sustainable earth walls to meet the building regulations. Energy Build. 37, 451–459.
19. Guettala, A., Abibsi, A., Houari, H., 2006. Durability study of stabilized earth concrete under both laboratory and climatic conditions exposure. Constr. Build. Mater. 20, 119–127.
20. Handa, O.C., 2002. Temple Architecture of the Western Himalaya: Wooden Temples. Indus Publishing Company, New Delhi.
21. Heathcote, K.A., 1995. Durability of earth wall buildings. Constr. Build. 9 (3), 185–189.
22. Houben, H., Guillaud, H., 1994. Earth Construction: A Comprehensive Guide. IT Publications, London.
23. Illampas, R., Ioannou, I., Dimos, C., 2014. Adobe bricks under compression: experimental investigation and derivation of stress–strain equation. Constr. Build. Mater. 53, 83–90.
24. IS: 2720-10, 1991.Determination of Unconfined Compressive Strength:Indian Standards.
25. IS: 2720-2, 1973. Determination of Water Content of Soils.
26. IS: 2720-3, 1980. Determination of specific gravity of soils.
27. IS: 2720-4, 1985.Grain Size Analysis: Indian Standards.
28. IS: 2720-5, 1985. Determination of Liquid and Plastic Limit: Indian Standards.
29. IS: 2720-7, 1985. Determination of Water Content-Dry Density Relation Using Light Compaction: Indian Standards.
30. Jaquin, P.A., Augarde, C.E., Gallipoli, D., Toll, D.G., 2009. The strength of unstabilized rammed earth materials. Ge´otechnique 59, 487–490.
31. John, G., Derek, C.C., George, J., 2005. Sustainable building solutions: a review of lessons from the natural world. Build. Environ. 40, 319–328.
32. Jua´rez, C., Guevara, B., Valdez, P., Dura´n-Herrera, A., 2010. Mechanical properties of natural fibers reinforced sustainable masonry. Constr.Build. Mater. 24, 1536–1541.
33. Kairamo, M., 1975. Vernacular architecture and its adaptation to the needs of modern life: the situation in Finland. Available from: http:// www.international.icomos.org/monumentum/vol15-16/vol15-16_9.pdf(Accessed on 12/01/2011).
34. Khosrow, G., Filho, R.D.T., Barbosa, N.P., 1999. Behavior of composite soil reinforced with natural fibers. Cement Concr. Compos. 21, 39–48.
35. Khosrow, G., Romildo, D., 1999. Toledo F, Normando P. Barbosa. Behavior of composite soil reinforced with natural fibers. Cem. Concr. Compos. 21, 39–48.
36. Kolop, R., Haziman, W.I.M., Eng, J.W., 2010. Properties of Cement Blocks Containing High Content of Oil Palm Empty Fruit Bunches (EFB) Fibers. Universiti Tun Hussein Onn, Malaysia.
37. Kumar, A., Pushplata, 2013a. Building regulations for environmental protection in Indian hill towns. Int. J. Sustainable Built Environ. 2,224–231.
38. Kumar, A., Pushplata, 2013b. Vernacular practices: as a basis for formulating building regulations for hilly areas. Int. J. Sustainable Built Environ. 2, 183–192.
39. Kumar, A., Walia, B.S., Mohan, J., 2006. Compressive strength of fiber reinforced highly compressible clay. Constr. Build. Mater. 20, 1063–1068.

40. Lima, S.A., Varum, H., Sales, A., Neto, V.F., 2012. Analysis of the mechanical properties of compressed earth blocks masonry using the sugarcane bagasse ash. Constr. Build. Mater. 35, 829–837.

41. Marı´n, C.G., Go´mez, C.R., Petric, J., 2010. Clay-based composite stabilized with natural polymer and fiber. Constr. Build. Mater. 24,1462–1468.

42. Miccoli, L., Mu¨ller, U.R.S., Fontana, P., 2014. Mechanical behaviour of earthen materials: a comparison between earth block masonry, rammed earth and cob. Constr. Build. Mater. 61, 327–339.

43. Middleton, G.F., 1992. Earth-Wall Construction, fourth ed. CSIRO National Building Technology Centre, Australia.

44. Millogo, Y., Morel, J.C., Aubert, J.E., Ghavami, K., 2014. Experimental analysis of pressed adobe blocks reinforced with hibiscus cannabinus fibers. Constr. Build. Mater. 52, 71–78.

45. Mohammed, A.E.S., 2004. Learning from tradition: the planning of residential neighborhoods in a changing world. Habitat Int. 28 (4), 625–639.

46. Morel, J.C., Pkla, A., 2002. A model to measure compressive strength of compressed earth blocks with the'3points bending test' Constr. Build. Mater. 16, 303–310.

47. Morel, J.C., Pkla, A., Walker, P., 2007. Compressive strength testing of compressed earth blocks. Constr. Build. Mater. 21 (5), 303–309.

48. Morela, J.C., Mesbaha, A., Oggerob, M., Walker, P., 2001. Building houses with local materials: means to drastically reduce the environmental impact of construction. Build. Environ. 36, 1119–1126.

49. Niroumand, H., Zain, M., Jamil, M., 2013. Various types of earth buildings. Procedia – Social Behav. Sci. 89, 226–230.

50. Nowamooz, H., Chazallon, C., 2011. Finite element modeling of a rammed earth-wall. Constr. Build. Mater. 25, 2112–2121.

51. Ogunye, F.O., Boussabaine, H., 2002. Diagnosis of assessment methods for weatherability of stabilized compressed soil blocks. Constr. Build. Mater. 16, 163–172.

52. Parisi, F., Asprone, D., Fenu, L., Prota, A., 2015. Experimental characterization of Italian composite adobe bricks reinforced with straw fibers. Compos. Struct. 122, 300–307.

53. Quagliarini, E., Lenci, S., 2010. The influence of natural stabilizers and natural fibers on the mechanical properties of ancient Roman adobe bricks. J. Cult. Heritage 11 (3), 309–314.

54. Quiteria, A.I., A´ ngeles, M.T., Vicente, G.L., Jose´, L.S., 2012. Traditional braces of earth constructions. Constr. Build. Mater. 30, 389–39.

55. Ramı´rez, R.A., Garcı´a, P.M., Reyes, J.M., Jua´rez, D.C.A., 2012. The use of sugarcane bagasse ash and lime to improve the durability and mechanical properties of compacted soil blocks. Constr. Build. Mater. 34, 296–305.

56. Ramı´rez, R.A., Garcı´a, P.M., 2012. Jacob, Jua´rez D.C.A., Gochi Y. The use of sugarcane bagasse ash and lime to improve mechanical properties of compacted soil blocks. Constr. Build. Mater. 34, 296–305.

57. Ranjan, G., Vasan, R.M., Charan, H.D., 1996. Probabilistic analysis of randomly distributed fibers-reinforced soil. J. Geotech. Eng-ASCE 122, 419–426.

58. Ranjan, G., Singh, Bhupal, Charan, H.D., 1999. Experimental study of soft clay reinforced with sand-fiber core. Ind. Geotech. J. 29 (4), 281–291.

59. Ren, K.B., Kagi, D.A., 1995. Upgrading the durability of mud bricks by impregnation. Build. Environ. 30 (3), 433–440.

60. Revised Development Plan, Hamirpur, 2012. Available from: http:// himachal.nic.in/tcp/DPHamirpur.pdf (Accessed on 25/12/2014).

61. Shukla, A., Tiwari, G.N., Sodha, M.S., 2009. Embodied energy analysis of adobe house. Renewable Energy 34, 755–761.

62. Singh, M.K., Mahapatra, S., Atreya, S.K., 2011. Solar passive features in vernacular architecture of North-East India. Sol. Energy 85 (9), 2011–2022.

63. Taylor, P., Luther, M.B., 2004. Evaluating rammed earth walls: a case study. Sol. Energy 76, 79–84.

64. Tiwari, P., Parh, K., Sharma, V., 1996. Performance evaluation of cost-effective buildings – a cost, emissions and employment point of view. Build. Environ. 31 (1), 75–90.

65. Torgal, F.P., Jalali, S., 2012. Earth construction: lessons from the past for future eco-efficient construction. Constr. Build. Mater. 29, 512–519.

66. Villamizar, M.C.N., Araque, V.S., Reyes, C.A.R., Silva, R.S., 2012. Effect of the addition of coal–ash and cassava peels on the engineering properties of compressed earth blocks. Constr. Build. Mater. 36, 276–286.

CHAPTER 9

AN ALTERNATIVE TO CLAY IN BUILDING MATERIALS: RED MUD SINTERING USING FLY ASH VIA TAGUCHI'S METHODOLOGY

Suchita Rai,[1,2] Dilip H. Lataye,[3] M. J. Chaddha,[1] R. S. Mishra,[1] P. Mahendiran,[1] J. Mukhopadhyay,[4] ChangKyoo Yoo,[5] and Kailas L. Wasewar[2,5]

[1]Jawaharlal Nehru Aluminium Research Development and Design Centre, Wadi, Amravati Road, Nagpur, India

[2]Advanced Separations and Analytical Laboratory, Department of Chemical Engineering, Visvesvaraya National Institute of Technology (VNIT), Nagpur, Maharashtra 440011, India

[3]Department of Civil Engineering, Visvesvaraya National Institute of Technology (VNIT), Nagpur, Maharashtra 440011, India

[4]Indian Institute of Technology, Gandhinagar, Gujarat, India

[5]Environmental Management & Systems Engineering Lab (EMSEL), Department of Environmental Science and Engineering, College of Engineering, Kyung Hee University, Seocheon-dong 1, Giheung-gu, Gyeonggi-do, Yongin-si 446-701, Republic of Korea

ABSTRACT

"Red mud" or "bauxite residue" is a highly alkaline waste generated from alumina refinery with a pH of 10.5–12.5 which poses serious environmental problems. Neutralization or its treatment by sintering in presence of additives is one of the methods for overcoming the caustic problem as it fixes nearly all the leachable free caustic soda present in red mud. In the present study, feasibility of reducing the alkaline nature of red mud by sintering using fly ash as an additive via Taguchi methodology and its use for brick production, as an alternative to clay, is investigated. The analysis of variance (ANOVA) shows that sintering temperature is the most significant parameter in the process. A pH of 8.9 was obtained at 25–50% of red mud and 50–75% fly ash with water and temperature of 1100◦ C. Alternatively 50% of red mud can be mixed with 50% of fly ash with water at temperature of 1200◦ C to get a pH of about 8.4. The mechanism of this process has been explained with also emphasis on chemical, mineralogical, and morphological analysis of the sintered red mud. The results would be extremely useful in utilization of red mud in building and construction industry.

INTRODUCTION

The Bayer process of extraction of alumina from bauxite remains the most economical process till date. In the Bayer process, the insoluble product generated after bauxite digestion with sodium hydroxide at elevated temperature and pressure to produce alumina is known as "red mud" or "bauxite residue" which is highly alkaline in nature with a pH of 10.5–12.5. Red mud is a mixture of compounds originally present in bauxite and of compounds formed during the Bayer cycle. The main reaction of aluminum oxide dissolution in the Bayer process is as follows:

$$2NaOH + Al_2O_3 \cdot 3H_2O \longrightarrow Na_2O \cdot Al_2O_3 + 4H_2O \quad (1)$$

Depending on the raw material processed, 1–2.5 tons of red mud is generated per ton of alumina produced [1]. An enormous quantity of red mud is generated worldwide (~75 million tons) every year posing a major environmental problem. In India, about 4.71 million tons/annum of red mud is produced which is 6.25% of world's total generation [2]. Red mud is disposed as semidry material in red mud pond or abandoned bauxite mines and as slurry having a high solid concentration of 30–60%. Problems associated with the disposal of red mud waste include its high pH, alkali seepage into underground water, safety in storage, alkaline air borne dust emissions, and the vast area of land required for disposal. Up to 2 tons

Citation: Suchita Rai, Dilip H. Lataye, M. J. Chaddha, et al., "An Alternative to Clay in Building Materials: Red Mud Sintering Using Fly Ash via Taguchi's Methodology," Advances in Materials Science and Engineering, vol. 2013, Article ID 757923, 7 pages, 2013. doi:10.1155/2013/757923.

of liquor with a significant alkalinity of 5–20 g/L caustic (as Na_2CO_3) accompanies every ton of dry mud. Neutralization of red mud will help to reduce the environmental impact caused due to storage activities of the residue and will also open opportunities for the reuse of the residue which till now has been prevented because of the high pH. It will also reduce degradation of clay or synthetic liners of red mud ponds. The overall risk of groundwater contamination would be reduced significantly by neutralizing red mud.

The literature shows that red mud can be treated with any of these methods to make it environmentally benign: acid leaching, CO_2 treatment, bioleaching, seawater neutralization, and sintering. A review of neutralization processes of red mud and its utilization is given by investigators [3].

A considerable research has been done on the utilization of red mud as a raw material for the production of a range of building products by sintering of red mud. Red mud can be used as a constructional/building material in bricks, blocks, light weight aggregates, roofing tiles, glass ceramics, cement industry as cements and special cements, and concrete industry. Red mud is made into useful ceramics articles by mixing 51–90% of the weight of red mud with 10–49% of the weight of at least one mineral and/or silicate containing material, shaping the mixture and firing it at a temperature of 950°–1250°C [4]. A patent [5] claims a process for manufacturing fired bricks wherein 50–90 wt% of red mud can be used along with clay and a water fixing agent. The raw bricks are dried with heated gases at a temperature below 70°C and subsequently fired at a temperature between 900°–1100°C. Though all these studies are based on sintering of red mud and its use, very little has been stated about the change in pH with the sintering temperature and other parameters. The ways in which these factors may interact in the neutralization or treatment process are poorly understood.

Hence the present study focuses on the study of change in pH in the sintering process due to a wide range of factors. The treatment process involves mixing of silicate material such as fly ash with red mud in presence of water. Fly ash is the by-product of burning of coal and is available from coal-based thermal power plants and other coal burning centers. An estimated availability of fly ash in India is about 80–100 million tons per year [6]. Every thermal power plant generating 1000 MW power on an average produces 3500 tons of fly ash every day [6]. It is estimated that an area of about 28,000 ha of land will be necessary to dump all the fly ash available in the country. The present utilization level of fly ash in India is less than 10% [6]. Fly ash contains 1 to 100 μm suspended particulates affecting plant growth besides being a health hazard [7]. Indian red mud is rich in titania and iron in alkali oxides while the fly ash is rich in silica and alumina. These features contribute to the optimum composition for brick making [7].

Hence use of fly ash has been made as a silicate material in sintering of red mud. In this way both the wastes can be made nonhazardous and can be reused.

Taguchi's experimental methodology has been employed to evaluate the pH values. Taguchi's design of experiments has been used extensively in product quality assessment and several other studies [8–12]. Taguchi's fractional factorial design of experiments has been used to examine the effect of significant parameters such as weight of red mud, weight of silicate material added, volume of water, and reaction temperature on the response characteristic (pH of red mud slurry). The average values and the signal to noise () ratio of the quality response characteristic for each parameter at 3 levels of their values have been calculated from the experimental data. The response curves have been graphically represented to reflect any change in the quality characteristic and *S*/*N* ratio with the variation in process parameters. These response curves are used to identify the effects of various parameters on the response characteristic. SSignificant parameters have been identified by using the analysis of variance (ANOVA) on the experimental data.

MATERIALS AND METHODS

Materials

Red mud from an alumina refinery situated at the eastern coast of India (18°49′9″N, 82°57′52″E) was ground to 100 mesh size and used for the study. Chemical composition of red mud is Al_2O_3 (16–17.5%), Fe_2O_3 (53.5–56%), SiO_2 (6-7%), TiO_2 (5-6%), Na_2O (4-5%), and CaO (2-3%). Mineralogically, red mud contains phases of undigested alumina, aluminosilicates, and phases of iron and titania. Alumina is in form of undigested gibbsite, alumogoethite, and sodalite along with silica and sodium (sodium aluminosilicates); iron is in form of hematite, alumogoethite, siderite, and ilmenite; titania is in form of anatase, rutile, and ilmenite; and calcium is in form of calcite. Average particle size of red mud is less than 10 microns with specific surface as 20.26 m^2/g.

The mineralogical composition of untreated red mud is gibbsite: $Al(OH)_3$, sodalite: $Na_2O\,Al_2O_3{\cdot}2SiO_2{\cdot}2H_2O$, hematite: Fe_2O_3, alumogoethite: FeAlOOH, anatase: TiO_2, rutile: TiO_2, ilmenite: $FeTiO_3$, siderite: $FeCO_3$, calcite: $CaCO_3$.

Sodium present in red mud (total caustic soda) is in two forms: free or soluble caustic soda and bound caustic soda. Free caustic soda is the entrained liquor in the red mud slurry which gets incorporated during digestion process and remains with red mud in spite of repeated washings. Free soda is in the form of NaOH, Na_2CO_3, $NaAlO_2$, and so forth. The pH of the red mud is due to the presence of these alkaline solids in red mud. Bound soda is in the form of sodalite complex which can be stated as "NAS" phases: $3(Na_2OAl_2O_32SiO_2)Na_2X$ ($X = CO_3^{2-}, 2OH^-, SO_4^{2-}, 2Cl^-$) (Kurdowski and Sorrentino, 1997) [13]. In red mud, about 20–25% is the free or soluble caustic soda while the rest is in the form of sodalite complex.

Fly ash has the following composition: 57-58% SiO_2, 34–36% Al_2O_3, and 4–6% Fe_2O_3. Mineralogically fly ash contains about 75–80% quartz (SiO_2) with remaining mullite ($2Al_2O_3SiO_2$), alumina in the form of mullite, and iron as hematite.

Experimental Setup

Red mud, fly ash, and water were mixed in different proportions as per the experiments designed by Taguchi's methodology and kept in a crucible at different temperatures in a muffle furnace (CHEMINCO, Kolkata). After sintering the mixture for a fixed duration, the crucible was cooled overnight and the mixture was taken in a beaker and 100 mL distilled water was added.

The mixture was stirred for 30 min using magnetic stirrer (Eltek M S 204). The pH of the slurry was measured after each test using LabX Light Titrator (Potentiometric Titrator), Mettler Toledo GmbH, Switzerland.

Taguchi's Design of Experimental Methodology

Dr. Taguchi of Nippon Telephones and Telegraph Company, Japan has developed a method based on "orthogonal array" experiments which gives much reduced "variance" for the experiment with "optimum settings" of control parameters. Taguchi methods [10] use orthogonal array distribution to design an experiment. A traditional experimental design to compare four independent variables at three different levels each requires a large number of individual experiments (eighty-one experiments). The logistical and resource implications of this experimental design make these experiments very difficult to carry out. By using L_9 orthogonal array with Taguchi method to design an experiment, a study involving 4 factors at 3 different levels can be conducted with only nine individual experiments. Taguchi method consists of 3 phases: designing the experiment, running and analyzing, and confirming and validating the experimental results. In an orthogonal array, each array can be identified by the form $L(B^C)$; the subscript of L, which is designated by A, represents the number of experiments that would be conducted using this design, and B denotes the number of levels within each column which denotes how many levels could be investigated, while the letter C indicates how many factors or variables could be included in the experiment [10].

L_9 orthogonal array consisting of four control parameters (factors), A, B, C, and D, is given in Table 1 and set at minimum, middle, and maximum value (i.e., levels 1, 2, and 3). The parameters were decided based on the preliminary experimentation carried out. In all nine experiments had to be conducted. Each experimental run was repeated thrice. The pH values (replicated thrice as R_1, R_2, and R_3) obtained for each experiment of the L_9 array are shown in Table 2. Subsequently analysis of variance (ANOVA) based on the Taguchi method was carried out using the values to determine the contribution of each parameter to the process of neutralization.

Table 1: Experimental layout of parameters and their levels

Factors	Parameters	Units	Level 1	Level 2	Level 3	Observed value
A	Weight of red mud	g	10	20	30	pH
B	Weight of silicate material	g	10	20	30	
C	Volume of water	mL	8	10	12	
D	Temperature	°C	900	1000	1100	

Table 2: table and observed values of pH

Run	Factors A Weight of red mud	Factors B Weight of silicate material	Factors C Volume of water	Factors D Sintering temp.	Final pH R_1	Final pH R_2	Final pH R_3	Ratios of pH R'_1	Ratios of pH R'_2	Ratios of pH R'_3	S/N ratio
1	20	30	8	1000	9.12	9.25	9.3	0.77	0.76	0.75	−2.40
2	20	20	12	900	10.7	10.65	10.54	0.65	0.66	0.66	−3.63
3	30	10	12	1000	9.65	9.62	9.59	0.73	0.73	0.73	−2.76
4	10	20	10	1000	9.58	9.32	9.31	0.73	0.75	0.75	−2.56
5	30	20	8	1100	9.10	9.15	9.18	0.77	0.77	0.76	−2.32
6	30	30	10	900	10.68	10.55	10.61	0.66	0.66	0.66	−3.62
7	10	30	12	1100	8.9	8.92	8.88	0.79	0.78	0.79	−2.09
8	20	10	10	1100	9.27	9.25	9.3	0.76	0.76	0.75	−2.43
9	10	10	8	900	10.69	10.6	10.582	0.65	0.66	0.66	−3.62

Treatment of Red Mud at High Temperature

The initial pH of the red mud slurry was 10.7 (50 gm of red mud was stirred with 100 mL of distilled water for 30 min). After each experimental run, designed as per Taguchi's experimental methodology, pH of slurry was measured. The slurry sample with the lowest pH value and close to 7.0 (of experimental run number 7) was filtered using Whatman paper no. 40. The filtered red mud was dried and analyzed.

Chemical constituents of red mud mix obtained after sintering at the optimized condition were analyzed by using Wet Chemical Method. Untreated red mud, fly ash, and sintered red mud were analyzed mineralogically for determination of phases using XRD, X-ray diffractometer (PANalytical X-Pert Pro), using Cu Kα radiation (λ = 1.54060 Å). Scanning electron microscopy (SEM) has been conducted to study the morphology of neutralized red mud using Electron Probe Analyzer (MAKE: JEOL, Japan, JXA-840A).

The slurry sample was prepared from red mud sintered at the optimized condition of parameters (of experimental run number 7) and 100 mL distilled water and steered thoroughly. pH rebound phenomena were studied for 2-week time duration with this sample of red mud slurry.

Analysis of Experimental Data

The experimental data generated was analyzed by analysis of variance (ANOVA). Taguchi excel sheet is used for all the analysis which is based on Taguchi method for determination of main effects of the process parameters.

R' values are the ratio of neutral pH (7) to experimental pH value. R'_1, R'_2, and R'_3, and are the values calculated from, R_1, R_2, and R_3. and . On this basis the "higher-is-better" quality characteristic was used in the analysis of experimental data. The plot of response curves and analysis of variance for raw data and S/N ratio data was used for the analysis of results. The mean of the response characteristic pH (Y_{opt}) at the optimal condition was estimated as

(2)

where T/N is the overall mean of the response, where T is the grand total of all results and N is the number of experiments, and A_{1avg}, B_{3avg}, C_{3avg}, and D_{3avg} represent average values of response at the first level of parameter A and third levels of parameters B, C, and D, respectively.

In order to confirm that the optimal parametric values using Taguchi's methodology are valid, selected confirmatory experiments were carried out under optimal conditions. The average of the results of the confirmation experiment is then compared with the anticipated average based on the optimal parameters and levels tested by Taguchi's methodology. The Y_{opt} value obtained is 0.78 and pH value of which comes out to be 8.969 which is close to the optimized experimental value of 8.8–8.92 in run 7. The deviation between actual and determined pH by Taguchi's methodology is much within 5%, and hence it is confirmed that the optimal parametric values determined by Taguchi's design are valid.

Utilization of Red Mud

Red mud was used as an alternative to clay in making bricks. Four standard size bricks were made by mixing suitable proportions of red mud, fly ash, and water. The bricks were made in commercial fly ash brick factory machine, that is, Brickman machine. They were pressed at ~200 kg/cm^2, dried overnight in an oven at 110°C, and sintered at 1100°C for 2 hours. The composition of the raw materials used and the properties of the bricks obtained are given in Table 6. The brick samples were tested for cold compressive strength (CCS), water absorption (WA), efflorescence and bulk density.

RESULTS AND DISCUSSION

Effect of the Process Parameters on pH

Table 3 provides the raw data for the average value ratios of pH and S/N ratio for each parameter at levels 1, 2, and 3. The sintering temperature from level 1 to level 2 has the greatest influence on value of pH. The difference between the levels also shows the same trend. Figure 1 shows the response curves of the individual parameters on the ratio of pH values, that is, R' values (7/experimental pH value) and S/N ratio.

Table 3: Average and main effects of ratios of pH values—Raw and S/N

Factors	Raw data, average value			Main effects		S/N data, average value			Main effects (S/N data)	
	L_1	L_2	L_3	L_2-L_1	L_3-L_2	L_1	L_2	L_3	L_2-L_1	L_3-L_2
A	0.74	0.73	0.71	−0.01	−0.02	−1.49	−2.80	−3.01	−1.30	−0.21
B	0.72	0.72	0.74	0.00	0.02	−2.92	−2.89	−2.70	0.03	0.19
C	0.72	0.73	0.72	0.01	0.00	−2.89	−2.79	−2.83	0.10	−0.04
D	0.66	0.75	0.76	0.09	0.01	−3.62	−2.52	−2.37	1.11	0.15

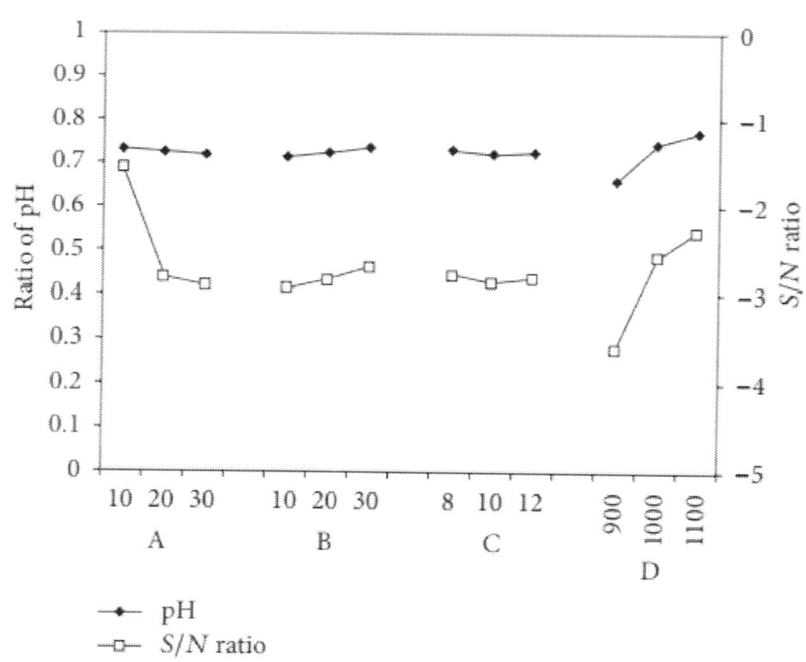

Figure 1: Main effects of each factor on ratio of pH along with S/N ratio of treated red mud slurry (ANOVA analysis). A: wt. of red mud (g); B: wt. of silicate material (g); C: vol of water (mL); D: temp. (°C).

From Figure 1, it can be seen that the ratio of pH decreases with the increase in the weight of red mud (factor A). It is due to the increase in free caustic soda in the slurry with the increase in quantity of red mud. There is a sharp decrease in the ratio of pH values when factor A changes from level 1 to 2 and also when the quantity of red mud changes from level 2 to 3. Sintering temperature (factor D) definitely decreases the pH of the slurry as a sharp increase in the ratio of pH is observed from level 1 to 2. Further increase in the ratio of pH values (i.e., decrease in pH) is observed from level 2 to level 3. Not much change in pH is observed at 900°C but the change is seen above this temperature. The increase in temperature decreases the pH substantially due to the fusion of free soda with the silica and alumina present in fly ash. The overall results convey that the pH of the slurry is mostly affected by the sintering temperature. The lowest pH is obtained with the weight of red mud at level 1, weight of silicate material at level 3, water at level 3, and temperature at level 3 ($A_1B_3C_3D_3$). The contribution of individual parameters was weighted as shown in Table 4 with analysis of variance (ANOVA) to see their effect on desired response characteristic (pH). The most significant parameters are found to be the sintering temperature with a very high contribution of 94.48% followed by weight of red mud (1.09%) and weight of silicate material (3.08%) and water (0.5%).

Table 4: ANOVA table

Factors	Total variance of each factor	Degree of freedom	Variance	Variance ratio	Pure sum of sq.	Percent contribution
A	0.001	2	0.000	10.5651	0.00062816	1.090
B	0.002	2	0.001	29.82904	0.00189326	3.079
C	0.000	2	0.000	4.087215	0.00020274	0.422
D	0.060	2	0.030	915.4204	0.06005186	94.480
Error	0.001	18	0.000	1	0.00032836	0.929
Totals	0.064	26	0.032			100.000

Confirmation Experiment

From Table 2, it can be seen that the experimental condition ($A_1B_3C_3D_3$) was the one in which the lowest value of pH value was achieved (8.88–8.9) which was close to the neutral value of 7 in all the 3 replications. As per the Guidelines of Australian and New Zealand Environment and Conservation Council (ANZEX) and Agriculture and Resource Management Council of Australia and New Zealand (ARMCANZ), the liquor being strongly alkaline with a high pH requires neutralization to a pH below 9 with an optimum value of 8.5–8.9 before becoming environmentally benign [14]. The optimized condition obtained from Figure 1 was $A_1B_3C_3D_3$having pH value of about 8.90 which is the lowest of all the values. Hence these conditions were fine-tuned by increasing the temperature to 1200°C and the pH achieved was 8.34 which is within the acceptable limit. Hence the final optimized conditions to achieve a pH of 8.9 were 25–50% of red mud and 50–75% silicate material, with water (10%) and a temperature of 1100°C. Alternatively 50% of red mud can be mixed with 50% of fly ash with water at temperature of 1200°C to get a pH of about 8.4. The results are in accordance with those obtained by investigators [4]. In absence of sintering, studies carried out at Jamaica [15] show that in neutralization of red mud with acidic fly ash, a large amount of fly ash is required with very slow neutralization rate (approx. 150 days to get a pH from 12.5 to 11).

Red Mud Analysis after Neutralization

Chemical Analysis

The chemical analysis of red mud after sintering is Al_2O_3: 24.50%, Fe_2O_3: 31.92%, SiO_2: 34.50%, TiO_2: 2.53%, Na_2O: 4.5%, and CaO: 1.09%. Sintered red mud has high alumina content, low iron content, and high silica content due to the addition of silicate material (fly ash) and subsequent sintering of the mixture as compared to the original red mud.

Formation of Different Phases

To investigate the phase structures of the red mud, the X-ray diffraction pattern was measured. Figure 2 shows the XRD pattern which shows a well crystallized sintered red mud powder. The peaks of the red mud show the phases as shown in Table 5. Alumina present in red mud in the hydroxide form has been converted into corundum which is alpha phase of alumina. Alumina has also reacted with silica from fly ash to form aluminosilicates (sillimanite) which impart strength to the fused material. Caustic soda has fused with alumina and silica to form sodium silicates and sodium aluminosilicates. Some of the free soda is still present in the sintered red mud in form of sodium aluminate. Hence pH value of above 8.9 is observed after the process. Calcium aluminosilicates and calcium silicates are also formed.

Table 5: Mineralogical phases in sintered red mud

Compound name	Chemical formula	JCPDS no.
Corundum	Al_2O_3	00-046-1212
Sillimanite	Al_2SiO_5	00-038-0471
Jadeite	$NaAlSi_2O_6$	00-022-1338
Burnt ochre	Fe_2O_3	00-033-0664
Aluminum iron oxide	$AlFeO_3$	00-030-0024
Brookite	TiO_2	00-015-0875
Calcium oxide	CaO	00-037-1497
Calcium silicate	Ca_2SiO_4	00-031-0299
Grossular	$Ca_3Al_2(SiO_4)_3$	00-039-0368
Sodium aluminum oxide	$NaAlO_2$	00-033-1200

Table 6: Composition and properties of bricks

	Brick composition		Properties of bricks				
	Red mud (%)	Fly ash (%)	CCS (kg/cm^2)	WA (%)	Efflorescence	Bulk density (g/cc)	Class of brick
Composition 1	50	50	31.7	24.0	Nil	1.47	Class 3.5
Composition 2	45	55	66.93	24.50	Nil	1.49	Class 5
Composition 3	40	60	86.73	23.8	Nil	1.50	Class 7.5
Composition 4	35	65	134.18	24.10	Nil	1.51	Class 12.5

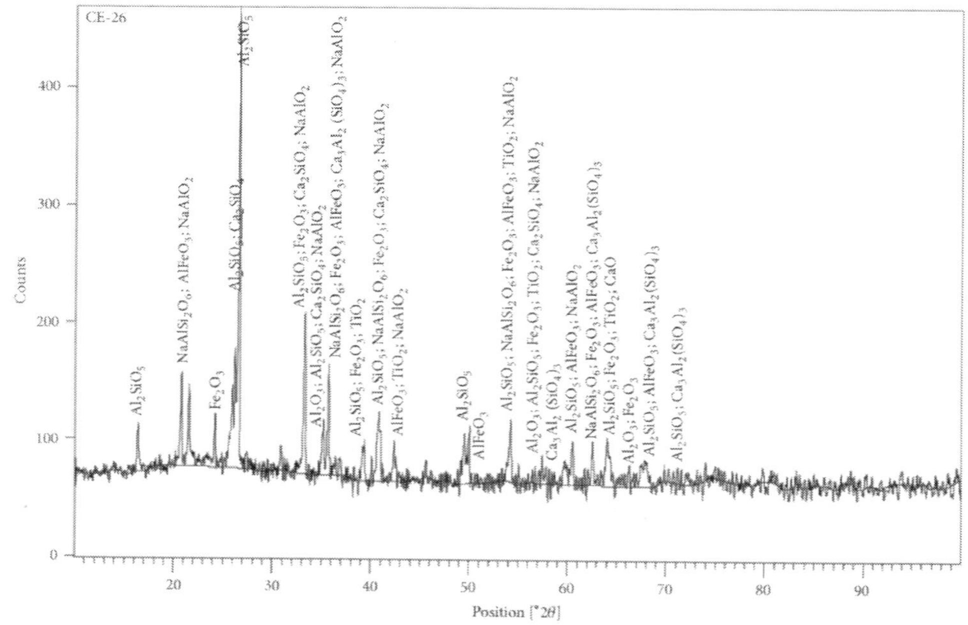

Figure 2: Mineralogical phases of sintered red mud mix with fly ash (XRD). Conditions: red mud: fly ash (1 : 3) with 10% water, 1100°C.

Some of the reactions taking place in the process are

$$Na_2CO_3 + SiO_2 \longrightarrow Na_2SiO_3 + CO_2 \quad (3)$$

$$2NaOH + SiO_2 \longrightarrow Na_2SiO_3 + H_2O \quad (4)$$

$$Na_2SiO_3 + Al_2O_3 \longrightarrow Al_2O_3 \cdot SiO_2 \cdot Na_2O \quad (5)$$

$$3Al_2O_3 + 2SiO_2 \longrightarrow 3Al_2O_3 \cdot 2SiO_2 \quad (6)$$

Morphology of Red Mud

Figure 3 shows the morphological structure of original (untreated) red mud as seen in SEM which indicates scattered fine particles of about 1 micron size. Globular particles with smooth surface which are formed due to the formation of sodium aluminosilicates are observed (Figure 4) in sintered red mud.

Figure 3: SEM photograph of untreated red mud.

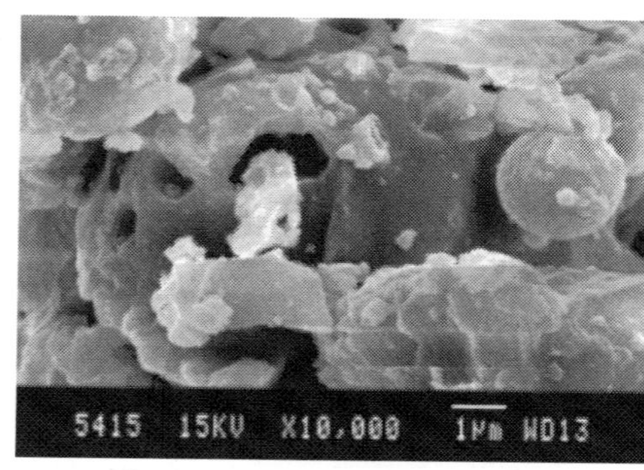

Figure 4: SEM photograph of sintered red mud.

Physical Characteristics of Red Mud

Physical characteristics of sintered red mud improve to a large extent as fast filtration of sintered red mud slurry is observed. Also the color of filtrate is clear as compared to filtrate obtained with untreated red mud.

Utilization of Sintered Red Mud

The chemical, mineralogical, and morphological properties of sintered red mud show that the material can be used in construction industry in place of clay. It can be utilized as a replacement of naturally available building material and for the production of new building materials. The developed bricks are having very good strength with CCS% ranging from 30 to 135 kg/cm^2 with slightly more value of WA% as shown in Table 6. The bricks are included in the class of 3.5 to 12.5. These promising results show that technically these bricks are suitable for construction of load-bearing walls. Red mud blocks can also be made from these mixtures which can also be used as a low cost housing material. Red mud in the range of 40–60% can be used for making good quality bricks. Efforts have been made at Central building Research Institute (CBRI), India, to produce burnt clay bricks by partially replacing the clay with red mud, lime, and fly-ash [16].

pH Rebound

The increase in pH and aluminium concentration after the neutralization of red mud is commonly known as reversion or rebound. No pH rebound was observed with the optimized sample as the pH remained the same even after two weeks. This means that there is no more caustic soda leaching after its initial leaching from sintered red mud.

CONCLUSION

The paper has discussed in detail the feasibility of sintering of red mud to deal with the caustic soda in it using Taguchi's methodology. The pH of red mud slurry reaches values within the acceptable limits after sintering red mud with a silicate material (fly ash) at a very high temperature of 1100–1200°C. The method of treating red mud with fly ash along with water can be a successful process to tackle the caustic soda problem. Also the resultant material can be reused as a construction material in place of clay. Taguchi's experimental methodology is found to be a valuable tool. The ANOVA analysis using Taguchi methodology shows that the most significant parameters are temperature of calcinations (94.48%) followed by weight of red mud (1.09%) and quantity of fly ash (3.079%). The optimized conditions to achieve a pH value of about 8.9 were 25–50% of red mud mixed with 50–75% of fly ash with water (10%) and a temperature of 1100°C. Alternatively about 50% of red mud with 50% of fly ash and 1200°C can be used where the pH obtained is within the permissible limit of 8.4.

The study shows that treatment of red mud with fly ash at a very high temperature lowers down the pH and hence alkalinity of red mud slurry permanently without pH rebound. This also modifies the physical characteristics as is evident from the morphology of red mud. Sintering of residue can be carried out to fix all leachable soda and to achieve high strength during the sintering process. Thus red mud can be used as an alternative to clay material in building and construction industry. Both red mud and fly ash can be reused. However, the cost would be very high due to the high energy consumption as an elevated temperature is required for sintering. But the process would certainly help in reducing pollution of soil, air, and water and would considerably save resources of mother earth.

REFERENCES

1. R. K. Paramguru, P. C. Rath, and V. N. Misra, "Trends in red mud utilization—a review," Mineral Processing & Extractive Metallurgy Review, vol. 26, no. 1, pp. 1–29, 2005. · ·
2. U. V. Parlikar, P. K. Saka, and S. A. Khadilkar, "Technological options for effective utilization of bauxite residue (Red mud)—a review," in International Seminar on Bauxite Residue (RED MUD), Goa, India, October 2011.
3. S. B. Rai, K. L. Wasewar, J. Mukhopadhaya, K. Chang, and H. Uslu, "Neutralization and Utilization of red mud for its better waste management," Archives of Environmental Science, vol. 6, pp. 13–33, 2012.
4. F. Puskas, "Process for the utilization in the ceramics industry of red mud from alumina plants," US Patent 4368273, 1983.
5. B. Garhard, "Method for producing bricks from red mud," US Patent 3886244, 1975.
6. K. G. K. Warrier, P. Krishna Pillai, P. Perumal, and C. L. Verma, "Liquid phase sintering of flyash to produce high volume flyash ceramics for a variety of applications," CSIR News, vol. 54, pp. 13–14, 2004.
7. Wealth from waste, special Report, "Construction World (Indian Edition)," 2002, http://www.constructionupdate.com/products/constructionworld/2002/jan2002/010.html.
8. D. M. Byrne and S. Taguchi, "The Taguchi approach to parameter Design," Quality Progress, vol. 20, no. 12, pp. 19–26, 1987. ·
9. P. J. Ross, Taguchi Techniques for Quality Engineering, McGraw Hill, New York, NY, USA, 1996.
10. G. S. Peace, Taguchi Methods: A Hands-on Approach, Addison-Wesley, 1993.
11. S. B. Rai, K. L. Wasewar, D. H. Lataye et al., "Neutralization of red mud with pickling waste liquor using Taguchi›s design of experimental methodology," Waste Management and Research, vol. 30, no. 9, pp. 922–930, 2012.
12. S. B. Rai, K. L. Wasewar, D. H. Lataye, and J. Mukhopadhyay,"Feasibilityofredmudneutralizationwith Seawater using Taguchi›s methodology," International Journal of Environmental Science and Technology, vol. 10, no. 2, pp. 305–314, 2013.
13. W. Kurdowski and F. Sorrentino, Waste Materials Used in Concrete Manufacturing, Edited by S. Chandra, William Andrew, Noyes, 1997.
14. C. Hanahan, D. McConchie, J. Pohl, R. Creelman, M. Clark, and C. Stocksiek, "Chemistry of seawater neutralization of bauxite refinery residues (red mud)," Environmental Engineering Science, vol. 21, no. 2, pp. 125–138, 2004. · ·
15. S. Khaitan, D. A. Dzombak, and G. V. Lowry, "Chemistry of the acid neutralization capacity of bauxite residue," Environmental Engineering Science, vol. 26, no. 5, pp. 873–881, 2009. · ·
16. Dass and S. K. Malhotra, "Lime-stabilized red mud bricks," Materials and Structures, vol. 23, no. 4, pp. 252–255, 1990. · ·

CHAPTER 10

EFFECT OF THE INSULATION BY THE MUD ON THE CONVECTION IN BUILDING IN THE SAHARA OF ALGERIA - CASE OF BECHAR CITY

Elhadj Benachour[1], Belkacem Draoui[1], Baachir Imine[2], Mohammed Hasnat[1], Lakhdar Rahmani[1]

[1] Laboratory of Energy in Arid Regions, University of Tahri Mohamed - Bechar, P.O. P 417. 08000 , Road Kenedza , Bechar, Algeria

[2] Department of Mechanical Engineering, University of Sciences and Technology of Oran-Mohamed Boudiaf, El-Mnaouar, Oran, Algeria

ABSTRACT

The thermal behavior of the buildings is a current problem which arouses the interest of many researchers. Indeed, the control of the loads of air conditioning or heating requires a thorough knowledge. Since the thermal quality of the buildings in the Maghreb is there generally very insufficient in particular in the Sahara of Algeria, the insulation proved that it is a very important parameter to minimize the thermal diffusion process inside on all when the insulator is a local material as the Mud which is in our region with a significant amount and almost free quantity. This work presents a study of numerical simulation aiming at the role and the influence of the insulation by the mud Reduced like a local material to the town of Bechar located at the south west of Algeria, or one is interested in a comparative study for pursued these goals. In this context, an analogy was used for the functions which are discretized by the finite difference method and integrated in the Fluent code which is based on the finite volume method. The validation of this procedure was confirmed while comparing some results. The results are presented in the form of distributions of the isotherms, the streamlines, local and average Nusselt of which the goal to study the influence on comfort.

INTRODUCTION

The thermal behavior of buildings is a current problem which arouses the interest of numerous and many researchers. Indeed, energy consumptions and greenhouse gas emissions are very high to the world. It is thus necessary to reduce the energy needs while improving the energy efficiency, in particular in the sector of the building. This one offers possibilities of considerable reduction of the energy consumption, in particular by improving the thermal insulation [1]. To reduce them, various strategies are organized, among which several aim at the improvement of the energy efficiency of buildings [2-5]. We indeed notice that numerous housing is not isolated yet. About 65 % of the walls of housing arrange no insulation. A significant portion and An important part of these not isolated external walls are massive solid brick walls or the thermal perceived in buildings was aimed by some research [6-9].Other works suggested studying the effect of the insulation on the energy consumptions and the energy performances in buildings. The insulation of the external walls is complex and several processes exist: insulation by filling of the slide, insulation by outside and insulation by the interior. In restoration, it very often arrives that the choice is restricted for technical reasons, economic and/or patrimonial reasons. Although it can generate many risks on the level of the hygrothermic and mechanical behavior of the walls, Then, since the thermal isolation is one of the investments

Citation: Elhadj Benachour, Belkacem Draoui, Baachir Imine, Mohammed Hasnat, Lakhdar Rahmani, Effect of the insulation by the mud on the convection in building in the Sahara of Algeria - case of Bechar City, DOIhttp://dx.doi.org/10.1051/epjconf/201611402003

of most profitable energy saving, in particular in the new building, Several numerical studies and experimental were realized[10-16]. The insulation by the interior then becomes often the only possible solution in particular with the buildings in the Sahara of Algéria. The hygrothermic parameters of the existing wall and reported materials must be well-known to design powerful and durable walls in time. Our study consists in studying the thermal transfer through walls. Two types of walls are studied:

- A not isolated wall,
- A wall isolated inside by the mud.

This work presents a study of numerical simulation aiming at the role and at the influence of the insulation by the mud as a local material at the city of Bechar situated in the south west of Algeria, or we are interested in a comparative study for aimed these goals. In this context, an analogy was used for the used functions which are discretized by the finite difference method and integrated in the Fluent code which is based on the finite volume method. The validation of this procedure was confirmed by comparing some obtained results. The results are presented in the form of distributions of isotherms, Stream lines and the local Nusselt number and average Nusselt number of whom (which) the purpose to study the influence on the comfort.

GEOMETRIC CONFIGURATIONS

By giving itself two configurations, the one being described the wall without insulation, and the other one with insulation by the mud. The studied configuration is represented on the figure 1. We chose the most widespread construction in the area. The studied room is a structure of a multi-house zone, but in is interested in only one, of dimension 16 m3 (4 x 4 x 4).To simplify the calculation we are interested in a phenomenon of two dimensions. The outer walls are simple concrete walls of 20 cm, with the mortar coating outside surface of negligible thickness compared to the thickness of the outside wall. The roof is insulated [supposed or presumedly adiabatic]. And on the bottom floor earth (ground) is a concrete slab with a wooden frame [supposed adiabatic]. Temperatures imposed on interior and exterior walls are captured experimentally in one day of a summer of August the Bechar city to 15h00.Then to model this phenomenon. This is a square cavity of length L and a height h. The left and right vertical walls of the cavity are heated to a constant temperature equal to Tc 48 ° C (Tc = 1 for the dimensionless case) Tf equal to 25 ° C (Tf = 0.5 for the dimensionless case), and the other walls are maintained adiabatic . One will exploit the existence and the absence of the interior insulation to facilitate the comparative study. The thickness of concrete wall e1 =L/20, and the thickness of the insulation interior by the Mud is e_2 =L/20. The factor of the form is A = 1.

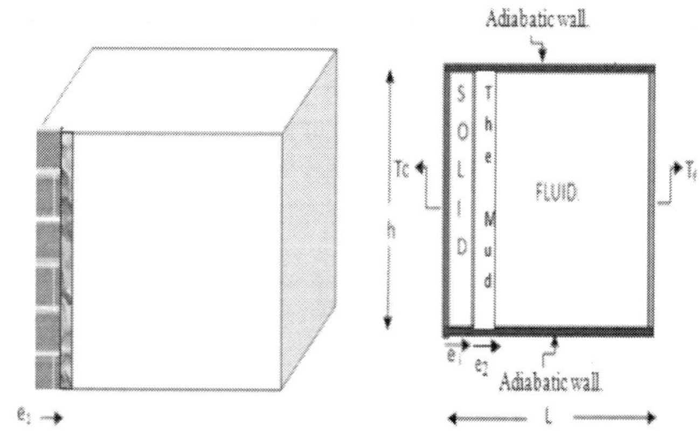

Figure 1: Schematic of the studied configuration

To simplify the problem, assume that:

- The fluid is Newtonian and incompressible.
- The heat dissipation by viscous friction is neglected.
- The Boussinesq approximation is considered

OBJECTS

We can express our aim for this study in the following points:

- Study the Modelling of the coupling of convection - conduction.
- Study of the effect of the distribution of the heat inside buildings on the convection.
- Comparative study between the construction of buildings with and without insulation.
- Study the effect of the Rayleigh number on the convection in building.
- Test the reliability of the insulation with the Mud as local material.

MATHEMATICAL MODEL

The fluid is assumed incompressible and obeys the Boussinesq approximation. In these cases, continuity in two dimensions and the equations governing the flow and energy is given by:

Continuity.

$$\frac{\partial U}{\partial X}+\frac{\partial V}{\partial Y}=0 \tag{1}$$

X-momentum.

$$\frac{\partial U}{\partial t}+U\frac{\partial U}{\partial X}+V\frac{\partial U}{\partial Y}=-\frac{1}{\rho}\frac{\partial P}{\partial X}+\upsilon\nabla^2 U \tag{2}$$

Y-momentum.

$$\frac{\partial V}{\partial t}+U\frac{\partial V}{\partial X}+V\frac{\partial V}{\partial Y}=-\frac{1}{\rho}\frac{\partial P}{\partial X}+\upsilon\nabla^2 V+g\beta\frac{\partial T}{\partial X} \tag{3}$$

Energy.

$$\frac{\partial T}{\partial t}+U\frac{\partial T}{\partial X}+V\frac{\partial T}{\partial Y}=\frac{\lambda}{\rho C_p}\nabla^2 T+\frac{1}{\rho C_p}\varphi \tag{4}$$

The derived equation of motion (2) over Y and the equation of motion (3) by contributing to x, then, after subtracting the two equations obtained, we obtain the equations dimensionless variables in writing Helmotz in terms of vorticity and stream function formulation are as follows:

$$\frac{\partial \omega}{\partial t}+U\frac{\partial \omega}{\partial X}+V\frac{\partial \omega}{\partial Y}=\Pr \nabla^2\omega+Ra\Pr\frac{\partial T}{\partial X} \quad (5)$$

$$\frac{\partial T}{\partial t}+U\frac{\partial T}{\partial X}+V\frac{\partial T}{\partial Y}=\frac{\partial^2 T}{\partial X^2}+\frac{\partial^2 T}{\partial Y^2}+\frac{\varphi}{\rho Cp}\frac{L^2}{a\Delta T} \quad (6)$$

$$\frac{\partial^2 \psi}{\partial X^2}+\frac{\partial^2 \psi}{\partial Y^2}=-\omega \quad (7)$$

The stream function and vorticity are related to the velocity components by the following expressions:

$$U=\frac{\partial \psi}{\partial Y},\ V=-\frac{\partial \psi}{\partial X}\ \ and\ \ \omega=\frac{\partial V}{\partial X}-\frac{\partial U}{\partial Y} \quad (8)$$

The dimensionless parameters in the equations above are defined as follows:

$$\begin{cases} X=\frac{x}{H}, & Y=\frac{y}{H}, \quad U=\frac{u}{u_i} \\ V=\frac{v}{u_i}, P=\frac{P}{\rho u_i^2}, & T=\frac{T-T_i}{T_c-T_i}, t^*=\frac{t}{L^2/a} \end{cases} \quad (9)$$

For solid, we are interested only in the following heat equation:

$$\frac{\partial T}{\partial t}=\frac{1}{a}\left(\frac{\partial^2 T}{\partial x^2}+\frac{\partial^2 T}{\partial y^2}\right) \quad (10)$$

PROCEDURE OF SIMULATION

The numerical calculation was conducted using the computer code "fluent" 6.2.13. The numerical procedure used in this work is that of finite volume. It involves the integration of differential equations of mathematical model on finite control volumes for the corresponding algebraic equations. The SIMPLE algorithm [17] was chosen for the coupling speed pressure in the Navier-Stokes equations on a staggered grid. The convective terms in all equations are evaluated using the schema upwind first order. The discretization of the time term is made in a totally implicit scheme. The convergence of the solution is considered reached when the maximum relative change of all variables (u, v, w, p, t) between two successive time is not less than 104 . With an aim of following well any variation of the fields thermal and hydrodynamic, we used a uniform grid of 14241 nodes and 14480 Elements in nonstationary mode.

VALIDATION OF THE CODE OF CALCULATION

Validation of the Mode

The computer code was validated on a natural convection problem of stale air in a square cavity with vertical walls differentially heated and adiabatic horizontal walls. Our results were compared with those obtained by De Vahl Davis (1983) [18] .The latter dealt with the same problem by adopting the finite difference method with the vorticity-stream function formulation (see Table 1).

Table1: Comparisons of the results of validation

	$Ra = 10^3$	$Ra = 10^4$	$Ra = 10^5$
Nu (average number) (hot wall)	**present study:** 1.2	**present study :** 2.257	**Present study :** 4.64
Nu (average number) (hot wall)	**De Vahl Davis [18]:** 1.12	**De Vahl Davis [18]:** 2.243	**De Vahl Davis [18] :** 4.52

RESULTS AND DISCUSSION

The structure of the flow, the temperature field and heat transfer through the hot wall are discussed in this section. In this study, to target the most important goal, we will show and studied the effect of the insulation by the Mud on dynamic and thermal behavior of the fluid in the cavity.

Isotherms

The isotherms are shown in Figures [2-9]. The heat distribution in the cavity is in accordance with the fluid circulation revealed by isotherms and iso currents. Indeed we find a heating fluid from the interface, if it causes the change of the heat distribution in the cavity (see figures) for different numbers of Rayleigh For a fixed value of the number of Prandtl equal 0.71. Gradually, as the Rayleigh number has increased, the isotherms become increasingly wavy and heat transfer increases, so the flow intensifies and natural convection is expanding and predominates (natural convection is predominant).

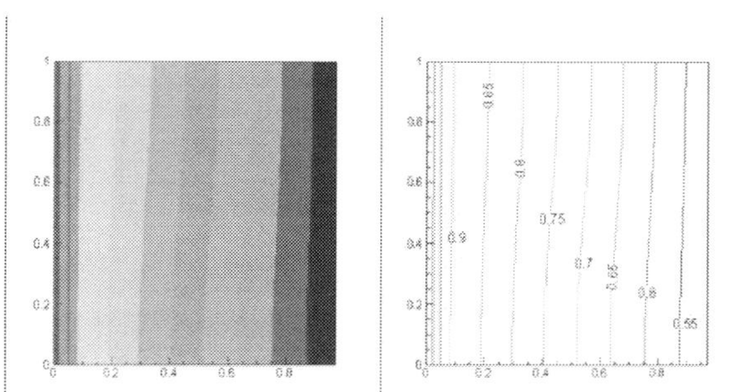

Figure 2: The Isotherms For uninsulated wall, Rayleigh Number Equal 10^3 , Pr=0.71

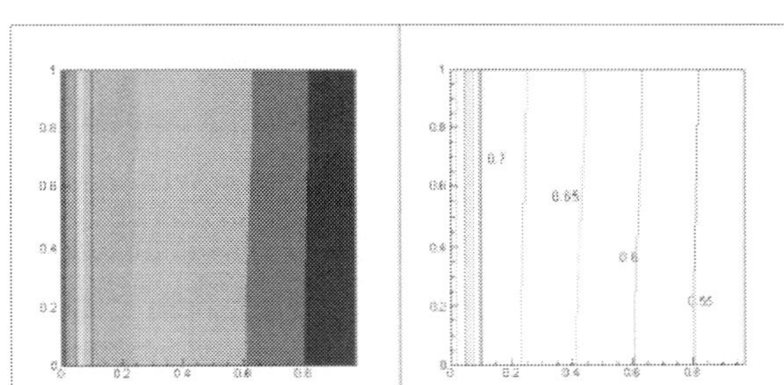

Figure 3: The Isotherms For Insulated Wall, Rayleigh Number Equal 10^3 , Pr=0.71

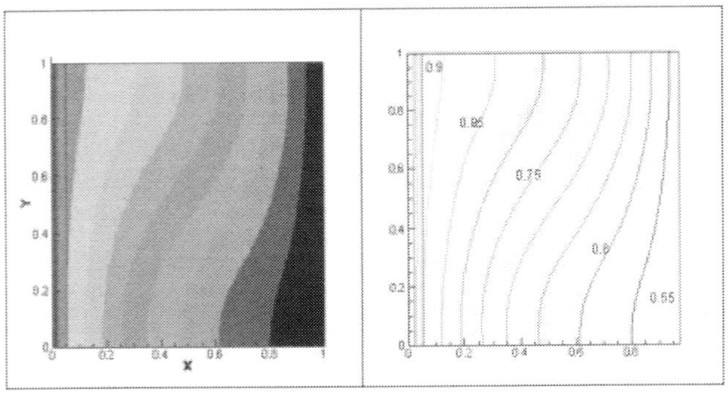

Figure 4: The Isotherms For uninsulated wall, Rayleigh Number Equal 10^4, Pr=0.71

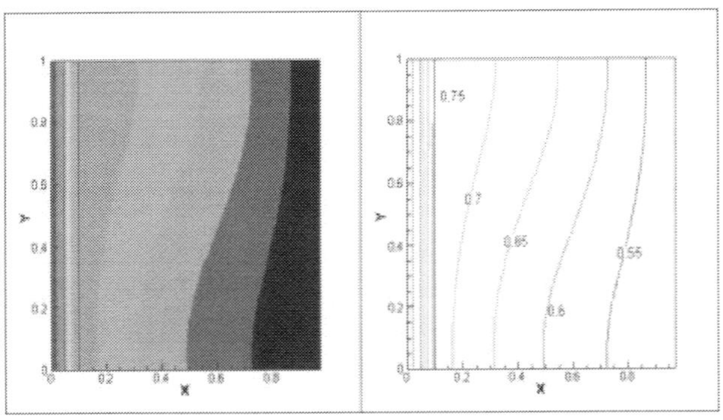

Figure 5: The Isotherms For Insulated Wall, Rayleigh Number Equal 10^4, Pr=0.71

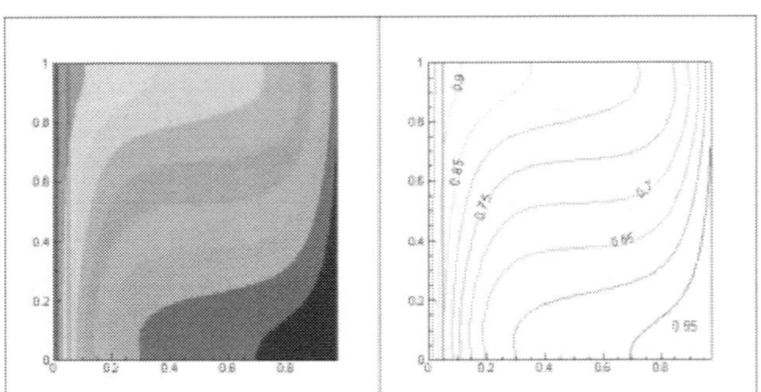

Figure 6: The Isotherms For uninsulated wall, Rayleigh Number Equal 10^5, Pr=0.71

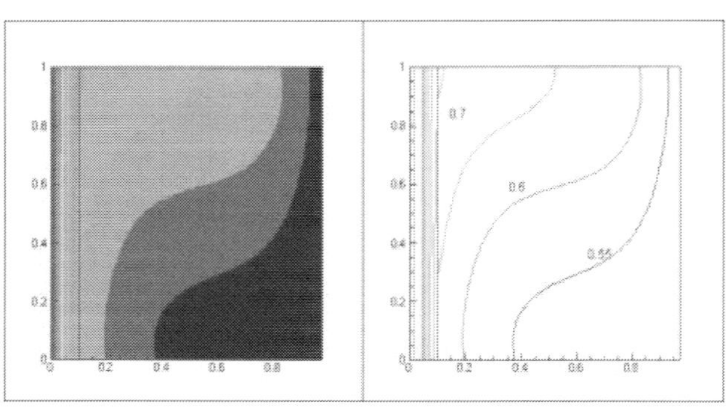

Figure 7: The Isotherms For Insulated Wall, Rayleigh Number Equal 10^5, Pr=0.71

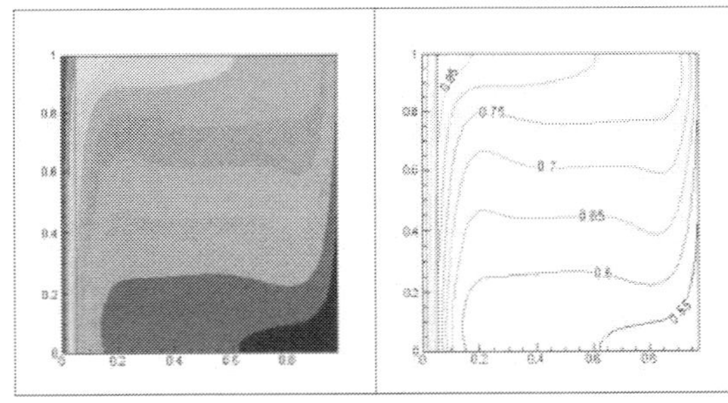

Figure 8: The Isotherms For uninsulated wall, Rayleigh Number Equal 10^6, Pr=0.71

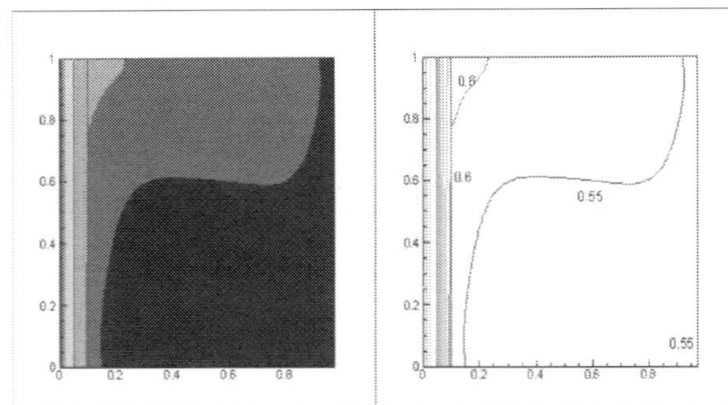

Figure 9: The Isotherms For Insulated Wall, Rayleigh Number Equal 10^6, Pr=0.71.

Streamlines

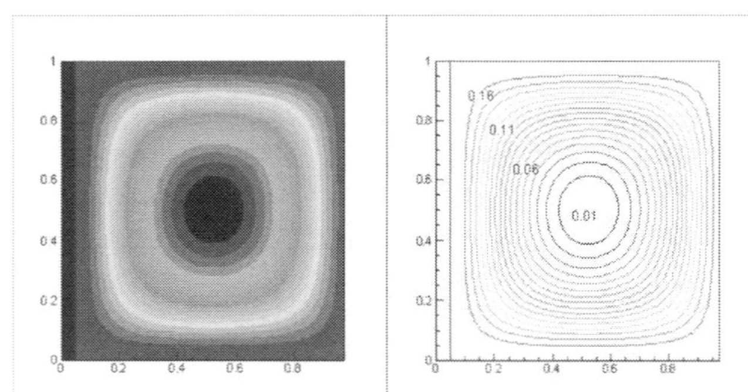

Figure 10: Streamlines obtained For uninsulated wall, Rayleigh Number Equal 10^3, Pr=0.71

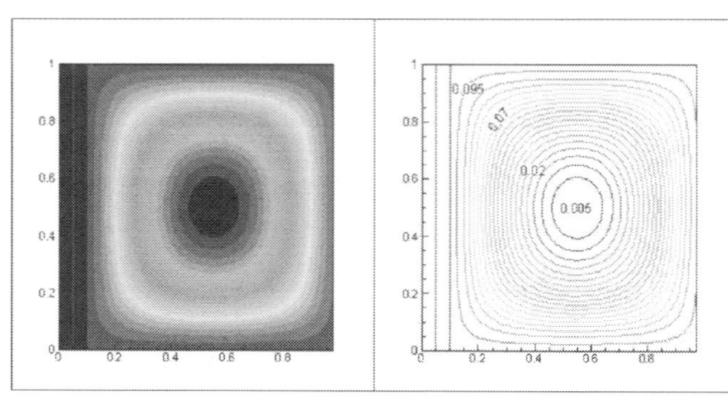

Figure 11: Streamlines obtained For Insulated wall, Rayleigh Number Equal 10^3, Pr=0.71

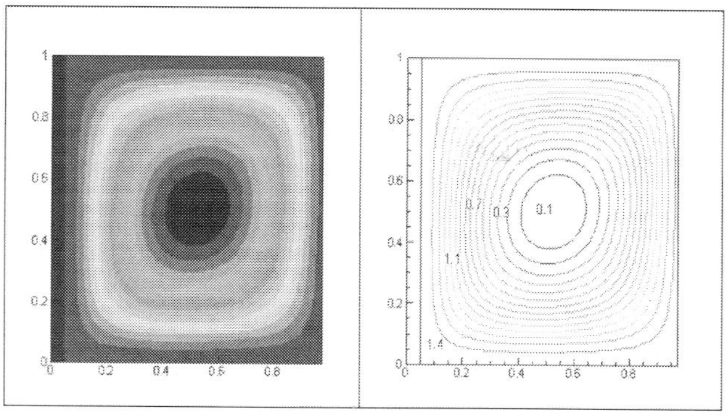

Figure 12: Streamlines obtained For uninsulated wall, Rayleigh Number Equal 104 , Pr=0.71

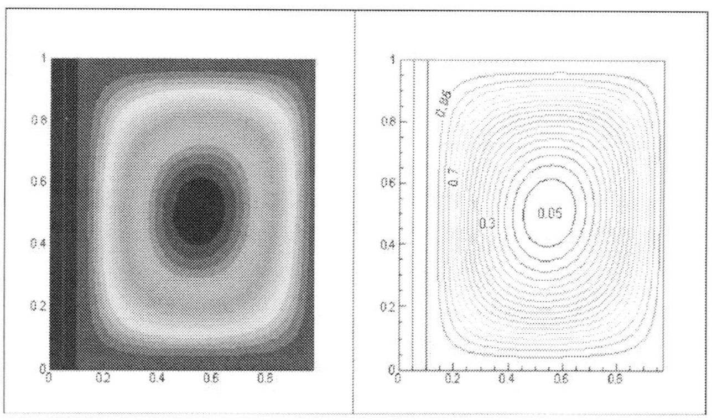

Figure 13: Streamlines obtained for Insulated wall, Rayleigh Number Equal 10^4 , Pr=0.71.

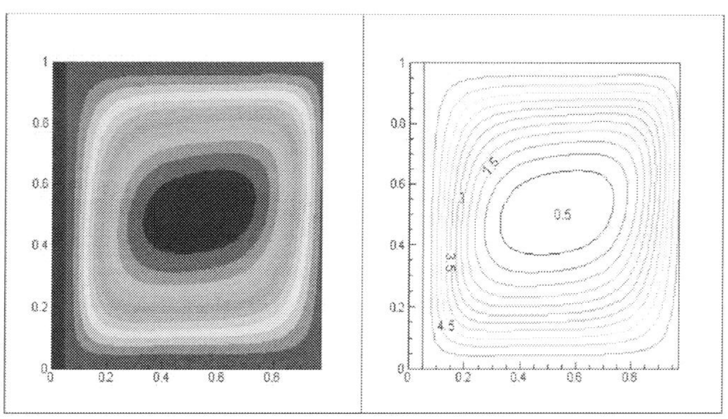

Figure 14: Streamlines obtained for uninsulated wall, Rayleigh Number Equal 10^5 , Pr=0.71

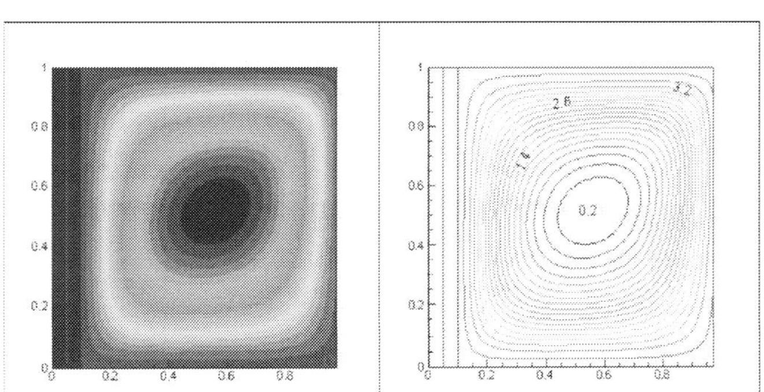

Figure 15: Streamlines obtained For Insulated Wall, Rayleigh Number Equal 10^5 , Pr=0.71.

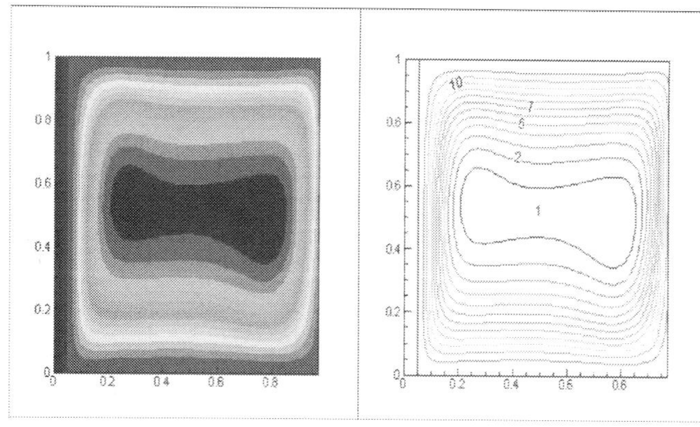

Figure 16: Streamlines obtained for uninsulated wall, Rayleigh Number Equal 10^6 , Pr=0.71

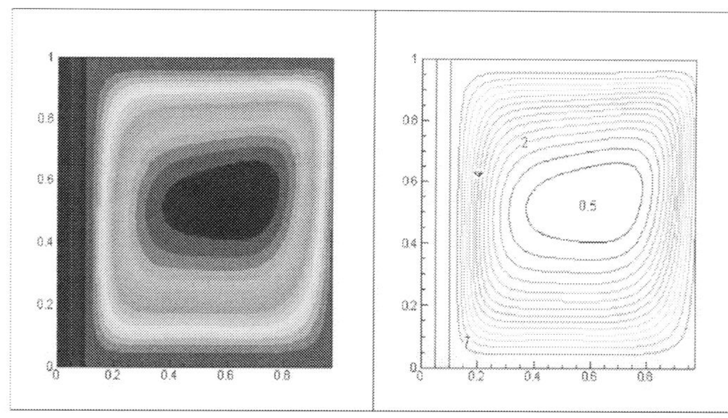

Figure 17: Streamlines obtained for Insulated Wall, Rayleigh Number Equal 10^6 , Pr=0.71

Figures [10-17] shows the current lines corresponding to Rayleigh numbers: 10^3 , 10^4 ,10^5 and 10^6 with a value of Prandt = 0.71 For uninsulated and Insulated Walls . When Ra = 10^3 , the flow is almost symmetric, we see that there are cells that are closed on itself, the fluid motion was a motion block, which justifies the existence of recirculation zones closer the center of the cavities, the flow lines are substantially parallel to the vertical walls. So also for the representation of flow lines corresponding to the Rayleigh number equal to 10^4 ,10^5 , we find that the current lines are tight at the vertical walls and in the middle, this is due to strong temperature gradient at the vertical walls and thus the flow is parietal which characterizes the large convective heat transfer to the vertical walls. We also observed the formation of two cells in the recirculation cavity for 10^6 same. It is also noted in the latter case, the expansion of two areas of recirculation of a cell in the clockwise direction and the other in the opposite direction, with the increase of Rayleigh.

Temperature Profiles

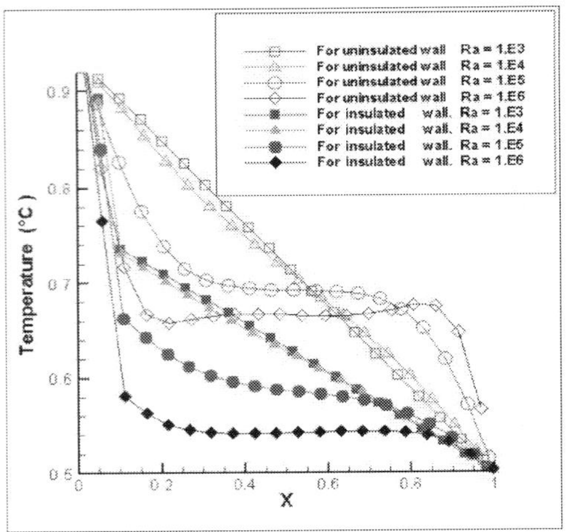

Figure 18: Dimensionless horizontal temperature profiles for different Rayleigh numbers for uninsulated and insulated walls, Pr = 0.71, Y = 0.5

Observation of the temperature profiles for different values of y = 0.5 (Figure 18) to Rayleigh leads us to confirm that the increase of temperature takes the form of the parabola is to say, this increase is condensed in the middle irrespective of the number Rayleigh, but it is sensitive, we can say there is a direct proportionality.

Nusselt Numbers

We report in Figure 20 the influence of the Rayleigh number on average Nusselt number. We found a significant influence of the Rayleigh number. The increase in average Nusselt number is accompanied with the increase of Rayleigh, but after the Rayleigh number equal to 105 , the heat transfer is almost the same, and then represents the better heat transfer in the cavity. The local Nusselt number is presented in Figure 19, it is clear that it becomes important for every increase of Rayleigh because of the change in the exchange ratio between the wall and the air, a difference of around 16% on average, between insulated and uninsulated walls to Ra = 10^5 , which can be doubled to Ra =10^6 .

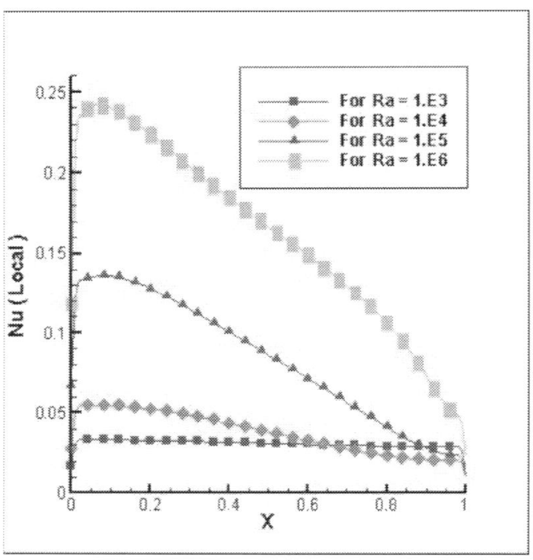

Figure 19: Profiles of Local Nusselt number, for different values of Ra=10^3 ,10^4 ,10^5 and 10^6 , for insulated walls, Pr = 0.71.

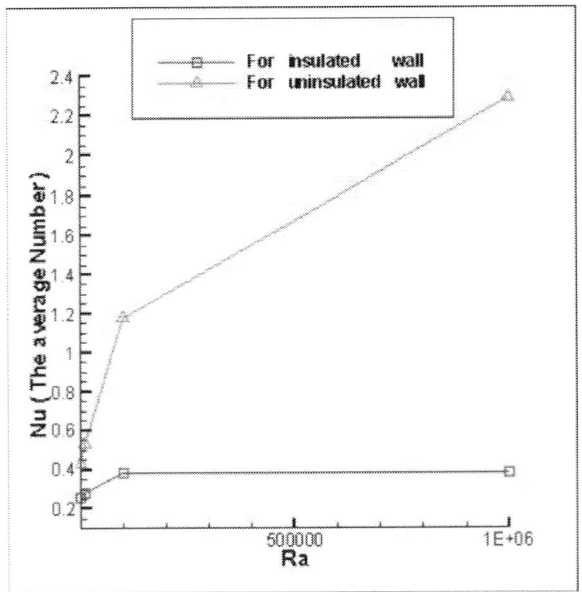

Figure 20: Profiles of average Nusselt number for different values of Ra=10^3 ,10^4 ,10^5 and 10^6 , for uninsulated and insulated walls, Pr = 0.71.

CONCLUSION

The idea of this work is based on actual events; this is just the underfloor heating in buildings arides. The objective of our work was to study the behavior of the air in enclosed with two-dimensional numerical study of natural convection or (where) laminar flow is considered on is interested by conduction or convection coupling. Solving equations governing natural convection written Helmontz's variables ψ and ω which are discretized by the finite difference method. We validated the work in the literature for the hydrodynamic and thermal studies in the cavities where he showed good consistency. For our simulation, we presented the streamlines, isotherms and the influence of key parameters such as positioning after the interface and the number of Rayleigh convection. We have shown by this simulation, the following:

1. Convection affects the structure of the isotherms.
2. The temperature of the fluid increases dramatically with increase in Rayleigh number.
3. The Interior thermal isolation reduces the losses of heat and minimizes the energy needs (request of heating) and thus reduced in fact, at the same time: the invoice, consumption and pollution.
4. The Mud is insulating effective which to guarantee the comfort and the air quality inside the building
5. The interior thermal insulation helps to preserve a constant temperature in the integrality of your housing winter like summer.
6. The Inside thermal insulation has a promising future provided that all actors, around the sector of construction are justified and motivated and conscious of the ecological and financial issues.

REFERENCES

1. R. Guechchati, M.A. Moussaoui, Ahm. Mezrhab and Abd. Mezrhab, Simulation of the effect of the thermal

insulation of buildings psychoeducational center Cas du centre psychopédagogique Case of SAFAA in Oujda, Renewable Energy Review, Vol. 13 N°2, 223 – 232, (2010).

2. R. Belarbi, M. Spérandio et F. Allard, Evaluation of the Building industries Provided with Systems of Passive Cooling. Application to the Case of Adiabatique' Evaporation, General Review of Thermics,Vol. 36, N°7, pp. 547 - 561, (1997).
3. C. lnard, P. Depecker et J. Roux, A Model Simplified for the Prediction of the Field of Temperature in the Building industries, General Review of Thermics, Vol. 36, N°2, pp. 113 - 123,(1997).
4. G.A. Florides, S.A. Tassou, S.A. Kalogirou, L.C. Wrobel, 'Measures Used to Lower Building Energy Consumption and Their Cost Effectiveness', Applied Energy, Vol. 73, N°3, pp. 299 - 328,(2002).
5. F.F. Al-Ajmi and V.I. Hanby, 'Simulation of Energy Consumption for Kuwaiti Domestic Buildings', Energy and Buildings, Vol. 40, N°6, pp. 1101 – 1109, (2008).
6. M.S. Mohsen and B.A. Akash, 'Some Prospect of Energy Saving in Building', Energy Conversion Management, Vol. 42, N°11, pp. 1307 - 1315, (2001).
7. L. Peeters, R. de Dear, J. Hensen and W. D'haeseleer, 'Thermal Comfort in Residential Buildings: Comfort Values and Scales for Building Energy Simulation', Applied Energy, Vol. 86, N°5, pp. 772 - 780, (2009).
8. E. Benachour and al, Effect of positioning the heating element on natural convection in a square cavity (habitat type), international review of mechanical engineering, vol.4N.5, July (2010).
9. A. Mecke, I. Lee, J.R. Baker jr., M.M. Banaszak Holl, B.G. Orr, Eur. Phys. J. E 14, 7 (2004).
10. U. Eicker, V. Fux, U. Bauer, L. Mei and D. Infield, 'Façades and Summer Performance of Buildings', Energy and Buildings, Vol. 40, N°4, pp. 600 - 611, (2008).
11. D. Medjelakh et S. Abdou, 'Impact of Thermal Inertia on Hygrothermic Comfort and the Energy consumption of the bâtiment', Review of Renewable energies, Vol. 11, N°3, pp. 329 - 341, (2008).
12. L. Peeters, R. de Dear, J. Hensen and W. D'haeseleer, 'Thermal Comfort in Residential Buildings: Comfort Values and Scales for Building Energy Simulation', Applied Energy, Vol. 86, N°5, pp. 772 - 780, (2009).
13. S.A. Klein, W.A. Beckman and J.A. Duffie, 'TRNSYS - A Transient Simulation Program', ASHRAE Transactions, Vol. 82, N°1, pp. 623 - 633, (1976).
14. S.A. Klein, W.A. Beckman, J.W. Mitchel, J.A. Duddie, N.A. Duffie and T.L. Freeman, 'TRNSYS Manual', University of Wisconsin, (2006).
15. N. Fezzioui, B. Draoui, M. Benyamine et S. Larbi, 'Influence of the Dynamic Characteristics of the Envelope of Building on Thermal Comfort in the South Algéria', Review of Renewable energies, Vol. 11, N°1, pp. 25 - 34, (2008).
16. S. V. Patankar, Numerical Heat Transfer and Fluid Flow, McGraw-Hill New York, (1980).
17. De Vahl Davis G, Natural convection of air in a square cavity. A bench mark numerical solution. (1983).

CHAPTER 11

EFFECT OF BANANA FIBERS ON THE COMPRESSIVE AND FLEXURAL STRENGTH OF COMPRESSED EARTH BLOCKS

Marwan Mostafa and Nasim Uddin

Department of Civil, Construction and Environmental Engineering, University of Alabama at Birmingham, Birmingham, AL 35294, USA

ABSTRACT

Sustainable development of the built environment in developing countries is a major challenge in the 21st century. The use of local materials in construction of buildings is one of the potential ways to support sustainable development in both urban and rural areas. Building with Compressed Earthen Blocks (CEBs) is becoming more popular due to their low cost and relative abundance of materials. The proposed Green-Compressed Earth Block (GCEB) consists of ordinary CEB ingredients plus Banana fibers, which will be the focus of this study. Banana fibers are widely available worldwide as agricultural waste from Banana cultivation. Banana fibers are environmentally friendly and present important attributes, such as low density, light weight, low cost, high tensile strength, as well as being water and fire resistant. This kind of waste has a greater chance of being utilized for different application in construction and building materials. This focused on the use of banana fiber and its effect on the compressive and flexural strength in CEB. The deflection at the mid-span of the blocks studied was calculated using the Linear Variable Differential Transformer (LVDT). The results of this study will highlight general trends in the strength properties of different soil mixes for CEBs. These efforts are necessary to ensure that GCEB technology becomes more widely accepted in the world of building materials and is considered a reliable option for providing low-cost housing.

INTRODUCTION

The demand for low cost sustainable building materials is growing as social, economic, and environmental issues evolve in today's society. The urgent need to develop suitable and affordable housing is born as a consequence of the fact that over one billion people in the world, most of whom live in developing nations, are either homeless or live in very poor housing. Today, 30% of our planet's people live in earth houses. The extensive use of earth is due to its low cost, availability, and its proven usefulness over time. Affordable housing for all income levels is an ever-present need. High material, labor and transportation costs have adversely affected the ability for some to reside even in developing communities. Natural building materials offer a number of environmental benefits, which are typically produced using simple, quick processes without the need for highly skilled labor, with low embodied energy, and by using raw materials from plant waste and construction materials. Compressed Earth Blocks (CEBs) are historically local masonry building materials, which are now in increasing demand by the sustainable building community. Building with CEB has become more popular due to it being an environmentally friendly process, with low cost, readily

Citation: Marwan Mostafa and Nasim Uddin, Effect of Banana Fibers on the Compressive and Flexural Strength of Compressed Earth Blocks, doi:10.3390/buildings5010282.

available raw materials, and making possible faster and easier construction [1].

CEBs offer a competitive alternative to conventional building materials because they utilize local resources that can be both cost effective and energy efficient, and closely follow existing masonry construction practices. Typically, CEBs are formed using a mixture (by weight) of angular sand aggregate (40%–70%), clayey soil (30%–60%), cement (4%–10%), and water (8%–12%) [2]. Compressive strength has become a basic and universally accepted unit of measurement to specify the quality of the masonry units. However, the CEB has limited compressive strength (2–3 MPa), which restricts their applications to only one-story buildings. Using traditional stabilizers, like cement or lime, with a high percentage increases, the CEB compressive strength significantly, but these additives also increase the material costs and use a lot of energy.

This study is going to show that these traditional binders can be replaced by environmentally friendly and sustainable alternatives from unutilized waste (*i.e.*, banana fibers). Over 10 million hectares of Banana plantation, with an average of 1500 plants per hectare, exist in more than 160 countries globally, creating tons of banana waste, which have been left over to decompose, emitting a huge amount of methane gas and carbon dioxide. These emissions have a negative impact on the environment, which increases global warming every year. Every ton of banana waste emits, on average, a half-ton of carbon dioxide per year. This study advances the utility of the CEBs by adding banana fibers, thereby creating an innovative Green Compressed Earth Block (GCEB) that further improves the mechanical properties of CEBs.

MATERIALS AND SPECIMENS PREPARATION

Materials

The materials used include local soil (from Cairo, Egypt), commonly available type I Ordinary Portland Cement (OPC), and banana fibers. The fiber is "stick-like" with an embossed surface to create deformations that provide mechanical anchorage. The lengths of the banana fibers were between 50 mm and 25 mm and cut from full-length fibers. These lengths were based on work done by Prochazka*et al.* (2010) [3].

Since banana fiber is acidic fruit material that has a pH value less than 7, and the CEB is an alkaline material compound that has a pH value more than 7, chemical pretreatment has to take place in order to increase the pH value of the banana fibers. Previous studies have shown that appropriate surface treatments enable mechanical bonding, and thereby improve matrix-reinforcement interaction. Alkali treatment of cellulosic fibers is the usual method to produce high quality fibers. Alkali treatment improves the fiber-matrix adhesion due to the removal of natural and artificial impurities. Therefore, the development of a rough surface topography and enhancement in aspect ratio offer better fiber-matrix interface adhesion and an increase in mechanical properties [4].

Characteristics of Banana Fibers

Banana fibers are generally lignocelluloses material, consisting of helically wound cellulose micro-fibrils in amorphous matrix of lignin and hemicelluloses. The cellulose content serves as a deciding factor for mechanical properties along with micro fibril-angle. A high cellulose content and low micro-fibril angle impart desirable mechanical properties for banana fibers. Lignins are associated with the hemicelluloses and play an important role in the natural decay resistance of the lignocelluloses material [5]. The composition of banana trunk obtained by elemental analysis, as given in Table 1.

Table 1: Composition of studied banana trunk fibers [5]

Constituents	Percentage
Cellulose	56%
Lignin	17%
Extractives	7%
Moisture	11%
Ashes	9%

Characteristics of Banana Fibers

Fiber Linear Density

The fiber diameter was evaluated from optical observations under microscope as the average of five diameter measurements taken at different locations along the fiber with a range of standard deviation from 0.05 to 0.1. Based on the diameters of the fiber, the whole fiber samples were divided into four broad categories. The diameter of the fiber was then measured at 100 different places along the length of four fibers. One hundred fibers were also taken at random from the sample and their diameter measured at 10 different places. The tex of the fiber was calculated assuming the density of banana fibers to be 1.4 g/cc, determined using a density gradient column prepared from xylene (0.865 g/cm^3) and carbon tetrachloride (1.595 g/cm^3) [5,6].

Fiber Stress-Strain Curves

Tensile testing was performed on four different strain rates of banana. Results of tensile testing revealed that strain rates played a significant role in the nature of the stress strain curves, the strength of the fibers and the nature of failure. The stress, as shown in Table 2, increased when the strain rate was increased to 0.5 min^{-1} but ultimately decreased with an increment in speed.

Table 2: Stress–strain average results and the effect of variation on strain rate [5,6]

Fiber No.	Strain Rate (min^{-1})	Stress (MPa)	Strain (%)
1	0.1	167.2	3.0
2	0.5	203.4	2.7
3	1	168.6	2.3
4	10	146.2	1.2

The representative stress-strain curves for banana fiber, with average values from 20 tests each on fibers of 7 tex, is shown in Figure 1. There was some initial compliance of the system. The averaged curves show the tendency of a dominantly brittle fracture for the fibers, except at the lowest strain rate of 0.1 min^{-1}. Some of the fibers showed verification for strain-hardening. This fact can be taken as a progressive reorientation of micro-fibrils, which occur for some of the fibers. Apparently from the stress–strain curves, higher strain rates resulted in higher apparent modulus values.

Banana fibers have shown high variability along the length and between fibers, which is a characteristic of natural fibers. The Stress of banana fibers was a function of the testing speed. At lower strain rate, an increase in strain facilitates the amorphous to crystalline sharing of load. At higher speeds, however, the faults dominate with catastrophic failure at the highest strain rates. Therefore, there is an inversely proportional relationship between the strain rate and the strain. Some of the stress strain curves showed signs of strain hardening.

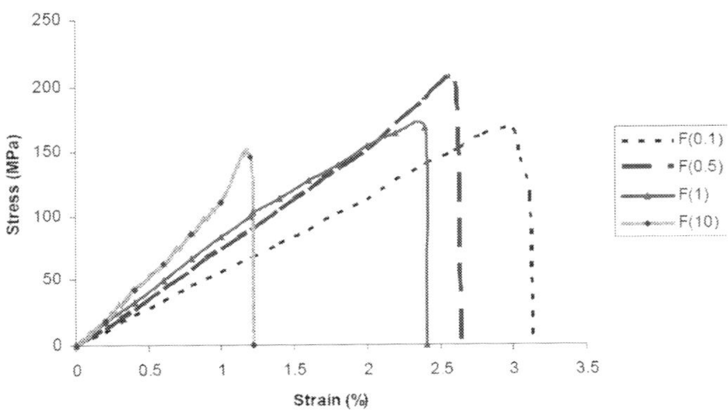

Figure 1: Representative stress–strain curves of banana fibers. Reproduced with permission from Journal of Engineered Fibers and Fabrics [5].

Banana Fibers Chemical Treatment

Alkali treatment increases surface roughness resulting in better mechanical bonding and the amount of cellulose exposed on the fiber surface. This increases the number of possible reaction sites and allows better fiber wetting. The possible reaction of the fiber and Sodium Hydroxide (NaOH) is represented in Equation (1) [7]:

$$\text{Fiber–OH} + \text{NaOH} \rightarrow \text{Fiber–O–Na}^{+} + H_2O \quad (1)$$

The banana fibers were cleaned and immersed in 6% NaOH solution for 2 h at room temperature as shown in Figure 2, and then thoroughly washed by immersion in a clean water tank to remove the non-reacted alkali until the fibers were alkali free. They were next rinsed under running water and filtered. The filtered fibers were then dried in an oven at 80 °C for 24 h [8].

Figure 2: Submerging banana fibers in NaOH solution.

Specimens Preparation

Dry soil, including sand and clay to be used for GCEB production were run through a manual sieve with a 3.40 mm^2 (0.00527 in^2) mesh size to remove lumps. The samples produced consisted of 10 plain CEBs (control) and 60 banana-fiber-reinforced GCEBs. Based on the typical CEBs mix ratios, GCEBs are formed using a mixture (by weight) of sand (35%), 5 mm aggregate (30%), clay (35%), cement (7%), and water (10%). With the exception of fiber length, all the identified factors were kept constant in all mixes to allow for observation of the influence of fiber length on strength of tested specimens. The development of strength properties of soil-cement-fiber mixes mostly depends on the formation of fiber-matrix adhesion, matrix-matrix cohesion and fiber-fiber cohesion. These identified bonds can be affected by dimensions, surface conditions and the quantity of fiber present [9]. Previous studies have recognized that reinforced CEB with 0.4% of the Polypropylene fibers recorded the highest compressive strength; also, these studies prove that increasing fiber content more than 0.4% reduces the strength significantly [10,11]. The designed mix is presented in Table 3.

Table 3: Seven design mixes of treated and untreated banana fibers

Mix Type	OPC %	Banana Fibers	Fibers Content (Weight Portion %)		
			25 mm	50 mm	Total
# 1 (control)	7%	No Fibers	0	0	0
# 2	7%	Untreated	0.35%	0	0.35%
# 3	7%	Treated	0.35%	0	0.35%
# 4	7%	Untreated	0.175%	0.175%	0.35%
# 5	7%	Treated	0.175%	0.175%	0.35%
# 6	7%	Untreated	0	0.35%	0.35%
# 7	7%	Treated	0	0.35%	0.35%

Mixing was continued until the matrix was observed to be consistent, and the fibers were added gradually while mixing. The dry mix was watered gradually in a uniform manner without stopping the mixing process. The mixing process was continued until the mix was visually deemed to be homogeneous and also satisfied the "drop test".

GCEBS PRODUCTION

GCEBs Mixing

Once the soil-cement-fiber matrix had satisfied the drop test and visual inspection, the mix was fed into the mold by an appropriate bucket from the manual compressed earth block's machine. The CEB's manual machine is designed to compress matrices from the bottom of the mold with the top plate flash by man force (this process is illustrated in Figure 3). After compression, the blocks were ejected from the mold, carefully moved, and placed on sieved sand outdoors as shown in Figure 4.

(a) (b)

Figure 3: (**a**) Filling the mold with the mix using a bucket; (**b**) Manual pressing using Compressed Earthen Block (CEB) machine.

(a) (b)

Figure 4: (**a**) Ejecting the Green-Compressed Earth Block (GCEB) from the mold; (**b**) Placing the blocks outdoors.

Curing Process

The GCEBs were covered with a plastic sheet after placement and sprayed lightly with water the next day. The blocks were stacked after two days but still kept under the plastic sheet and moisture-cured daily for a minimum of seven days. The relative humidity was about 100% under the plastic sheet. The purpose of the initial curing done during the first two days was to minimize the occurrence of cracks and the breaking off of pieces of the block during stacking [12]. The temperature was about 26 °C. The blocks were tested 28 days after production.

EXPERIMENTAL WORK

Compressive Strength Test

The nominal dimensions of blocks produced were 12 cm × 12 cm × 9 cm for compressive test. The specimens for the compressive strength tests were tested according to American Society for Testing and Materials (ASTM) international C67-07 [13]. A total of 35 GCEBs were tested for compressive strength; 5 blocks for each of the 7 mix designs. The blocks were tested under uniaxial compression load using a COMTEST Impact 2000 KN block (Impact, North Ayrshire, Scotland), cube and cylinder compression machine with a maximum load capacity of 2000 KN. The rate of compression was set at 300 N/s (67 lbs/s) until failure.

Compressive Test Results and Discussion

Reinforced blocks with randomly distributed natural banana fibers yielded higher compressive strength results compared to the unreinforced blocks, as shown in Table 4 and Figure 5. At 0.35% banana fiber content by weight, the percentages of increase in compressive strength compared to the unreinforced specimens, #1, are presented in Table 5.

Table 4: Average compressive strength results of GCEBs

Sample / Mix	# 1	# 2	# 3	# 4	# 5	Average Compressive Strength (MPa)	Standard Deviation
# 1 (control)	3.21	3.26	3.47	3.36	3.39	3.33	0.10
# 2	3.91	4.69	4.56	4.03	4.70	4.37	0.38
# 3	4.08	4.83	4.77	5.68	5.45	4.96	0.63
# 4	4.29	3.92	3.29	3.49	3.74	3.76	0.39
# 5	4.34	4.01	3.85	3.81	4.14	4.03	0.22
# 6	5.63	4.92	5.12	5.16	6.13	5.36	0.49
# 7	5.96	5.66	6.18	5.47	6.30	5.92	0.35

A research study attributed similar observations made in compressive strength of Polypropylene (PP) fiber reinforced concrete specimens to the confinement provided by the PP fiber bonding [14]. There was a noticeable increase of 17% in the strength of the treated fibers by 6% NaOH concentration relative to the untreated fibers. In this study, the matrix reinforced with only 50 mm fibers (mix # 7) recorded the highest compressive strength with value of 5.92 MPa. A research study also observed that longer PP fibers (12 mm) performed better in compressive strength at 28 days compared to shorter fibers (6 mm) [15]. Cracks typically formed before peak load was reached during testing. This observation was true for both plain and fiber reinforced specimens.

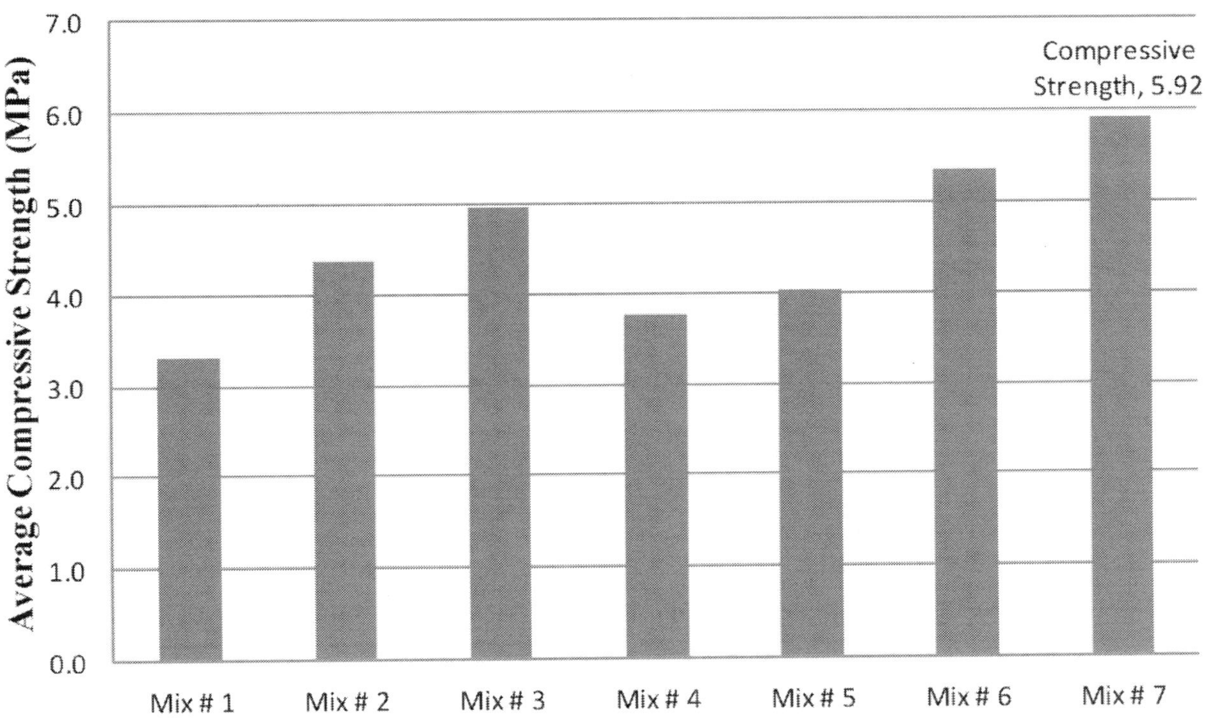

Figure 5: Average compressive strength results for GCEB mixes.

Table 5: Percentage increase in compressive strength

Mix Type	% Increase
# 2	31%
# 3	49%
# 4	13%
# 5	21%
# 6	61%
# 7	78%

Failure Mode of Plain and Reinforced Blocks

The failure mode of the plain and reinforced CEBs are very close to the failure mode of the typical concrete blocks. However, there is a slight difference in failure mode between the plain and the reinforced blocks as shown in Figure 6. Soil spalling is observed in the plain CEB compression failure leading to a void in the block, while in the reinforced GCEB compression failure sample, fibers held the soil, preventing a void.

Figure 6: (**a**) Plain CEB failure mode; (**b**) Reinforced GCEB failure mode.

Flexural Tensile Strength Test

Flexural strength testing was done according ASTM C67-07 [13]. The loading scheme was modified from third-point loading to center-point loading. A total of 5 samples were tested for each of the seven mix designs. A Universal Hydraulic UH Series Shimadzu universal testing machine with a maximum load capacity of 1000 KN was used. The machine was set up with a Linear Variable Differential Transformer (LVDT) displacement sensor to record mid-span deflection as shown in Figure 7. The values for flexural strength were computed using Equation (2):

$$S = 3WL/2bd^2 \tag{2}$$

where:

- S = Modulus of rupture of the block at the plane of failure, PSI (Pound per Square Inch) (MPa);
- W = Maximum load indicated by the testing machine, lbs (N);
- L = Span length, in (mm);
- b = Average width of the block at the plane of failure, in (mm);
- d = Average depth of the block at the plane of failure, in (mm).

Figure 7: (**a**) Leveling the two supports and centering Linear Variable Differential Transformer (LVDT); (**b**) Flexural strength test set-up.

Using center point loading, flexural strength is determined assuming simple, pure bending. In this case the span to depth ratio was approximately 2. Failure of the tested blocks did not always occur at the mid-point. During testing, all the unreinforced blocks experienced sudden failure, while none of the fiber reinforced blocks did, as shown in Figure 8. The reinforced GCEBs with treated 50 mm of banana fibers (mix #7) yielded the highest average modulus of rupture (MOR) with a value of 0.95 MPa; also, the average flexural strength and the maximum displacement of the blocks are shown in Table 6. Table 7 presents the percentage of increase in flexural strength compared to mix # 1 with a value of 0.49 MPa.

(a) (b)

Figure 8: (**a**) Plain CEB flexural failure; (**b**) Reinforced GCEB flexural failure.

Table 6: Average flexural strength and maximum mid-span displacement results of GCEBs

Sample Mix	# 1	# 2	# 3	# 4	# 5	Avg. Flexural Strength (MPa)	Standard Deviation	Maximum Displacement (mm)
# 1	0.57	0.62	0.40	0.41	0.57	0.49	0.10	0.8
# 2	0.61	0.59	0.53	0.70	0.70	0.63	0.07	1.5
# 3	0.82	0.75	0.74	0.69	0.84	0.77	0.06	2.8
# 4	0.46	0.73	0.43	0.66	0.58	0.57	0.13	3.0
# 5	0.79	0.71	0.77	0.73	0.70	0.74	0.04	2.0
# 6	0.81	0.64	0.73	0.77	0.68	0.73	0.07	2.3
# 7	0.88	0.96	0.99	1.00	0.94	0.95	0.05	3.4

Table 7: Percentage increase in flexural strength

Mix Type	% Increase
# 2	28%
# 3	58%
# 4	17%
# 5	51%
# 6	49%
# 7	94%

Fibers are known to oppose crack formation in step with increasing stress and also bridge micro cracks from expanding [15,16]. The block geometry did not provide enough span to allow for failure to occur at the center of the blocks, which would have allowed for a better assessment of the influence of the fibers. Also, the center point is not always the weakest point in the block, therefore values reported using center-point loading could be higher compared to third-point loading results.

Most of the 35 tested samples' failures occurred at the mid-point. Each tested sample exhibited linear elastic characteristics prior to initial crack, which typically occurred at peak load. The load-deflection responses of the fiber-reinforced samples were different from the unreinforced ones. Typical load-deflection curves of the tested samples are presented in Figure 9. In general, the fiber-reinforced blocks performed better in post-initial crack behavior compared to the plain matrix. The results also suggest that the fibers affect the brittle behavior of the matrices. The unreinforced samples exhibited sudden failure in all instances. Most of the fiber-reinforced matrices experienced complete failure at more than 2 mm deflection, while the plain samples experienced complete failure at less than 1 mm displacement. It was also observed that some of the fibers' cracks bridged. That could explain the cause of gradual failure as shown in Figure 10.

The fibers at the crack zone bear the tensile stress transferred from the rupture section. Previous studies established a relationship between increased aspect ratio of fibers and the ability of fibers to bridge micro cracks. In this study, the 50 mm fibers had a higher aspect ratio and produced a higher tensile strength in matrices compared to the 25 mm fibers, suggesting that their embedded length was insufficient to develop full tensile capacity. Observations during testing showed that the strength rebound occurred when fibers kicked in. The maximum post-initial crack load for the matrices reinforced with treated 50 mm fibers (mix #7) was the highest peak load recorded for all matrices.

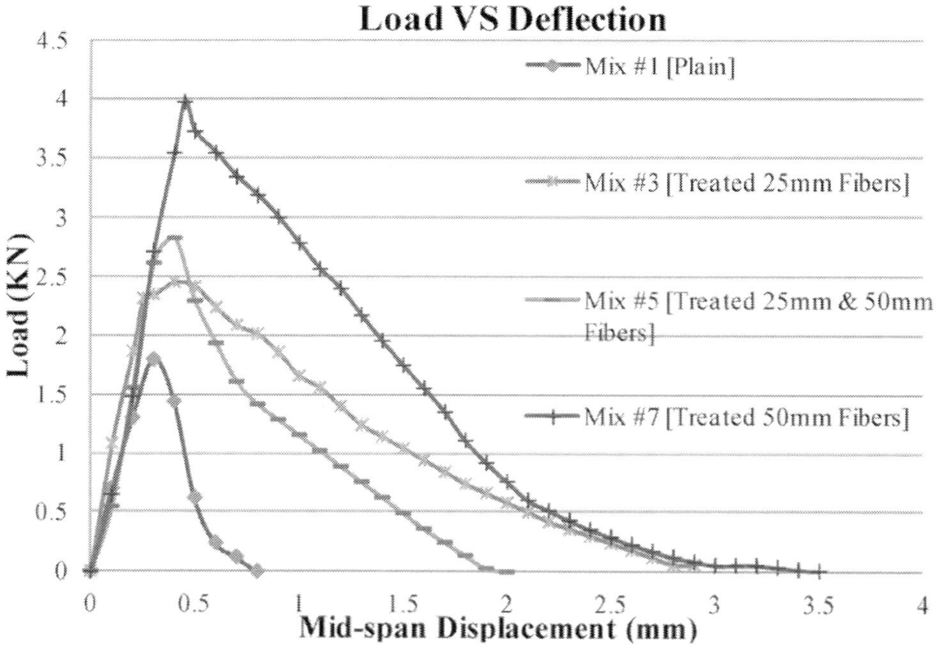

Figure 9: Typical load-deflection curves for GCEBs Mixes.

Figure 10: Banana fibers bridging the GCEB cracks during failure.

The findings point out that extensible banana fibers did not totally pull out of matrices when specimens reached peak strength. It was also noted that gradual fiber slipping and stretching resulted in a high post-peak strength even at high deformation levels. Comparing the results from the different matrices in this study, matrices with different proportions of the shorter fibers (25 mm) did not sustain as much fiber slippage as matrices with only 50 mm fibers. Subsequently, more gradual failure, after initial crack of all the fiber-reinforced matrices, suggests an improved performance in ductility that can be attributed to the banana fibers. Also, both fiber pullout and breakage were observed in GCEBs.

Water Absorption Test

The specimens from the flexural tests were used for the water absorption test. The water absorption test is basic in nature but may be the most useful in assessing the durability of GCEBs. After 28 days curing, GCEBs were weighed and recorded. This was the dry weight (W_d). The weighted samples were then totally immersed in a water tank for 24 h. They were then taken out, wiped with a cloth, and reweighed. The weight of each specimen was recorded. This was the wet weight (W_s). The water absorption of the samples was calculated using Equation (3):

$$W_A\ (\%) = 100 \times (W_s - W_d)/W_d \qquad (3)$$

where:

- W_A = Block water absorption;
- W_s = Saturated weight of the block after submersion in cold water;
- W_d = Dry weight of the block.

The absorption measures the unit's total capacity to absorb moisture. The measure of plain-CEB recorded approximately 7.4% water absorption, while the water absorption of the GCEB with banana fibers recorded an average of 10.6%. The final density of the block is 128 PCF (Pound per Cubic Feet). Blocks with banana fibers have a slightly higher water permeability or absorption than plain blocks. The higher water absorption capacity of fiber-reinforced blocks may be attributed to the amount of water absorbed by the cellulose fiber. It can be concluded that fibers increase block permeability to water. The resulting data, however, shows that all the processed GCEBs were either in the 10%–20% water absorption range or below. Therefore, with respect to this parameter, all the GCEBs met the minimum requirements for their use in building construction [17].

Relationship between Flexural Strength, Compressive Strength, and Fiber Length

Table 8 and Figure 11 summarize the average compressive and flexural strength results in each of the seven GCEB mixes. The ratios between the values of such strengths are also presented. The recorded compressive strength values were, on the average, 5.45 to 7.35 times higher than the flexural strength values. Mix # 5 recorded the lowest compressive strength to flexural strength ratio of 5.45, while mix # 6 recorded the highest ratio of 7.35. The average compressive strength to flexural strength ratio for the banana fiber reinforced blocks was 6.5.

Table 8: Ratio and average values of compressive and flexural strength

Mix \ Average	Compressive Strength (MPa)	Flexural Strength (MPa)	Ratio
# 1	3.33	0.49	6.80
# 2	4.37	0.63	6.94
# 3	4.96	0.77	6.44
# 4	3.76	0.57	6.60
# 5	4.03	0.74	5.45
# 6	5.36	0.73	7.35
# 7	5.92	0.95	6.23
Average	**4.53**	**0.70**	**6.54**

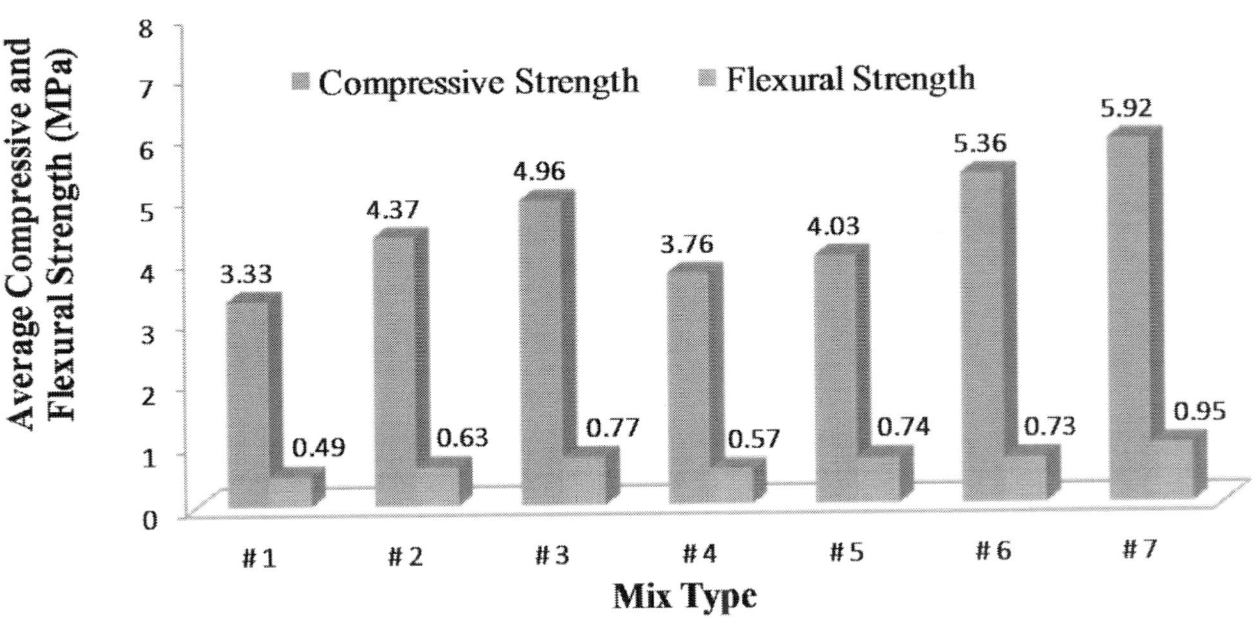

Figure 11: Average compressive and flexural strength.

SUMMARY AND CONCLUSIONS

The meaning of this experimental study was to evaluate the influence of banana fiber length on the compressive and flexural strength of GCEBs. The results presented in this study provide important insights into the influence of fiber length on the flexural and compressive strengths of GCEBs. In GCEB matrices, soil constituents, fibers content, level of chemical stabilization (OPC content), compaction effort, and curing conditions all contribute to high mechanical properties. All these variables were kept constant throughout the study to enable observation of the influence of the banana fibers on the flexural strength, compressive strength, and post-initial crack behavior of fiber-reinforced GCEBs.

On average, specimens reinforced with 50 mm fibers performed better, in both flexural and compressive strength, compared to the unreinforced specimens and specimens reinforced with different variations of 25 mm fibers. The highest flexural and compressive strength values were recorded at reinforcement with 50 mm fibers and 0.35% fiber content by weight. The results made plain that the treated 50 mm fibers were the better performing fiber in this study. Compared to the unreinforced specimens, specimens reinforced with 50 mm fibers at 0.35% by weight, were 94% and 77% higher, in terms of flexural and compressive strength, respectively. The incorporation of fibers into the matrices prevented sudden failure of the tested samples during the MOR test.

Material properties of banana reinforced GCEB are largely influenced by fiber type, fiber volume, fiber geometry and length (aspect ratio), fiber surface conditions, method of production, and composition of matrices. All these factors were considered when selecting the engineered natural banana fibers used. The findings show that with an adequate understanding of GCEBs specimens, such fibers can enhance the lateral load performance of earthen masonry. Consequently, experimental work and sample size will be increased to validate these initial findings.

AUTHOR CONTRIBUTIONS

Marwan Mostafa and Nasim Uddin conceived and designed the experiments; Marwan Mostafa performed the experiments; Marwan Mostafa and Nasim Uddin analyzed the data; Nasim Uddin contributed reagents/materials/analysis tools; Marwan Mostafa wrote the paper.

REFERENCES

1. Walker, P.J. Strength and erosion characteristics of earth blocks and earth block masonry. *J. Mater. Civ. Eng.* 2004, *16*, 497–506.
2. Minke, G. The Building Material Earth in Modern Architecture. In *Earth Construction Handbook*; WIT Press: Southampton, UK, 2000.
3. Prochazka, P.; Trčková, J.; Dolezel, V. Mixture Reinforced by Polypropylene Fibres at Higher Temperature Conditions. In Proceedings of the 35th Conference on Our World in Concrete and Structures, Singapore, 25–27 August 2010.
4. Bessadok, A.; Roudesli, S.; Marais, S.; Follain, N.; Lebrun, L. Alfa fibers for unsaturated polyester composites reinforcement: Effects of chemical treatments on mechanical and permeation properties. *Compos. A* 2009, *40*, 184–195.
5. Samrat, M.; Raul, F.; Yusuf, A. Banana fibers variability and fracture behavior. *J. Eng. Fibers Fabr.* 2008, *3*, 39–45.
6. Sandeep, K.; Varma, K. Degradation studies of polycaprolactone in banana fibers reinforced thermoplastics composites. *J. Macromol. Sci. B* 2006, *45*, 153–164.
7. Girisha, C.; Sanjeevamurthy; Guntiranga, S. Effect of alkali treatment, fiber loading and hybridization on tensile properties of sisal fiber, banana empty fruit bunch fiber and bamboo fiber reinforced thermo-set composites. *Int. J. Eng. Sci. Adv. Technol.* 2012, *2*, 706–711.
8. Lina, H.E.; Pillay, S.; Vaidya, U. Fiber Composites for Automotive and Transportation Applications. In Proceedings of the 8th Annual Automotive Composites Conference and Exhibition, Troy, MI, USA, 16 September 2008.
9. Khedari, J.; Watsanasathaporn, P.; Hirunlabh, J. Development of fibre-based soil-cement block with low thermal conductivity. *Cem. Concr. Compos.* 2005, *27*, 111–116.
10. Prasad, C.K.; Nambiar, E.K.; Abraham, B.M. Plastic fiber reinforced soil blocks as a sustainable building material. *Int. J. Adv. Res. Technol.* 2012, *1*, 42–45.
11. Elenga, R.G.; Mabiala, B.; Ahouet, L.; Goma-Maniongui, J.; Diris, G.F. Characterization of clayey soils from congo and physical properties of their compressed earth blocks reinforced with post-consumer plastic wastes. *J. Sci. Res. Geomater.* 2011, *1*, 88–94.
12. Bland, D.W. In-Plane Cyclic Shear Performance of Interlocking Compressed Earth Block Walls. Master's Thesis, California Polytechnic State University, San Luis Obispo, CA, USA, April 2011.
13. ASTM. *Standard Test Methods for Sampling and Testing Brick and Structural Clay Tile*; ASTM International: West Conshohocken, PA, USA, 2007.
14. Patel, P.A.; Desai, A.K.; Desai, K.A. Evaluation of engineering properties for polypropylene fiber reinforced concrete. *Int. J. Adv. Eng. Technol.* 2012, *3*, 42–45.
15. Bagherzadeh, R.; Pakravan, H.R.; Sadeghi, A.; Latifi, M.; Merati, A.A. An investigation on adding polypropylene fibers to reinforce Lightweight Cement Composites (LWC). *J. Eng. Fibers Fabr.*2012, *7*, 13–21.
16. Tang, C.; Gao, W.; Fengjun, C.; Cai, Y. Strength and mechanical behavior of short polypropylene fiber reinforced and cement stabilized clayey soil. *Geotext. Geomembr.* 2007, *25*, 194–202.
17. Ugwuishiwu, O.; Mama, O.; Okoye, M. Effects of natural fiber reinforcement on water absorption of compressed stabilized earth blocks. *IJSR Int. J. Sci. Res.* 2013, *2*, 165–167.

CHAPTER 12

DEVELOPMENT OF CLAY TILE COATINGS FOR STEEP-SLOPED COOL ROOFS

Anna Laura Pisello , Franco Cotana, Andrea Nicolini and Lucia Brinchi

CIRIAF–Interuniversity Research Center on Pollution Control and Environment Protection "Mauro Felli", University of Perugia, Via Duranti 67, Perugia 06125, Italy

ABSTRACT

Most of the pitched roofs of existing buildings in Europe are covered by non-white roofing products, e.g., clay tiles. Typical, cost effective, cool roof solutions are not applicable to these buildings due to important constraints deriving from: (i) the owners of homes with roofs visible from the ground level; (ii) the regulation about the preservation of the historic architecture and the minimization of the visual environment impact, in particular in historic centers. In this perspective, the present paper deals with the development of high reflective coatings with the purpose to elaborate "cool" tiles with the same visual appearance of traditional tiles for application to historic buildings. Integrated experimental analyses of reflectance, emittance, and superficial temperature were carried out. Deep analysis of the reflectance spectra is undertaken to evaluate the effect of different mineral pigments, binders, and an engobe basecoat. Two tile typologies are investigated: substrate-basecoat-topcoat three-layer tile and substrate-topcoat two-layer tile. The main results show that the developed coatings are able to increase the overall solar reflectance by more than 20% with acceptable visual appearance, suitable for application in historic buildings. Additionally, the effect of a substrate engobe layer allows some further contribution to the increase of the overall reflectance characteristics.

INTRODUCTION

The widespread use of materials with high reflectivity to the solar radiation and high spectral emissivity, *i.e.*, cool materials for building envelope applications, is considered as one of the most effective techniques to reduce energy requirements for cooling [1] and also to mitigate urban heat islands, which are related to higher temperatures registered in urban areas with respect to the rural surroundings [2]. The urban heat island effect [3], in particular, is emphasized by the high density of constructions and paving, together with the lack of green areas, that are responsible for higher absorption of the solar radiation and for the increase of anthropogenic generated heat [4].

The research issue concerning high reflective building envelopes has been investigated through important international research projects [5] which have contributed to encourage the adoption of "cool roof" options for several building typologies, in different climate conditions [6]. Numerical and experimental studies on tens of buildings in the United States and in Europe [7,8,9] have shown the decrease of the cooling peak load by up to 70% and of the indoor free-floating temperature up to 3 °C in the thermal zones adjacent to the roof, produced by cool roof implementation [10].

Although cool roofs represent acknowledged solutions for energy saving, their application is still limited to high reflective coatings for non-sloped roofs, mainly concerning industrial,

Citation: Anna Laura Pisello , Franco Cotana, Andrea Nicolini and Lucia Brinchi, Development of Clay Tile Coatings for Steep-Sloped Cool Roofs, doi:10.3390/en6083637.

commercial, or office buildings [11]. Additionally, these building typologies are characterized by higher internal gains, which tend to increase the energy requirement for cooling and the consequent cool roof efficacy [12]. Nevertheless, especially in European historical cities of the Mediterranean Basin, cool roof applications should also concern many buildings located in historical centers and urban architecture heritage contexts, which are largely responsible for the summer peak of electricity requirement for cooling, and where more invasive retrofit interventions are almost impossible to implement [13]. These contexts are mostly characterized by historic buildings located in very dense urban areas, which roofs are sloped and covered by traditional clay tiles [14], where typical cool roof coatings and membranes are impossible to apply for cultural heritage preservation reasons [15]. Therefore, given the acknowledged role of the envelope and roof performance for building energy saving [16,17,18,19], the development of effective solutions specifically focused on possible application in historic European urban contexts has an important role for the optimization of: (i) building thermal-energy performance [20] at inter-building scale [21,22,23] and (ii) urban comfort and reduction of pollutants' concentration [24].

MOTIVATION AND PURPOSE OF THE WORK

The outlined perspective shows how the investigation of specific coatings for clay represents an important research issue for cool roof development with the purpose to optimize the solar reflectance of traditional buildings' roofs without affecting their appearance [25]. In particular, this work concerns an investigation of the thermal-optical properties of pigments and binders for clay coatings, which should have equivalent visible characteristics with respect to traditional tiles. In fact, the purpose of this work is the elaboration of high solar reflectance coatings with the same spectral response of traditional clay tiles in the visible region, but high reflectance in other solar spectrum range [26], which includes more than half of the solar radiation [15].

The chosen mineral based coatings are specifically optimized in order to elaborate covering layers for tiles representing non-impacting passive solar techniques which contribute to building energy efficiency and indoor thermal comfort during the cooling season, for possible applications in Italian and European historic buildings [27].

In the choice of materials—tiles, binders, and pigments—traditional and natural building materials and techniques are preferred. In fact, parallel to the concept of cool materials optimization [28], the appreciation of traditional techniques and natural local materials and technical elements, *i.e.*, clay tiles, contributes to the purpose of providing important advantages from a global environment point of view [29,30,31], such as: low embodied energy, ability to regulate indoor humidity, reduced toxicity, and relatively low price [14].

In order to optimize the final performance of the cool roof, two kinds of tiles are compared: (i) the substrate-topcoat two-layer tile; and (ii) the substrate-basecoat-topcoat three-layer tile (Figure 1). Accordingly, new samples are prepared by applying the same coatings upon two different tile substrates: a natural tile and a novel white-engobe clay tile [14]. Therefore, current investigation is aimed at systematically estimate and optimize the optical and thermal behavior of these two sets of clay tiles, in order to provide a better understanding of factors that play major roles in the performance of cool roof technologies.

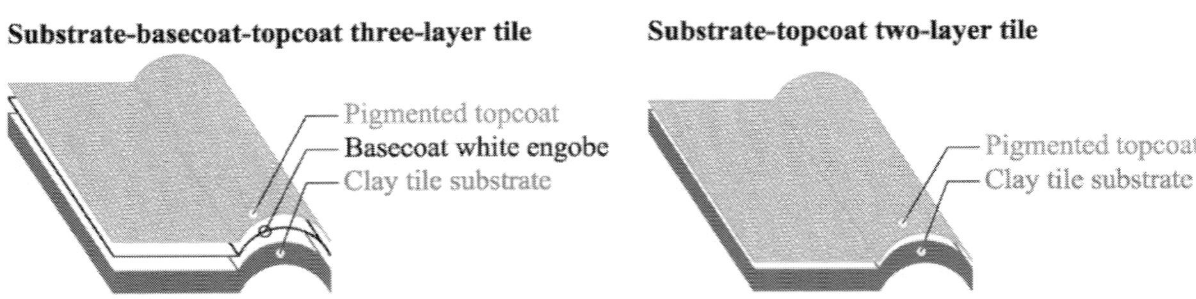

Figure 1: The two investigated tile typologies.

MATERIALS AND METHODS

Clay Tiles

Clay tiles are the most diffuse roof covering element used in Italian buildings. In particular, clay tiles represent the traditional technique used to direct rain water towards the drainpipes. A typical clay tile is about 42 cm × 25 cm is size, and weighs around 3 kg. The natural colors of the tiles are light brown, red or beige, depending on varying clay pit location (Figure 2a,b), but also different finishes are available in the market in order to reproduce fake ancient effects or other colors (Figure 2c,d).

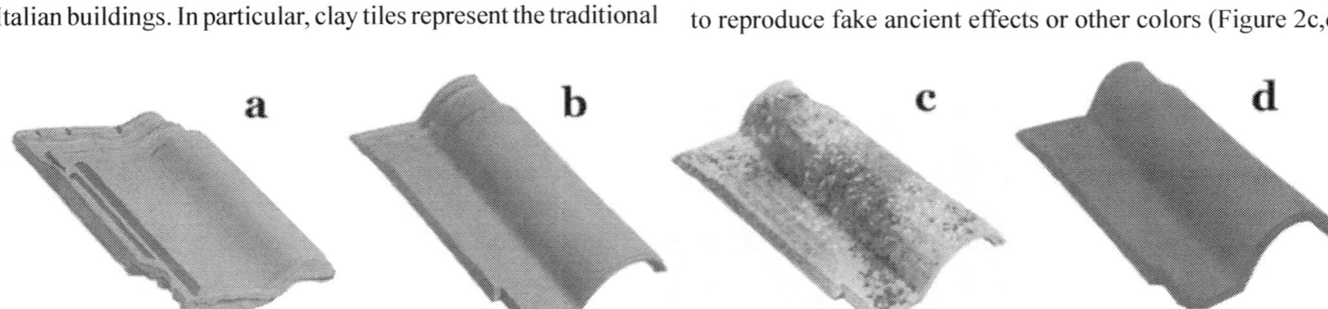

Figure 2: Typical commercialized clay tiles. (**a**,**b**) Natural "terracotta" tiles; (**c**) Clay tile with old-looking effect; (**d**) Brown engobe tile.

Coatings

The chosen pigments and the binder used are easily commercially available at low price. A mineral- based binder with high permeability to aqueous vapor is chosen. The selected white opaque stabilized potassium silicate-based binder addresses DIN 4108.3 and DIN 18363 requirements. It contains TiO_2pigment: a strongly scattering, weakly absorbing, stable, non-toxic, inexpensive and hence extremely popular white pigment, which also exhibits photocatalytic activity [32]. As regards colored pigments, natural mineral earths are preferred, because of their high stability under weathering conditions, light oxidation and corrosion [33]. Iron oxide-based yellow (encoded with *Y* in Table 1), red (encoded with*R*) and brown (encoded with *B*) are used as pigments.

Preparation of Samples

Small coupons of 7 cm × 7 cm are cut from tiles and covered with the experimental coatings. The sample name and composition of the coatings are reported in Table 1.

Table 1: Codenames and composition for the elaborated samples

Sample code	White basecoat	Experimental coating			
		Y (g)	*R* (g)	*B* (g)	Opaque binder (g)
N0	no	-	-	-	-
N1	no	-	-	-	100
N1A	no	3.6	-	-	100
N1B	no	7.2	-	-	100
N1C	no	7.2	0.6	-	100
N1D	no	7.2	1.4	-	100
N1E	no	7.2	1.4	1.0	100
N1F	no	7.2	1.4	1.6	100
N1G	no	11.3	1.4	1.6	100
W0	yes	-	-	-	-
W1	yes	-	-	-	100
W1A	yes	3.6	-	-	100
W1B	yes	7.2	-	-	100
W1C	yes	7.2	0.6	-	100
W1D	yes	7.2	1.4	-	100
W1E	yes	7.2	1.4	1.0	100
W1F	yes	7.2	1.4	1.6	100
W1G	yes	11.3	1.4	1.6	100

Colored coatings are prepared in the lab by mixing the opaque white binder with water (as suggested by manufacturer) and with different weighted amounts of pigments, obtaining different colors. The various coatings are prepared in such a way that two consecutive coatings in Table 1 differ in only one added component; using this method of changing one variable at a time, useful comparisons between various coatings are carried out.

The coatings are applied upon the tiles immediately after preparation by means of a paint-brush. The same coating, referred to as "experimental coating" in Table 1, is applied as the only coating in natural tile (encoded with N in Table 1) and as topcoat over a white engobe basecoat in white tiles (encoded with W in Table 1), as described in Figure 1. The coated tiles are allowed to dry at room temperature for 2 weeks before measurements described in Section 2. The experimental analysis of the samples concerns the evaluation of natural brick tiles and coated tiles, which roughness could be considered as equivalent, in order to evaluate the effect of the pigment in terms of visual appearance (color) and reflectance performance.

EXPERIMENTAL PROCEDURE

The analysis of the thermal and optical properties of the clay tile samples is carried out by an integrated experimental analysis consisting of three main phases: (i) the solar reflectance measurement (Figure 3a) and (ii) the thermal emittance measurement (Figure 3b). Additionally, (iii) infrared thermography is used to evaluate the absorbed heat by the tile samples in the same boundary conditions. The used infrared camera is a FLIR B360 (FLIR Systems, Inc., Wilsonville, OR, USA) which characteristics are reported in Table 2.

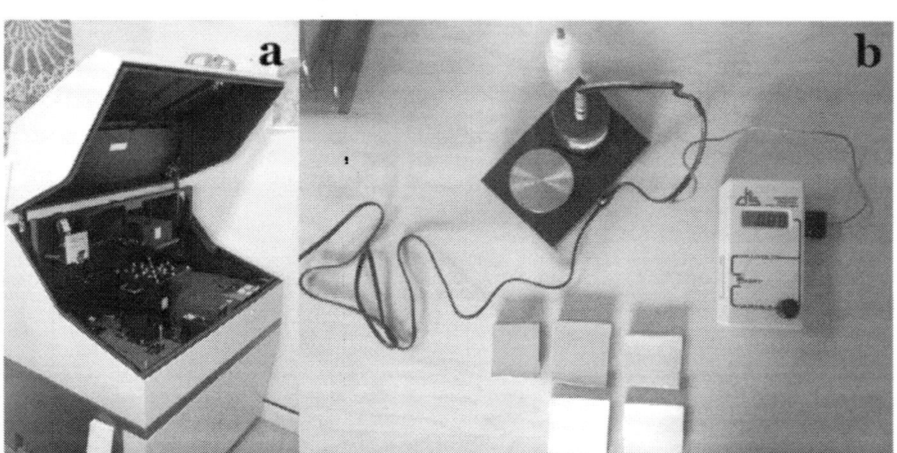

Figure 3: (**a**) Spectrophotometer; (**b**) emissometer instruments for the in-lab experimental analysis.

Table 2: Characteristics of the infrared camera

Field of view	25°/19°	IR resolution	320 × 240
Minimum focus distance	0.4	Spectral range	7.5 ÷ 13 μm
Thermal sensitivity	0.06 °C	Object temperature range	−20 °C to +120 °C
Detector type	focal plane array	Accuracy	±2% of reading

The spectrophotometer with integrating sphere used in order to characterize the solar reflectance of the samples is a Shimadzu SolidSpec 3700 (Shimadzu Corporation, Kyoto, Japan), which characteristics are reported in Table 3. The solar hemispherical reflectance of the samples is determined by following the procedure described in [34], with reference to the solar spectrum reported in [35].

Table 3: Characteristics of the spectrophotometer

Spectral bandwidth	UV/VIS: 0.1–8 nm (8 steps) NIR: 0.2–32 nm (10 steps)	IR resolution	320 × 240
Spectrum interval	240 ÷ 2600 nm	Noise	<0.0002 Abs (500 nm, SBW 8 nm), <0.00005 Abs (1500 nm, SBW 8 nm) determined under conditions of RMS value at 0 Abs and 1 s response
Resolution	0.1 nm	Photometric range	−6 to 6 Abs
Wavelength accuracy	UV/VIS: ±0.2 nm, NIR: ±0.8 nm	Accuracy	±2% of reading
Wavelength repeatability	UV/VIS: ±0.08 nm, NIR: ±0.32 nm		

The emittance measurements are carried out AE1 RD1 emissometer with scaling digital voltmeter: the guidelines reported in [36] are followed. The instrument is composed by a differential thermopile radiant energy detector, a heater, a heat sink with a flat surface, two sets of reference standards for the calibration, and the samples to be tested. The samples are prepared in order to be slightly larger (7 cm × 7 cm) than the outer dimensions of the emissometer measuring head, as suggested by [36] international standard. The emittance measurement has a repeatability of ±0.01 emittance units and it approximates total hemispherical emittance at 65 °C. The detector responds to the radiation heat transfer and it produces a voltage output that is linear with emittance.

RESULTS AND DISCUSSION

Visual Appearance of the Samples

As previously mentioned, the elaboration of the coating consists of a progressive increase of the percentage of colored natural mineral earths into the white opaque mineral based binder. As reported in Figure 4 and in Table 1, the colored pigments are applied step-by-step following this order:

- -Yellow pigment: W1A, N1A, W1B, N1B samples;
- -Red pigment: W1C, N1C, W1D, N1D samples;
- -Brown pigment: W1E, N1E, W1F, N1F, W1G, N1G samples.

Therefore, the darkest samples, *i.e.*, N1G and W1G, present the overall mix of these pigments, which purpose is to obtain a visual appearance of the tile as close as possible to the ancient traditional clay tiles found in Italy (Figure 4). The application of such coatings shows that: (i) the appearance of the tiles is very close to that of natural clay tiles of ancient buildings starting from the C series of samples to the G series, which color is equivalent to the natural clay color; (ii) the final appearance of the tiles does not show any perceivable difference between the natural tile (N series) and the tile with the white engobe (W series) as substrate. This implies that the visible feature of the tiles is mainly attributable to the chosen pigments and to the mineral binder. Therefore, no perceivable effect of the engobe in the visible part of the reflected spectrum is expected, while the purpose is to increase mainly the infrared reflectance capability of the proposed tiles, where there is about the half of the solar radiation, which represents a key cooling capability to be developed [10].

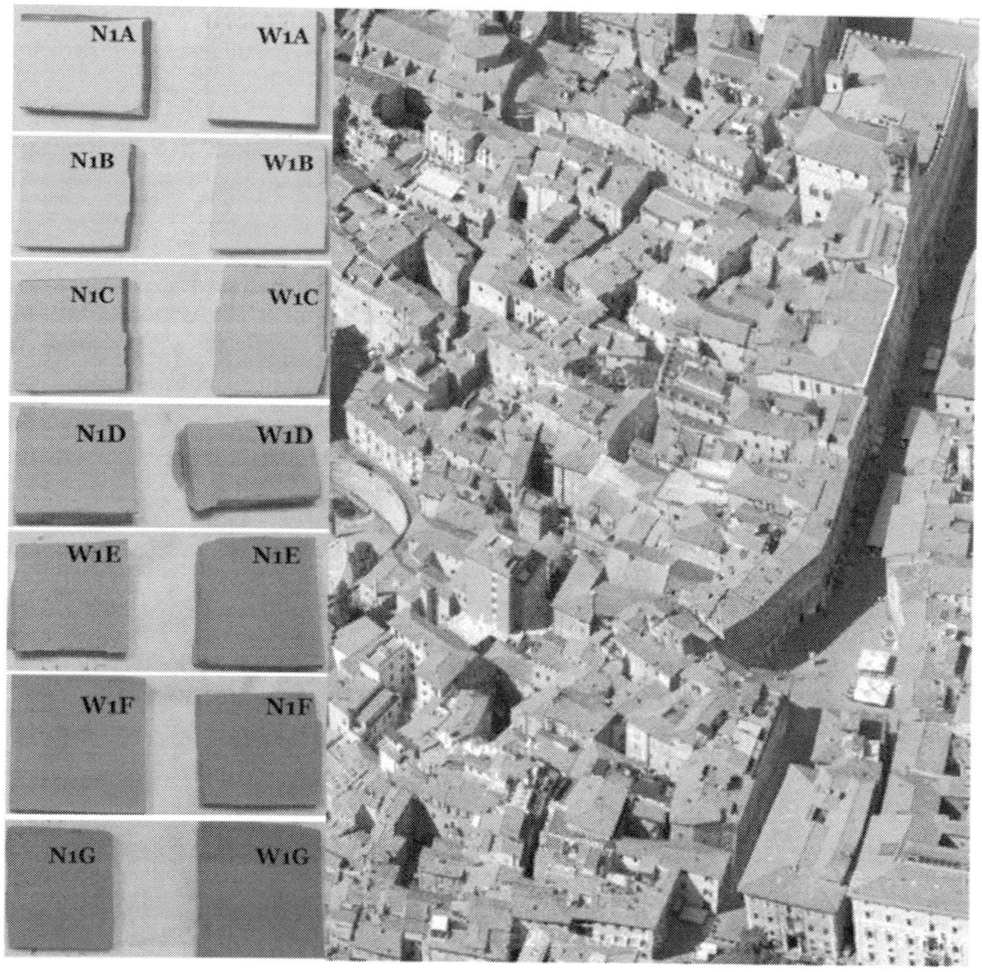

Figure 4: Elaborated samples and comparison of the visual appearance with respect to the traditional roof tiles of the historic center in Perugia, Italy.

Infrared Thermography

Infrared thermography was performed on 14 June 2013. This day was chosen for the hot weather conditions, typical of the summer period in most of Italian territory (Figure 5). All the samples were exposed to the solar radiation at about 9:00 a.m. and they were tested under the same weather and exposure conditions all morning long. The overall photography campaign was carried out at the same time (2:00 p.m.) in order to obtain reliable comparative assessment of the superficial temperature of the tiles. Figure 5 represents the daily profile of the main weather parameters impacting the superficial temperature of the samples, and the relative cool roof performance, which have been continuously monitored during that day for the purpose of this study: outdoor dry bulb temperature, wind velocity, global solar radiation of the measurement site. Although the thermography analysis is a preliminary screening test, it allows us to describe the comparative effect on temperature rise in the sun of the tested samples. The infrared images illustrated in Figure 6 show that the surface of the original natural red tile (N0) is hotter than the other developed clay tile. In particular, the N1 and W1 tiles are colder than the red tile by about 12 °C, while the N1C and W1C tiles, which present less impacting color appearance, present lower (by about 7 °C) superficial temperatures. Finally, N1G and W1G samples are colder with respect to the natural red tile by about 4 °C, despite the equivalent visual appearance. Additionally, the engobe samples seem to be slightly colder than the N-based samples but the precision level of the thermography and the variability of the human measurement position are not reliable enough to outline a definite behavior of the W-samples with respect to the N-samples, which is detailed in Section 4.3.

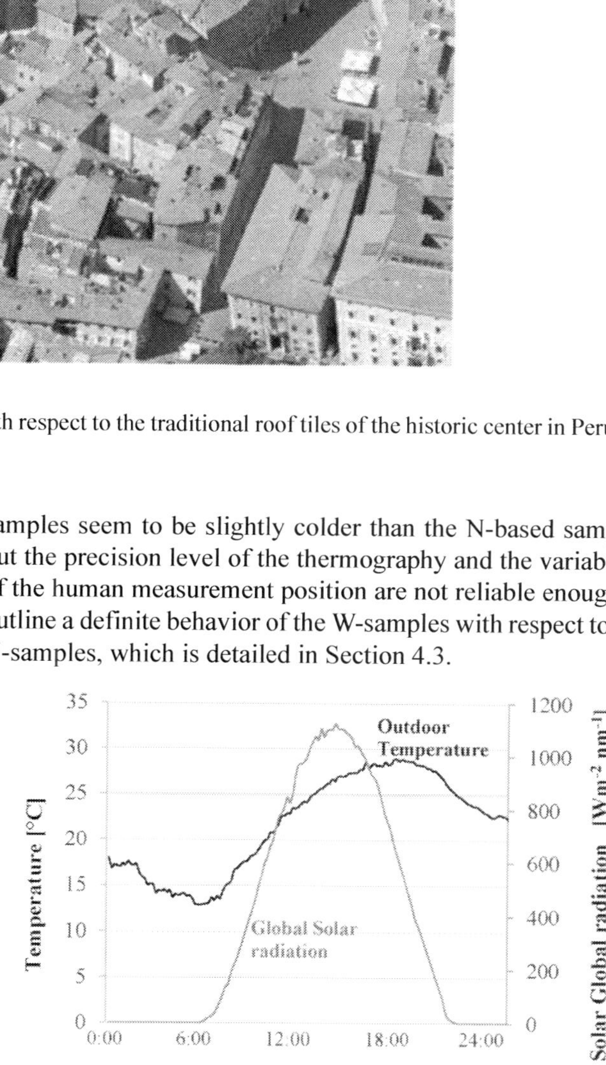

Figure 5: Main weather parameters continuously monitored during the thermography analysis.

Figure 6: Thermography images of the developed samples.

Solar Reflectance Spectra

Reflectance spectra of the various samples are reported in Figure 7 and Figure 8. The values of solar reflectance calculated following the procedure detailed in Section 3, by spectrophotometer with integrating sphere, are reported in Table 4. A gradual decrease of reflectance with increasing pigment content in both sets of tiles is clear from reflectance spectra; such decrease is particularly important in the VIS region, as expected, because the pigments are added to intentionally modify the visual appearance. On the other hand, all 14 of the experimentally colored tile samples exhibit rather good solar reflectance, *i.e.*, $R > 53$, and nine samples have $R > 65$. In order to have a better understanding of the role played by the various components, comparisons are carried out, as shown in Figure 7 andFigure 8.

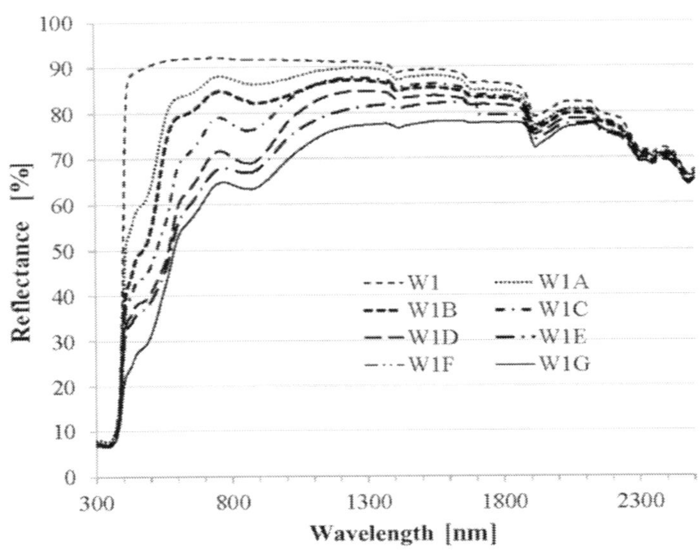

Figure 7: Solar reflectance of the substrate-basecoat-topcoat 3-layer tiles.

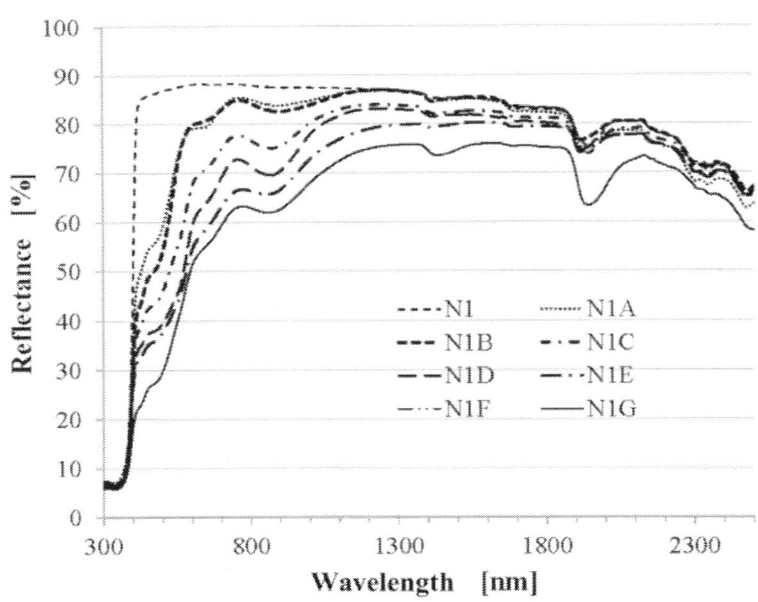

Figure 8: Solar reflectance of the substrate-topcoat 2-layer tiles.

Table 4: Solar reflectance and thermal emittance of the various samples

Sample codenames	R_{UV} Reflactance (%) 300–380 nm	$R_{VISIBLE}$ Reflectance (%) 380.5–780 nm	R_{NIR} Reflectance (%) 781–2500 nm	R_{SOLAR} (R) Reflectance (%) 300–2500 nm	IR emittance Emittance (%)
N0	8.8	31.1	65.9	45 [1]	0.88
N1	8.5	85.2	86.2	83	0.89
N1A	8.7	69.7	84.2	74	0.89
N1B	8.0	67.8	83.9	73	0.90
N1C	7.9	58.5	79.6	66	0.89
N1D	7.5	52.1	76.9	61	0.90
N1E	7.7	48.5	73.3	58	0.90
N1F	7.1	45.2	70.3	55	0.90
N1G	6.7	43.4	69.1	53	0.90
W0	45.3	74.1	82.8	77 [2]	0.90
W1	8.6	88.6	89.9	87	0.89
W1A	9.1	73.7	86.8	77	0.89
W1B	7.9	67.5	83.8	73	0.89
W1C	8.1	59.1	81.8	67	0.88
W1D	8.2	52.3	77.2	62	0.88
W1E	7.7	48.4	73.1	58	0.88
W1F	7.6	46.8	72.0	56	0.89
W1G	7.3	44.7	71.0	55	0.88

[1] Our value is consistent, although slightly better, than lit. Values of solar reflectance for terracotta ceramic tiles which are in the 25%–40% range [2]. [2] This value is consistent, although slightly better, than the lit. Values for white clay tile, which are in the 60%–75% range [2].

Effect of the Binder

Figure 9 reports the solar reflectance of the tile W0 with the white engobe, the natural red terracotta tile N0, and the same two samples (W0 and N0) with the white opaque binder experimental coating, respectively named N1 and W1 (for codenames see Table 1).

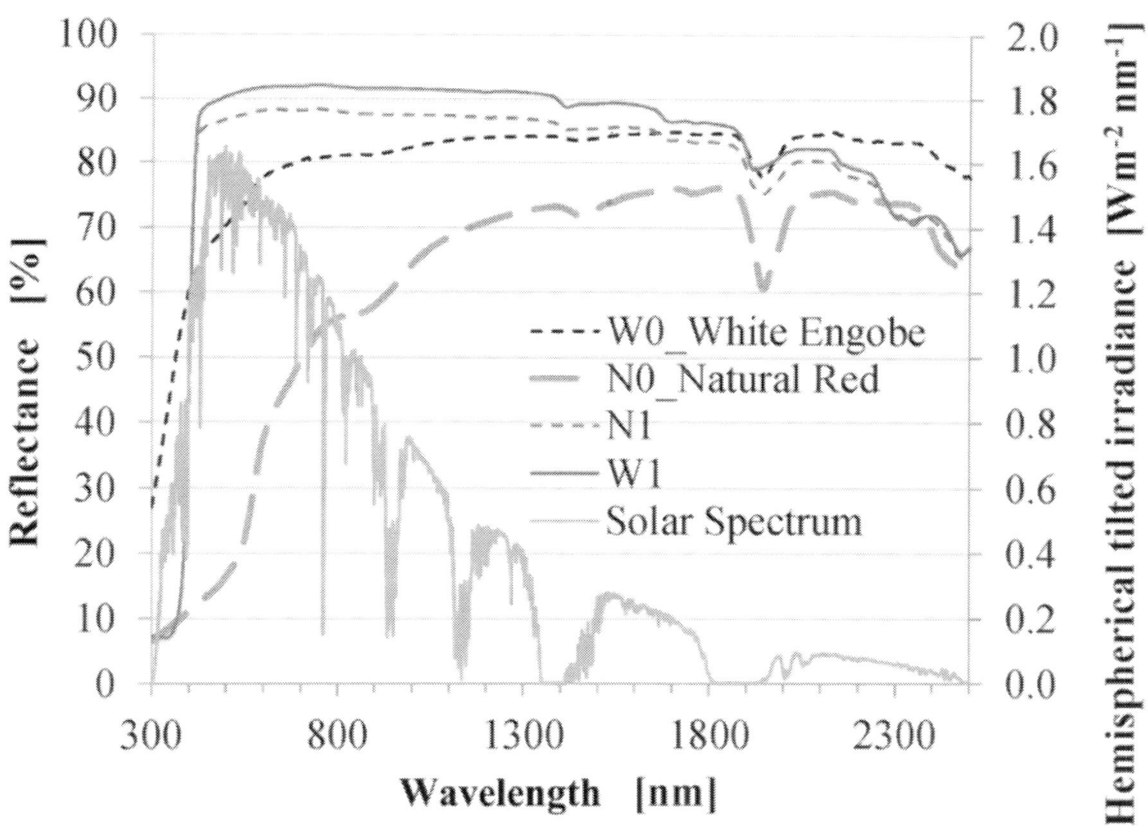

Figure 9: Comparison between the solar reflectance of samples to investigate the role of the binder.

By comparing W0 and N0, it is evident that the engobe is able to produce an important increase of solar reflectance, especially in the most powerful zone of the solar radiation spectrum (grey line), up to 1500 nm, also given the much lighter color of the engobe. In fact, R_{UV} and R_{VIS} values of the W0 sample are higher by 36% and 43%, respectively, with respect to the N0, while in terms of R_{NIR} the difference is 17%. Accordingly, the value of R (Table 4) is much higher for W0 (77%) than for N0 (45%). It is noteworthy that the reflectance of W0 is slightly higher than the values given in the literature for white clay tiles [2].

By comparison of the white coated tiles, N1 and W1, with the engobe tile, W0, Figure 9 clearly shows how the white coating produces a significant increase in reflectance values in the interval 400/1900 nm, and also a brief decrease of reflectance between 300 and 400 nm and beyond 1900 nm. These two contributions to reflectance can be ascribed, respectively, to the content increase of TiO_2 and to absorbance of the binder, including the little amount of organic stabilizers present in it. Therefore, the selected binder has a relatively low impact on the overall solar reflectance capability, with a better performance than organic binders such as acrylic matrix [10]. The overall effect of application of the experimental white coating corresponds to a significant increase in the solar reflectance, R, with values increasing up to 83% and 87% for N1 and W1 respectively, from base values of 45% and 77%, for N0 and W0 respectively (Table 4).

Effect of the Colored Coatings

Selected differential spectra are evaluated [37,38] and reported in Figure 10. In Figure 10a the differences of solar reflectance of experimental coatings 1 and 1A, for both sets of tiles encoded as N and W, are plotted *vs.* the wavelength. This plot clearly shows the effect of addition of 3.6 g of iron oxide yellow pigment to 100 g of white opaque binder upon the reflectance of the experimental coating. Similarly, in Figure 10b the differences of reflectance *vs.* the wavelengths of coatings 1B and 1D are reported, showing the effect of addition of 1.4 g of iron oxide red. Additionally, the differences of reflectance between 1D and 1F are reported in Figure 10c, showing the effect of addition of 1.6 g of iron oxide brown pigment.

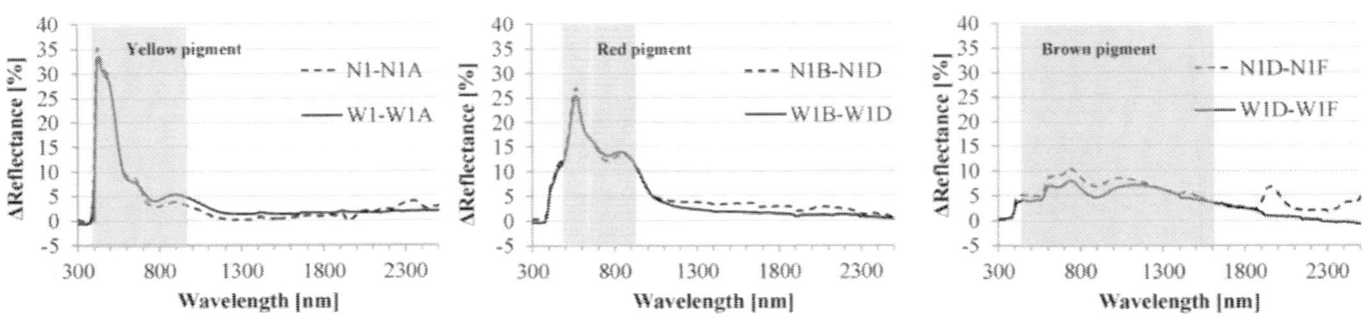

Figure 10: Evaluation of the effect of colored pigments through differential spectra: (**a**) Yellow; (**b**) Red; (**c**) Brown.

Due to the different amounts used of the various pigments, given the main purpose to reproduce equivalent visual appearance with respect to the traditional "terracotta" tile, and the complexity of the coatings, the comparisons do not pretend to give rigorous insight into the behavior of each pigment. Nevertheless they are aimed at outlining the role played by pigments upon the overall observed reflectance of the complex systems. It is noteworthy, however, that especially for the simplest of the systems investigated in Figure 10a, the differential spectrum is consistent with the absorption coefficient variation with wavelength of a similar iron-oxide based yellow pigment obtained by a rigorous treatment [25]. In more complex system, *i.e.*, with red and brown pigments, the differences found between the spectra measured in this paper with respect to the literature are more important. Despite that fact, the analyses carried out for the purpose of this work in terms of differential spectra are of some utility, as specified below.

As regards the yellow pigment, Figure 10a shows that spectra are basically similar for the two sets of tiles and that important absorbance is registered in the VIS region around 400–600 nm, while low absorption is registered in the NIR region, with just a small peak around 900 nm. Accordingly, the values of solar reflectance, *R*, for addition of the yellow pigment (Table 4, samples with codenames 1A and 1B compared with coating 1 for both sets of tiles N and W) decrease by up 17%–18% in the VIS region, but only by about 2% in the NIR region. The overall solar reflectance values are rather good, in the range 73%–77%, which makes this pigment a "cool" coating. However, red and brown addition seems necessary to give the tile a visible appearance similar to the natural terracotta color, as evident from the samples reported in Figure 4. As regards the red, it is clear from Figure 10b that the iron oxide based pigment gives an important contribution to the decreased reflectance observed both in VIS and in NIR regions, especially around 500–1000 nm, whereas its behavior is cooler at wavelengths above 1000 nm. Its addition leads to a decrease in the values of reflectance by up to 22% in VIS and 7% in NIR, but the overall values of R are still good, in the range 61%–67% (Table 4). As regards the brown, the differential spectra in Figure 10c give an indication of a relatively small contribution of this pigment in the VIS region (if compared to yellow and red pigments), and this can be rationalized as related to a leveling effect. Anyhow, brown pigment contributes to decrease of reflectance in the wavelength range up to 1600 nm, and R decreases to values in the range 53%–56% (Table 4). These values can be considered as a good result for a relatively dark colored tile, in particular if compared to the natural red tile, which solar reflectance corresponds to 45%, with the same visible appearance of N1G and W1G (Figure 4).

Effect of the White Enrobe Basecoat

The effect of the basecoat, *i.e.*, engobe layer, is now assessed in order to analyze its possible role in emphasizing the solar reflectance when topcoat colored pigments are applied. Figure 11 reports reflectance spectra of both kinds of tiles (N and W) with some of the opaque experimental coatings used, for a rapid visualization of the effect of the engobe. The solar spectrum is also reported in order to gain immediately the idea of the relative importance of the spectral region in terms of solar irradiance. It is clear that the engobe produces a beneficial effect on the overall performance of all the tiles investigated. The major effect is observed in white tiles, with higher values of reflectance in the region 400–1900 nm (Figure 11), and higher values of R by about 4% points (Table 4) for W1 with respect to N1. On addition of the pigments, the beneficial effect of the engobe is less important, but it is still present although it does not give any effect in terms of visual appearance. In fact, in presence of the engobe basecoat, reflectance is higher in the region 400–2300 nm for yellow (Figure 11, N1A, W1A), and in 900–2200 nm for red (Figure 11, N1C, W1C), and in 900–2500 nm for brown (Figure 11, N1G, W1G). For colored tiles the values of R are generally higher in presence of the engobe basecoat by 1%–4% points (Table 4).

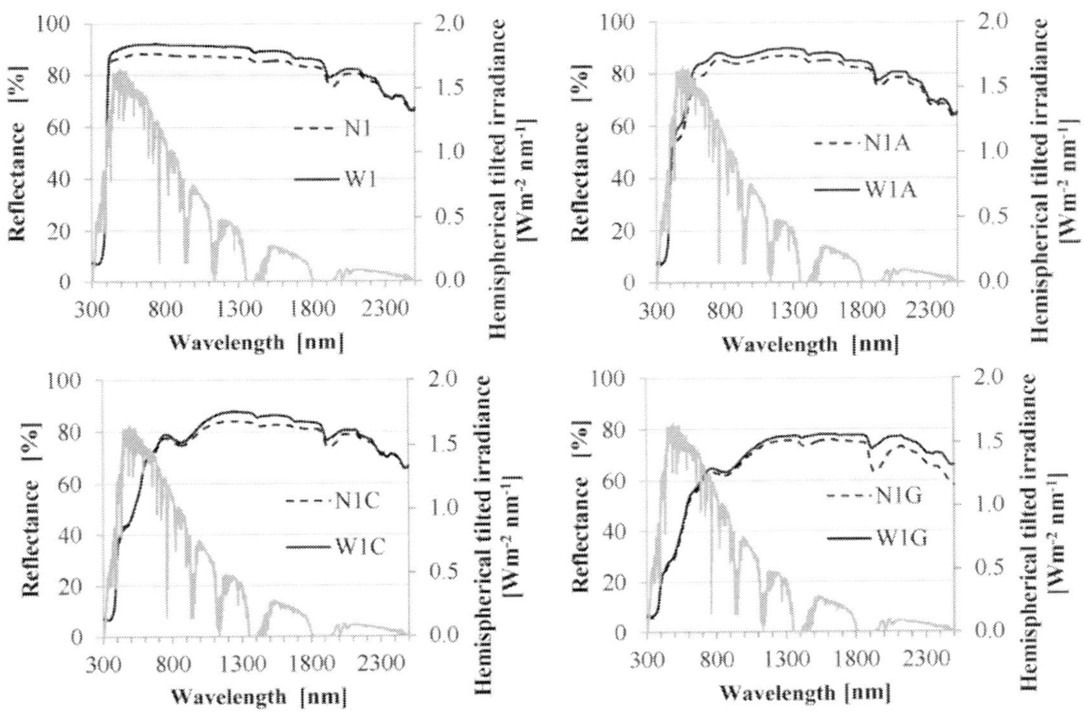

Figure 11: Solar reflectance comparison between samples with and without the basecoat engobe.

CONCLUDING REMARKS AND PERSPECTIVES

The current work examined the behavior of new tiles developed in order to evaluate possible innovative cool roof applications specifically fitting with the traditional Italian and European context, in particular for possible application to historic buildings, which energy efficiency retrofits are much restricted by specific constraints due to the preservation of the architectural heritage. In fact Italian regulation imposes heavy limits about the implementation of systems and technologies impacting ancient buildings and environmentally protected areas. These limits represent the boundary conditions of this research and the objective of the experimental study at the same time. The development of these innovative tiles is effectively aimed at elaborating high performance tiles in terms of cool roof potential, but with the same visual appearance of traditional "terracotta" tiles.

To this aim, new tile prototypes were prepared, consisting of traditional clay tiles where colored opaque coatings were applied with a visual appearance essentially the same of the existing traditional "terracotta" tiles. Two kinds of tiles are compared: (i) the substrate-topcoat two-layer tile; and (ii) the substrate-basecoat-topcoat three-layer tile. Accordingly, new samples were prepared by applying the same coatings upon two different tile substrates, a natural tile and a novel white-engobe clay tile. The experimental methodology consisted of: a screening test by thermography, a spectrophotometer and emittance analysis, a final deep analysis of the solar reflectance spectra profile for different wavelength bands. In the same monitored weather conditions, the thermography allowed to find out different superficial thermal behavior of the samples, which presented colder surface temperature with increasing visual clarity of the appearance. Moreover, even the darkest prototyped tile presents lower superficial temperature by about 4 °C than natural red tile. The spectrophotometer analysis allowed to characterize the solar reflectance spectra of the samples and to identify the influence of: (i) the binder; (ii) the several colored pigments; (iii) the engobe basecoat layer.

First results showed that, if a white tile can be used, tile W0 with $R = 77\%$ is an optimal choice for a new roof, with a good ratio performance/cost because the engobe can be directly applied at the factory, following the same procedure of industrially colored tiles, at low cost. Additionally, the durability performance of the engobe has the same guarantee of the tile element itself, which represents an important benefit in terms of the maintenance of the cool roof effect for decades. If a layer of the experimental coating is added, the value of the global solar reflectance, R, goes up to 87%, with the relative benefits in terms of the overall performance of the cool roof system. On the other hand, for existing roofs, application of the experimental coating 1 over traditional tiles is an effective and simple option, with high performance (R = 83%) and a rather significant improvement with respect to natural terracotta tiles, although the labor cost should be included into the overall evaluation.

Finally, in historical centers where white tiles cannot be used, a weakly light red coating with relatively high increase of solar reflectance, is represented by the 1C coating, with R = 66%–67%. If the dominant color of the area is even darker, a still effective coating for tile is 1F or the darker 1G, with R = 53%–56%. These two colored coatings would lead to an important improvement in cooling while maintaining basically the same visual aspect of traditional "terracotta" tiles. In fact, the values of solar reflectance, R, are increased with respect to natural tile N0 by a factor of up 50% for 1C, and up to 22% for the darkest tile 1G. Additionally, in the case of colored tiles, the presence of the white engobe basecoat gives an overall improvement of 1–4 percentage points in terms of solar reflectance. Therefore, specific case study analyses should investigate the cost-effectiveness of this solution with different boundary conditions. In this work, the preliminary assessment of the cost-effectiveness of the proposed tiles is carried out by considering several published results [11] where the effect of increased reflectance is evaluated in different climates, when cool roofs are applied to residential buildings. In particular, the estimated effect in Italian climate corresponds to about 25% and 17% of reduction of discomfort hours in summer, in Palermo and Rome, Italy, respectively. [11]. The primary energy saving for cooling is 33% and 30% in Palermo and Rome, when a cool roof with the characteristics of the proposed 1C coating is applied, corresponding to an increase of solar reflectance by about 40% with respect to the base case. It was also demonstrated that the corresponding winter penalty is much less important than the cooling benefit for Mediterranean climates. Additionally, these year-round benefits are even higher for low or non-insulated constructions, which represent the majority of Italian historic buildings.

For all the new tiles presented in this current work, experiments about weathering are in due course. The effect of the tile roughness will be also evaluated, and the weathering performance of the proposed tiles will be compared with existing rougher tiles. On the other hand, other tiles are currently under preparation, with the aim to further increase the performance, with increased amount of the TiO_2 white pigment in the basecoat.

Future developments of this research will deal with the analysis of new pigments, through mixing metals' oxides, and new binders, developed in-lab, where also the dimension of the particles will be investigated, in order to develop innovative nano-based cool coatings. Finally, the application to a real case study building will be examined, with the final purpose to investigate both the thermal-energy and the economic implications of such an application in Italian historic buildings.

ACKNOWLEDGMENTS

Authors' acknowledgments are due to Gloria Pignatta for having contributed to the in-lab measurements, and to Leandro Lunghi who prepared the clay tile substrate samples.

REFERENCES

1. Synnefa, A.; Saliari, M.; Santamouris, M. Experimental and numerical assessment of the impact of increased roof reflectance on a school building in Athens. *Energy Build.* 2012, *55*, 7–15.
2. Santamouris, M.; Synnefa, A.; Karlessi, T. Using advanced cool materials in the urban built environment to mitigate heat islands and improve thermal comfort conditions. *Sol. Energy* 2011,*85*, 3085–3102.
3. Akbari, H.; Konopacki, S. Energy effects of heat-island reduction strategies in Toronto, Canada.*Energy* 2004, *29*, 191–210.
4. Oke, T.R.; Johnson, D.G.; Steyn, D.G.; Watson, I.D. Simulation of surface urban heat island under "ideal" conditions at night—Part 2: Diagnosis and causation. *Bound. Layer Meteorol.* 1991,*56*, 339–358.
5. Synnefa, A.; Santamouris, M. Advances on technical, policy and market aspects of cool roof technology in Europe: The Cool Roofs project. *Energy Build.* 2012, *55*, 35–41.
6. Synnefa, A.; Santamouris, M.; Apostolakis, K. On the development, optical properties and thermal performance of cool colored coatings for the urban environment. *Sol. Energy* 2007, *81*, 488–497.
7. Bozonnet, E.; Doya, M.; Allard, F. Cool roofs impact on building thermal response: A French case study. *Energy Build.* 2011, *43*, 3006–3012.
8. Zinzi, M.; Agnoli, S. Cool and green roofs. An energy and comfort comparison between passive cooling and mitigation urban heat island techniques for residential buildings in the Mediterranean region. *Energy Build.* 2012, *55*, 66–76.
9. Pisello, A.L.; Cotana, F. The thermal effect of an innovative cool roof on residential buildings in Italy: Results from two years of continuous monitoring. *Energy Build.* 2013. submitted for publication.
10. Libbra, A.; Tarozzi, L.; Muscio, A.; Corticelli, M.A. Spectral response data for development of cool coloured tile coverings. *Opt. Laser Technol.* 2011, *43*, 394–400.
11. Synnefa, A.; Santamouris, M.; Akbari, H. Estimating the effect of using cool coatings on energy loads and thermal comfort in residential buildings in various climatic conditions. *Energy Build.*2007, *39*, 1167–1174.
12. Santamouris, M.; Papanikolaou, N.; Livada, I.; Koronakis, I.; Georgakis, C.; Argiriou, A.; Assimakopoulos, D.N. On the impact of urban climate to the energy consumption of buildings.*Sol. Energy* 2001, *70*, 201–216.
13. Pisello, A.L.; Bobker, M.; Cotana, F. A building energy efficiency optimization method by evaluating the effective thermal zones occupancy. *Energies* 2012, *5*, 5257–5278.
14. Pisello, A.L.; Rossi, F.; Cotana, F. On the Impact of Cool Roofs in Italian Residential Buildings: Experimental Assessment of Summer and Winter Performance. In Proceedings of SEBUA-12 ICHMT International Symposium on Sustainable Energy in Buildings and Urban Areas, Kusadasi, Turkey, 14–20 July 2012.

15. Libbra, A.; Muscio, A.; Siligardi, C.; Tartarini, P. Assessment and improvement of the performance of antisolar surfaces and coatings. *Prog. Org. Coat.* 2012, *72*, 73–80.
16. Asdrubali, F.; Cotana, F.; Messineo, A. On the evaluation of solar greenhouse efficiency in building simulation during the heating period. *Energies* 2012, *5*, 1864–1880.
17. Buratti, C.; Moretti, E. Lighting and energetic characteristics of transparent insulating materials: Experimental data and calculation. *Indoor Built. Environ.* 2011, *20*, 400–411.
18. Lai, C.M.; Wang, Y.H. Energy-saving potential of building envelope designs in residential houses in Taiwan. *Energies* 2011, *4*, 2061–2076.
19. Berardi, U.; Albino, V. Green buildings and organizational changes in Italian case studies. *Bus. Strategy Environ.* 2012, *21*, 387–400.
20. Pisello, A.L.; Goretti, M.; Cotana, F. A method for assessing buildings' energy efficiency by dynamic simulation and experimental activity. *Appl. Energy* 2012, *97*, 419–429.
21. Pisello, A.L.; Taylor, J.E.; Xu, X.; Cotana, F. Inter-building effect: Simulating the impact of a network of buildings on the accuracy of building energy performance predictions. *Build. Environ.*2012, *58*, 37–45.
22. Xu, X.; Taylor, J.E.; Pisello, A.L.; Culligan, P. The impact of place-based affiliation networks on energy conservation: An holistic model that integrates the influence of buildings, residents and the neighborhood context. *Energy Build.* 2012, *55*, 637–646.
23. Xu, X.; Pisello, A.L.; Taylor, J.E. Simulating the Impact of Building Occupant Peer Networks on Inter-Building Energy Consumption. In Proceedings of the 2011 Winter Simulation Conference, Phoenix, AZ, USA, 11–14 December 2011; pp. 3373–3382.
24. Santamouris, M. Heat island research in Europe—The state of the art. *J. Adv. Build. Energy Res.*2007, *1*, 123–150.
25. Levinson, R.; Berdahl, P.; Akbari, H. Solar spectral optical properties of pigments—Part II: Survey of common colorants. *Sol. Energy Mater. Sol. Cells* 2005, *89*, 351–389.
26. Ferrari, C.; Libbra, A.; Muscio, A.; Siligardi, C. Energy performance of opaque building elements in summer: Analysis of a simplified calculation method in force in Italy. *Energy Build.* 2013, *64*, 384–394.
27. Levinson, R.; Akbari, H.; Reilly, J.C. Cooler tile-roofed buildings with near-infrared-reflective non-white coatings. *Build. Environ.* 2007, *42*, 2591–2605.
28. Kolokotsa, D.; Maravelaki-Kalaitzaki, P.; Papantoniou, S.; Vangeloglou, E.; Saliari, M.; Karlessi, T.; Santamouris, M. Development and analysis of mineral based coatings for buildings and urban structures. *Sol. Energy* 2012, *86*, 1648–1659.
29. Dall'O', G.; Speccher, A.; Bruni, E. The green energy audit, a new procedure for the sustainable auditing of existing buildings integrated with the Leed protocols. *Sustain. Cities Soc.* 2012, *3*, 54–65.
30. Cotana, F.; Rossi, F.; Nicolini, A. Evaluation and optimization of an innovative low-cost photovoltaic solar concentrator. *Int. J. Photoenergy* 2011, *2011*.
31. Rossi, F.; Nicolini, A. Ethanol reforming for supplying molten carbonate fuel cells. *Int. J. Low-Carbon Technol.* 2013, *8*, 140–145.
32. Bettoni, M.; Brinchi, L.; Del Giacco, T.; Germani, R.; Meniconi, S.; Rol, C.; Sebastiani, G.V. Surfactant effect on titanium dioxide photosensitized oxidation of 4-dodecyloxybenzyl alcohol. *J. Photochem. Photobiol. A* 2012, *229*, 53–59.
33. Bikiaris, D.; Daniilia, S.; Sotiropoulou, S.; Katsimbiri, O.; Pavlidou, E.; Moutsatsou, A.P.; Chryssoulakis, Y. Ochre-differentiation through micro-Raman and micro-FTIR spectroscopis: Application on wall paintings at Meteora and Mount Athos, Greece. *Spectrochim. Acta A* 2000,*56*, 3–18.
34. *Standard Test Method for Solar Absorptance, Reflectance, and Transmittance of Materials Using Integrating Spheres*; ASTM E903-96; American Society for Testing and Materials: West Conshohocken, PA, USA, 1996.
35. *Standard Tables for Reference Solar Spectral Irradiances: Direct Normal and Hemispherical on 37° Tilted Surface*; ASTM G173-03; American Society for Testing and Materials: West Conshohocken, PA, USA, 2012.
36. *Standard Test Method for Determination of Emittance of Materials Near Room Temperature Using Portable Emissometers*; ASTM C1371-04a; American Society for Testing and Materials: West Conshohocken, PA, USA, 2010.
37. Brinchi, L.; Germani, R.; Savelli, G.; di Michele, A.; Onori, G. Premicelles of cetyltrimethylammonium methanesulfonate: Spectroscopic and kinetic evidence. *Coll. Surf. A*2009, *336*, 75–78.
38. Di Michele, A.; Germani, R.; Pastori, G.; Spreti, N.; Brinchi, L. Effects of temperature on micellar-assisted bimolecular reaction of methylnaphthalne-2-sulphonate with bromide and chloride ions. *J. Colloid Interface Sci.* 2013, *402*, 165–172.

CHAPTER 13

LIME AS AN ANTI-PLASTICIZER FOR SELF-COMPACTING CLAY CONCRETE

Gnanli Landrou [1], Coralie Brumaud [1], Frank Winnefeld [2], Robert J. Flatt [3] and Guillaume Habert [1]

[1]Institute of Construction and Infrastructure Management, Chair of Sustainable Construction, Swiss Federal Institute of Technology (ETH Zurich), Stefano-Franscini-Platz 5, Zürich 8093, Switzerland

[2]Laboratory for Concrete/Construction Chemistry, Empa, Swiss Federal Laboratories for Material Science and Technology, Überlandstrasse 129, Dübendorf 8600, Switzerland

[3]Institute for Building Materials, Physical Chemistry of Building Materials, Swiss Federal Institute of Technology (ETH Zurich), Stefano-Franscini-Platz 3, Zürich 8093, Switzerland

ABSTRACT

This paper focuses on the modification of clay properties with inorganic additives to deflocculate and flocculate inorganic soil for the development of a material that would be as easy to use as the current concrete products, but with a much lower environmental impact. Considering that the rheological behaviour of clays is controlled by their surface charge, we first introduce potential determining ions to deflocculate the clay particles and to reduce the yield stress of the earth material. Their efficiency is characterized using zeta potential measurements and rheological tests. We then achieve the flocculation of clay particles by using natural minerals that slowly dissolve in the interstitial liquid and ultimately precipitate calcium silicate hydrate (C–S–H). The precipitation products are identified by X-ray diffraction and the consequences of this delayed precipitation are followed by oscillatory rheometric measurements. Finally, it is suggested that in this process, C–S–H precipitation is not used as a binding vector but as an anti-plasticizer that removes the inorganic dispersant additives.

INTRODUCTION

Traces of earthen architecture date to 10,000 years ago, and this building material technique is still used in most climates and societies [1]. Without transport and with infinite recycling possibilities, earth is among the building materials which have the lowest environmental impact [2,3] and very efficient temperature and moisture regulation properties for indoor living spaces [4]. Earth construction is currently under strong development, likely due to environmental concerns. However, this development is limited because the conventional earth construction techniques are both time-consuming and labour-intensive. In contrast, cement is an incredibly easy-to-use material but has a significant environmental impact [5]. Substantial engineering and scientific efforts have been invested to improve the understanding and processing of cement-based concrete, but no or very little engineering improvement has been made in the case of earth.

Recent attempts have been made to fluidify earth material and cast it using the same techniques that are used for concrete. Actually, experiments were conducted with castable earth in which cement plasticizers are used to reduce the yield stress at the fresh state and 5 to 8 wt % of cement is added to allow setting [6]. The technique works, but the mechanisms controlling clay behaviour are not well understood, and a large amount of cement is still used. In another experiment of castable earth [7] the product developed is able to flow and set. However, the amount of lime that has to be used is close to 10 wt %, which is a reduction of only approximately 30% compared

Citation: Gnanli Landrou, Coralie Brumaud, Frank Winnefeld, Robert J. Flatt and Guillaume Habert, Lime as an Anti-Plasticizer for Self-Compacting Clay Concrete, doi:10.3390/ma9050330.

with conventional cement-based concrete for a compressive strength that does not exceed 15 MPa. Thus, the objective of this study is to use the concrete technologies to transform earthen architecture by providing a material that is as easy and cheap to use as the current concrete products, with the addition of a minimum amount of cement. To achieve this, our approach is to transfer knowledge from concrete technology to earthen construction, recognizing analogies between the two materials. In particular, colloidal interactions and adhesion forces are present in both materials, but there are also differences since the cohesion forces between particles are much weaker for clay particles [8] due to the difference in the constituted binder (no hydraulic reaction occurs).

To improve the material workability, a careful control of the rheology of the clays requires a better understanding of the colloidal interactions between particles [9,10,11] and knowledge transfer from the fundamental physics of grains and colloids to civil engineering [12,13].

Clay, a phyllosilicate mineral based on an octahedral (O) and tetrahedral (T) sheet combine in either a 1:1 or 1:2, respectively, to form an anisotropic TO or TOT layer [14] is an inorganic charged particle. It shows a lamellar morphology with a permanent negative charge at the surface and pH-dependent positive charge at the edge. As a unique binder in earth, clay can have its surface interaction changed with the help of organic dispersants from the cement and ceramic industries. Among the deflocculants commonly used in ceramic processing, sodium hexametaphosphate, a hexamer of composition $(NaPO_3)_6$, guarantees an efficient dispersion effect for natural clay materials [15,16,17,18,19,20,21]. Its deflocculating action is a combination of two mechanisms. First, it increases the overall negative surface charge because of the adsorption of phosphate anions, particularly at the edges of the clay mineral particles, and second by complexing the dissolved flocculant alkaline earth cations and replacing them with Na^+ cations, thereby increasing the thickness of the electrical double layer [15,16,17,18,19]. The second deflocculant is sodium silicate Na_2SiO_3, which is widely used in raw clay and kaolin dispersion as a deflocculant because of its high efficiency [17,18,19,20,21,22,23,24,25,26,27]. Similar to sodium hexametaphosphate, studies have shown that it acts as deflocculant by the adsorption of a polyanion that modifies surface charge as well as by increasing the pH of the solution [22,28]. Other important phenomena, with respect to dispersion stability and rheological behavior, are the dissolution of the clay mineral constituents [29,30,31] and the reactions between the sodium silicates and the dissolved or desorbed ions from the particles [18].

In this paper, we first assess the efficiency of these two clay dispersants and determine how valuable they could be for the development of a self-compacting clay concrete. Then, it is of fundamental importance to be able to flocculate earth material again to remove the formwork and allow for the complete drying of the material. This second stage is currently achieved through the addition of cement, which provides additional strength [1,6,7]. In this study, to develop a material without cement use while transferring technology from cement chemistry and ceramic science, we focus on a different technique and explore the potential role for Ca^{2+} ions coming from the dissolution of natural minerals that contain calcium. This allows for the formation of insoluble salts between the dispersing anions (phosphate and silicate), as well as the restoration of calcium as a flocculating agent between the particles.

MATERIALS

Earth

A commercially available mineral earth material for plastering with a specific density of 2.72 g/cm^{3}and specific surface area S_{BET} = 46.9 m^2/g was used in this study. The particle size distribution (PSD) was measured using a laser particle size analyzer (Partica LA-950, Horiba, Munich, Germany). For the measurements, the earth injected dry in the machine sample bath was dispersed with 0.1 wt % of sodium hexametaphosphate (Fisher Chemical, Reinach, Switzerland). Between each measurement, an automatic ultrasonic treatment in the sample bath was applied during 1 min (repeated 10 times) to avoid agglomeration between particles. The studied clay earth contained 95% fine particles (<63 μm) (Figure 1), including clays (36.6 wt %) and silt (57.9 wt %). Its chemical composition, obtained through X-ray fluorescence spectrometry (XRF), is given in Table 1. The X-ray powder diffraction technique revealed, thanks to Rietveld methods [32,33], that the main mineralogical components are smectite (24 wt %), quartz (41 wt %), a lower amount of kaolinite (7.5 wt %), and traces of carbonate (0.4 wt %). The complete mineralogical composition is given in Table 2. When prepared as a suspension, earth develops a pH of the order of 9.

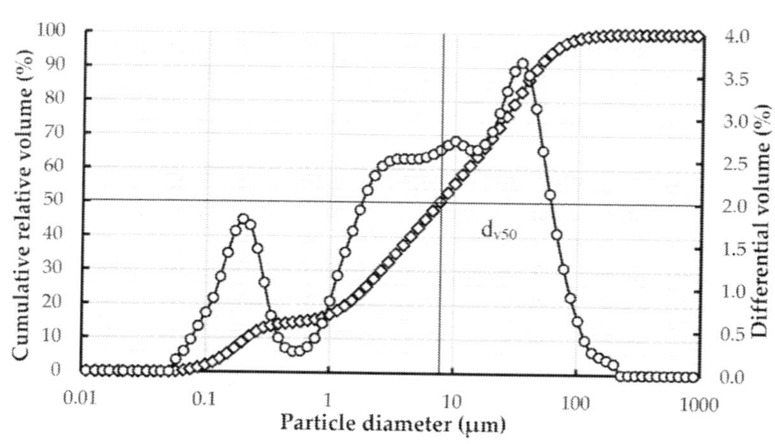

Figure 1: Earth particle size distribution and differential volume.

Table 1: Chemical composition of the earth studied by X-ray fluorescence analysis. Values are given in mass %

SiO_2	TiO_2	Al_2O_3	Fe_2O_3	MnO	MgO	CaO	Na_2O	K_2O	P_2O_5	Cr_2O_3	Ignition Loss
65.8	1.2	14.9	7.4	0.1	1.1	1.1	0.5	2.4	0.3	0.02	5.02

Table 2: Mineralogical composition of the earth studied by X-ray diffraction

Quartz	Smectite	Illite	Kaolinite	Muscovite	Plagioclase/Albite	Others	Total
40.9	23.8	5.2	7.5	7.7	7.1	7.7	99.9

Dispersants-Deflocculants

Three different dispersants used in the cement and ceramic industries were investigated.

The first one was a high-range water reducer agent (HRWRA), a comb co-polymer that is commonly used in concrete and mortar applications to improve flow material properties by deflocculating cement particles [34]. It is based on a polycarboxylate ether (referred to here as PCE) composed of polyethylene oxide side chains grafted with a methacrylic acid backbone.

The two other dispersants were inorganic deflocculants taken from the ceramic industry: high purity sodium hexametaphosphate $(NaPO_3)_6$ (>99.0% pure) (referred to in this paper as NaHMP) in powder form sourced from Fisher Chemical (Reinach, Switzerland) and sodium silicate solution (noted here NaSil) from Sigma Aldrich (Buchs, Switzerland) with a composition of 10.6% Na_2O and 26.5% SiO_2, the rest being water.

All dispersant are added as a solution at the same time as the water.

Calcium Products

Three different calcium products were used as additives: calcium hydroxide ($Ca(OH)_2$, >96% pure), calcium carbonate ($CaCO_3$, >99.0% pure), and calcium chloride ($CaCl_2$, >93% pure), all sourced from Fluka—Sigma Aldrich (Buchs, Switzerland). These products have different Ca^{2+} release rates in the interstitial solution because of their different solubilities.

MEASUREMENT PROCEDURES

Rheology Measurements

Rheology measurements were performed on earth pastes to study the effect of dispersants on the rheological properties of the suspension. To compare the results obtained from different measurements, the water to solid ratio (W/S) for earth pastes was kept constant (0.5). Only the nature and dosage of the dispersant were varied from one sample to another: these variations are discussed in the relevant sections of this manuscript. In this paper, all dosages are expressed as the percentage of the mass of solids in the system, and water is deionized. The tested earth pastes were prepared using the following mixing procedure: the dispersant in powder or liquid form was added to the required amount of mixing water to ensure its dissolution before introduction into the solid phase. Water (with or without dispersant) was mixed with earth for 3 min at 365 rpm with a mechanical stirrer equipped with a four-bladed mixing tool (Heidolph, Dietikon, Switzerland).

Yield Stress Measurements

The rheology measurements were performed using a MCR501 Anton Paar (Buchs, Switzerland) stress-controlled rheometer equipped with Vane geometry [35] at room temperature (23 °C ± 0.1 °C). The Vane geometry was a four-bladed paddle with a diameter of 22 mm. The outer cup diameter was 25 mm, and its depth was 60 mm. Twenty minutes after the first contact of the constituents (including the mixing phase), the cup of the rheometer was filled and covered to limit evaporation, and the sequence was started. An increasing shear rate ramp from 0.1 to 300 s^{-1} (with a logarithmic distribution of shear rates) was then applied for 300 s followed by a decreasing shear ramp from 300 to 0.1 s^{-1} for 300 s. Only the decreasing ramps are analyzed for the value of the yield stress (Figure 2), extrapolated at low shear rates according to the Bingham model $\tau=\tau0+\eta p\gamma^{\cdot}$, where τ is the shear stress, $\gamma^{\cdot}$ the shear rate, τ_0 the yield stress and η_p the plastic viscosity.

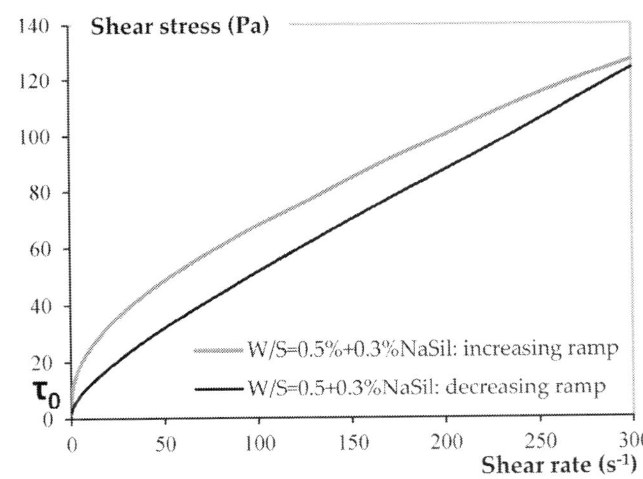

Figure 2: Flow curve of a clay paste prepared with 0.3% of NaSil.

Viscoelastic Properties

The limits of the linear viscoelastic behavior (LVR), where the rheological properties of the materials are not strain- or stress-dependent, were first determined by identifying the critical value of stress τ_c through a stress sweep test. A stress amplitude (from 0.1 to 3000 Pa) at a constant frequency of 1 Hz was applied and the end of the linear elastic regime was defined as when the elastic modulus falls to 90% of the plateau value [36]. The dynamic viscoelastic properties of the materials can be quantified in terms of the storage modulus G' within the linear viscoelastic region (LVR). The viscoelastic behaviour of the pastes was thus investigated in the time domain by applying for six hours a constant shear stress of 1 Pa (lower than τ_c) and a constant frequency of 1 Hz to the material, and by recording the subsequent strain. This test is used as an indication of the time-dependent structural changes within the material.

Zeta Potential Measurements

The ζ-potential, which is the electrokinetic potential in colloidal systems and, thus, the key indicator of the stability of dispersions, of the concentrated suspensions of earth was measured with a ZetaProbe (Colloidal Dynamics, Hofheim, Germany) [37], which is based on the electro-acoustic method. A high-frequency alternating electric field is applied and causes charged particles to oscillate. The motion of the particles generates a sound wave, which is recorded and delivers the dynamic mobility of the suspended particles. The software calculates the ζ-potential from the dynamic mobility [38,39]. The ζ-potential experiments were conducted on suspensions of solid volume fractions of approximately 6 wt %. The sedimentation issues and resulting measurement artefacts are avoided by blade stirring. The dispersant dosages were kept the same as those used in the rheology measurements. The single point measurement performed enables to follow the ζ-potential over time at constant dosage in the mineral soil. The reported value was recorded once no variation is observed in surface charge.

Mineral Characterization

To understand the role played by calcium, the dispersants and the calcium source have been mixed with a magnetic stirrer at 900 rpm. After mixing 3 min or 3 h, the samples are dried at 60 °C and 50 mbar. This drying process preserves the microstructure of weak crystallized reaction products compared with higher temperature oven-drying techniques (classically 105 °C). It is similar to the pre-treatment used to prepare concrete samples for mercury intrusion porosimetry in [40]. The measurement was done without clay as it is assumed that, even if clay particles in the whole system may react in presence of NaHMP or NaSil under alkaline conditions, the studied reaction is the same as the concentration of released ions during the reaction is low in such a diluted system [18]. The impact of the presence or not of clay for the identification of reaction products was thus negligible. The mineralogical composition of the formed reaction products was determined on randomly oriented powder specimens with X-ray diffraction (XRD). A frontloading XRD specimen was prepared using two blades to minimize the preferred orientation. XRD measurements were performed using Bragg–Brentano geometry (Bruker AXS D8 advance, CuKα radiation, Karlsruhe, Germany). The powder samples were step-scanned at room temperature from 2° to 80° 2θ (step width 0.02° 2θ, counting time 4 s). Finally, the reaction products were also observed with an environmental scanning electron microscope (SEM) (FEI quanta 200 3D).

Adsorption Measurements

A total organic carbon (TOC) analyzer was used in this work for samples prepared with PCE, only organic dispersant used in this study. The analysis technique involves a two-stage process commonly referred to as TC–IC. It allows for the measurement of both the amount of inorganic carbon (IC) by acidification of the sample and the amount of total carbon (TC) in the sample. TOC is calculated by subtraction of the IC value from the TC of the sample. A blank value was first done with an earth paste sample not containing any PCE in order to take into account the amount of organic carbon in earth as a reference. Furthermore, another measurement of the organic carbon in a reference solutions of PCE was done. The earth pastes with different amounts of PCE (0%, 0.5%, 1%, and 2%) were prepared according to the same procedure as for rheological measurement (W/S = 0.5). After mixing, the pore solution was collected through a syringe equipped with a nylon filter 0.45 μm by air pressure filtration. The extracted liquid was then tested with the TOC analyzer Sievers 5310C GE Analytical Instruments (Basel, Switzerland). By subtracting the TOC value measured on the extracted interstitial fluid (and corrected from the baseline value of the earth) from the value obtained in the reference PCE solution, the amount of PCE adsorbed or absorbed on earth particles was computed.

EXPERIMENTAL RESULTS AND DISCUSSION

Effect of Dispersants on Earth Products

Rheology

In Figure 3, the yield stress is plotted as a function of dispersant dosage for the earth pastes prepared with NaHMP and NaSil. The water to solid ratio is equal to 0.5, which allows for an initial material that is sufficiently thick to evaluate the efficiency of admixtures. As expected, when the dispersant is added, the yield stress of the earth paste is significantly reduced. The addition of only 0.1% NaHMP significantly decreases the yield stress of the reference paste by a factor of 20. This can be attributed to the deflocculating action of the inorganic dispersant, which modifies the clay surface [16,18,19]. Actually, as the clay's rheological behavior is controlled by the surface charge, increasing the surface charge with charged inorganic additives deflocculates the clay particles, which considerably reduces the yield stress of the earth material. This observation will be confirmed by ζ-potential measurement in the next section. It can also be noted in Figure 3 that the yield stress becomes constant from 0.2% NaHMP and 0.3% NaSil. We estimates that this is related to the saturation dosage of each dispersant. Beyond these respective dosages, and for this W/S ratio, it is expected that all particle surfaces are covered, and an additional dispersant is no longer efficient. Finally, it is worth noting that the NaHMP deflocculant is more efficient than NaSil, but it is less robust. A small variation of the dosage, which can easily occur in practice, leads to a yield stress close to zero. Since the yield stress of the constitutive paste controls the stability of the material [13,41,42], this very low yield stress could lead to static segregation issues. To prevent these issues, the yield stress of the paste must be sufficiently high to support particles in suspension. According to [42] and considering the maximum particle size of the earth material of 100μm (*cf.* Figure 1), the minimum yield stress needed is of the order of 9 Pa.

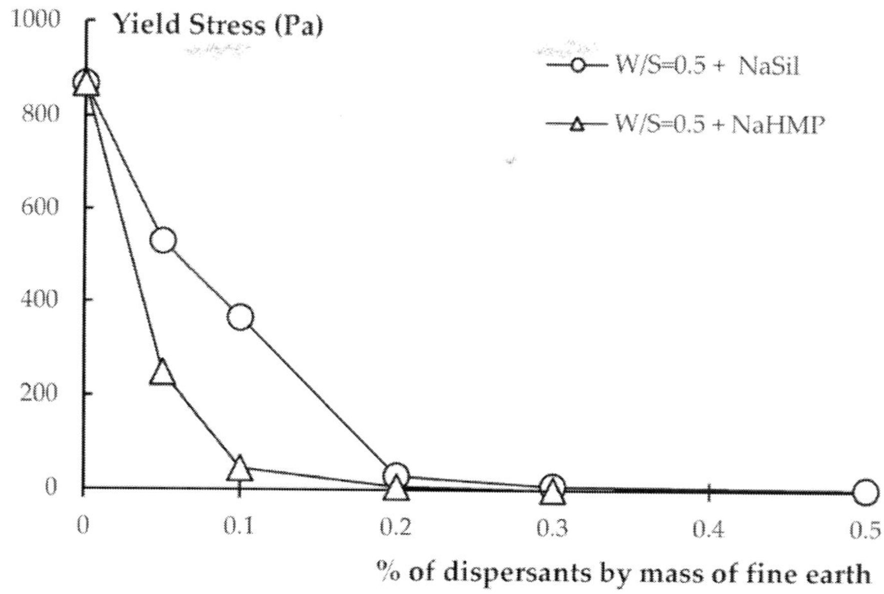

Figure 3: Effect of NaHMP and NaSil on the yield stress of earth paste.

In Figure 4, earth pastes containing 0.2% NaHMP and 0.3% NaSil are compared with the earth paste containing PCE as dispersant. Adding 2% PCE, which is a high dosage for this product, does not decrease the yield stress of the suspension. On the contrary, a negative effect is observed given that the yield stress of the earth paste prepared with 2% PCE is higher than the yield stress of the reference earth paste. In cement applications, the dispersing effect of PCE is primarily due to adsorption on cement particle surfaces (positively-charged) with a negatively-charged backbone. This induces predominantly steric repulsive forces between particles since cementitious systems have a high ionic strength that basically suppresses electrostatic repulsion [43,44]. As no effect is observed with clay particles, we can suppose that the side chains of PCEs are "trapped" between the aluminosilicate layers of smectite, leading to a large polymer consumption but no dispersion. This phenomenon is well known [45,46,47,48] and will be confirmed with TOC and ζ-potential measurement in the next section.

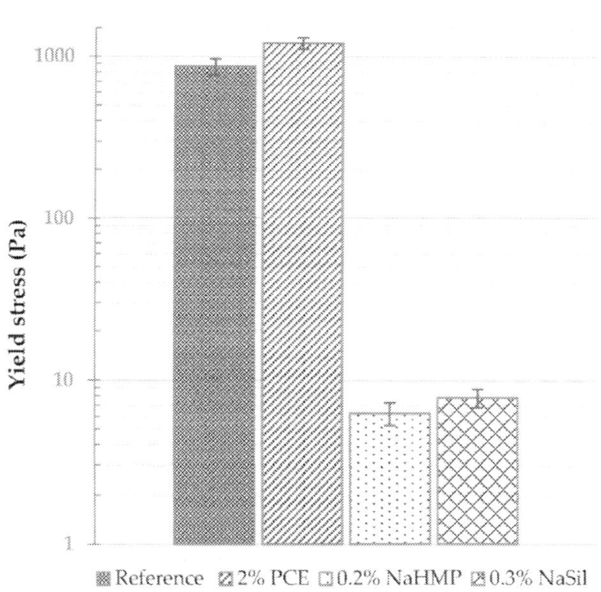

Figure 4: Comparative yield stress of earth pastes mixed with different dispersants.

Zeta Potential

The ζ-potential provides an indication of the surface charge of the clay particles in suspension. The magnitude of this parameter is often used as a measure of the strength of the repulsive interactions between similarly charged particles in suspensions. The same approach has been applied to evaluate the electrostatic forces between our clay particles. ζ-potential measurements were performed, and the particle charge values for the reference paste and those prepared with different types and amounts of dispersants are gathered in Table 3. In deionized water, the ζ-potential of the earth sample is negative because of its negative surface charge. The addition of 0.2% NaHMP increases its magnitude by a factor close to three. A similar observation can be made for 0.3% NaSil, though this increase has a lower intensity (increase by a factor close to two). These results confirm the interpretation that these two inorganic dispersants deflocculate clay particles by increasing the repulsive forces between particles. NaHMP is more efficient than NaSil in terms of deflocculation, which is correlated with higher negative charge of particles. On the contrary, the ζ-potential of earth material does not change with the addition of PCE, which indicates that the PCE molecules does not affect the surface charges of clay particles. This is no surprise as these polymers are not highly charged, that they are known to induce only minor changes to the zeta potential in addition to dispersing through steric hindrance rather than electrostatic repulsion. More important information with regard to the PCE behavior can be obtained by looking at its adsorption isotherm measured with TOC analysis and plotted with respect to PCE dosage [49]. Indeed, Figure 5 shows that almost 100% of the polymer introduced is adsorbed and that no plateau is observed. This (most probably) indicates that the introduced polymer is captured by the clay minerals with at least some of its side chains being trapped into clay layers, so that its dispersing efficiency is lost. This sorption of extremely large amount of PCEs has been well described in literature for the clay mineral types present in the studied material (smectite) [45,46,47,48]. No saturation of the clay particles is observed which could suggest that at a higher dosage PCEs could be active as dispersants. However, 2% PCE is already a very high dosage which means that reaching the saturation level will not be cost competitive.

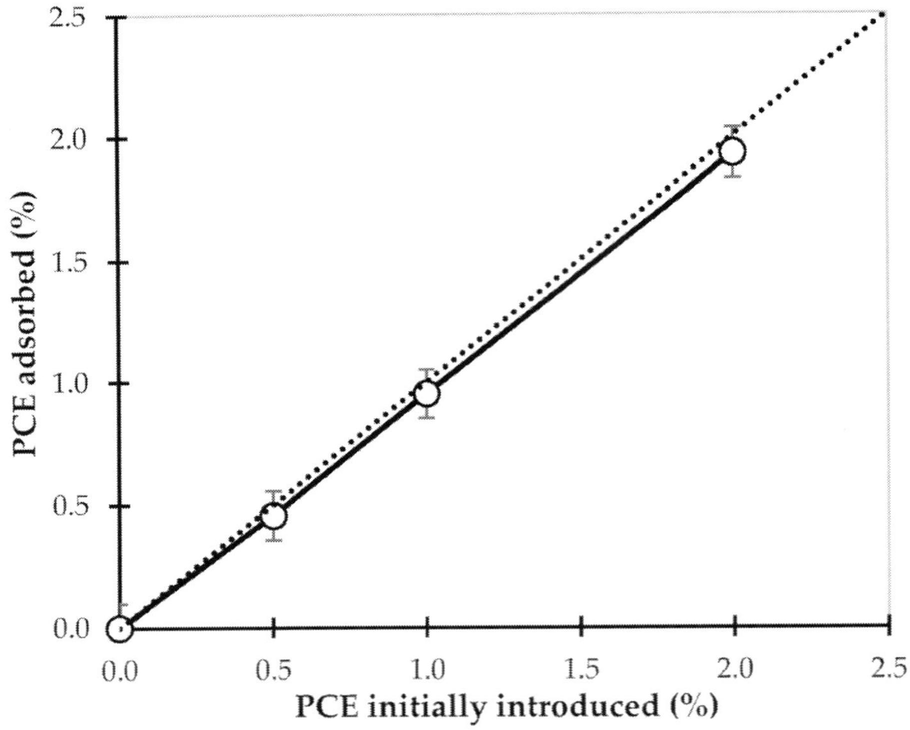

Figure 5: Adsorption isotherm for the PCE molecules reported with respect to the added PCE. The dotted line indicates full consumption.

Table 3: ζ-potential values for the reference earth, the earth prepared with 2% PCE, the earth prepared with 0.2% NaHMP and the earth prepared with 0.3% NaSil.

Parameter	Reference	2% PCE	0.2% NaHMP	0.3% NaSil
Average ζ (mV)	−4.6	−4.5	−13.4	−9.7
SD (mV)	±0.1	±0.1	±0.4	±0.2

Effect of Releasing Calcium

Various types of calcium-containing compounds were introduced into the mixed earth/dispersant to study their effect on the rheology of the material and their potential ability to flocculate particles. The dosages were calculated in order to introduce the same amount of calcium in the system (regardless of solubility). Depending on the dispersant used, the ratios of Ca/Si (when NaSil is used as dispersant) and Ca/P (for NaHMP) were fixed, respectively, at 1.3 and 1.5. Note that these amount are quite small in comparison to earth material as the ratio of 1.3 to 1.5 is calculated in relation to the dispersant which is, itself, calculated in relation to the earth content (0.2 wt %). In comparison to previous studies [6,7], we are then using ten times less calcium salts.

Calcium Ions and Structuration over Time

In Figure 6 and Figure 7, the variation of the storage modulus $G' - G_0$ (*i.e.*, the storage modulus of the earth paste prepared with calcium and NaSil or NaHMP G' minus the storage modulus of earth paste containing only dispersant G_0) is plotted as a function of time. The elastic modulus is an indicator of the time-dependent structural changes within the material and is an indication of the strength of the particle network at a given time. As a general observation, for all tested earth pastes, the elastic modulus increases with time. This behavior is entirely ascribed to the effect of calcium addition because, by referencing to the neat paste, we account for the potential contribution of water absorption and evaporation that can occur during the test and thereby modify the microstructure. This first observation shows that the use of calcium modifies the internal structure and cohesion of the material. Adding calcium flocculates the system. In the case of NaHMP, some authors have described this phenomenon as a complexation between phosphate anions and calcium ions [16,50] and an interaction of cations with the surface of clay minerals. Andreola *et al.* [15] reports that dissolved Ca^{2+}ions have deleterious effects on dispersing clay materials by (i) adsorbing onto the clay surface, thereby reducing the thickness of the electrical double layer; and (ii) decreasing the amount of phosphate anions that are available to be adsorbed on clay edges through soluble complex formation. By complexation with calcium ions Ca^{2+}, a network of interaction is created between particles, which gives rise to cohesion forces within the system. If we consider each earth paste behavior in detail, for the three calcium containing compounds introduced into the earth paste containing 0.2% NaHMP, the evolution is not similar, although the elastic modulus increases with time. First, it can be noted that the earth paste prepared with $CaCl_2$ displays a high elastic modulus at the start of the test followed by a low rate of increase (Figure 6).

For $Ca(OH)_2$ and $CaCO_3$, the elastic modulus at the beginning of the test is relatively low and increases rapidly during the test. The effect is more pronounced for the earth paste prepared with $Ca(OH)_2$, where the elastic modulus tends to ultimately reach the same high elastic modulus value as the earth paste prepared with $CaCl_2$. The same observation can be made for the earth pastes initially prepared with NaSil as deflocculant and calcium (Figure 7), apart from the elastic modulus of the earth paste containing $CaCO_3$. The curve can be decomposed in three parts: a plateau is observed at the beginning of the test, followed by an increase and then, after approximately 20 min, a plateau. Considering that the modification of the cohesive network within the material is linked to the interaction between calcium ions and mineral anions (same amount for each set of experiments), the different viscoelastic behaviors can be explained by the accumulated amount of calcium released in solution at a given time or, more accurately, the amount of calcium that has reacted with the mineral anions (Section 4.3). These three calcium products do not present the same dissolution rate, and calcium ions are slowly or quickly released in the interstitial solution depending on the type of calcium containing compound used. The calcium ions coming from $CaCl_2$ are released almost instantaneously in the interstitial fluid, which leads to the high initial elastic modulus displayed and, thus, to an initially-structured state, whereas the calcium ions coming from $Ca(OH)_2$ are slowly released over time, which delays the material structuration or flocculation. It can be suggested that, in the given time, the total amount of $CaCO_3$ cannot be dissolved in the interstitial fluid: the calcium ions coming from this mineral are, thus, not entirely available for the reaction, leading to a slowed down structuration rate.

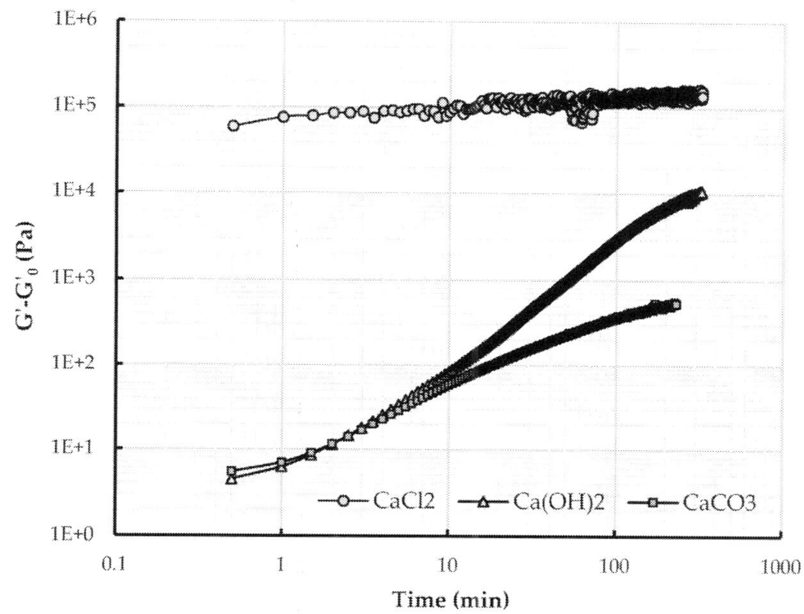

Figure 6: Variation of the storage modulus for earth pastes prepared with 0.2% NaHMP and different calcium products. Values are reported as a difference with respect to G_0, the modulus of the paste without addition of calcium compounds at each measuring time.

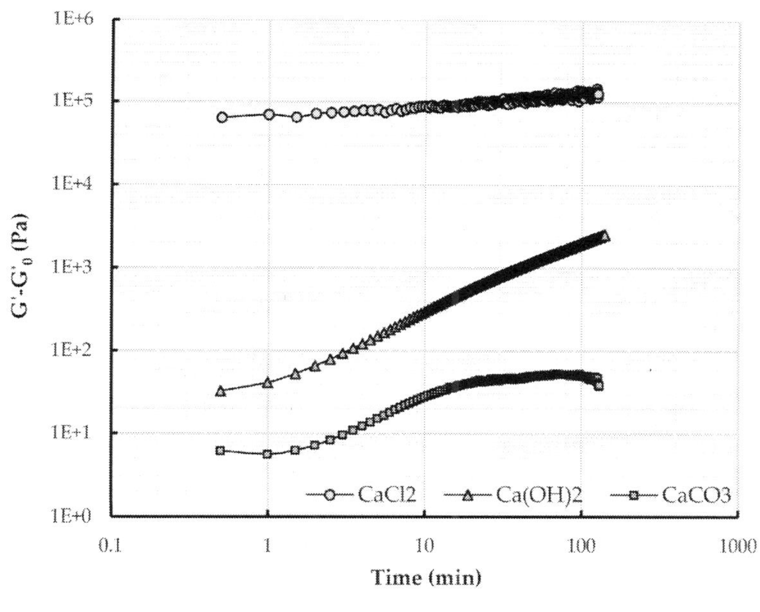

Figure 7: Variation of the storage modulus for earth pastes prepared with 0.3% NaSil and different calcium products. Values are reported as a difference with respect to G_0, the modulus of the paste without addition of calcium compounds at each measuring time.

Structuration and Nature of Interactions

In Figure 8, the elastic modulus is plotted as a function of strain for the reference earth paste, the paste prepared with 0.3% NaSil and the same earth paste containing $Ca(OH)_2$ obtained from stress sweep test to determine the limit of the elastic linear regime. As previously described, adding 0.3% NaSil decreases the elastic modulus, higher the critical strain of the reference earth paste and modifies the structure of the material by flocculation.

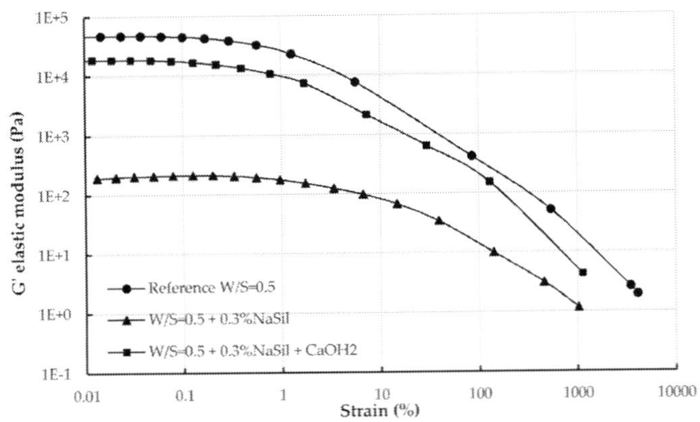

Figure 8: Elastic modulus as a function of strain for the reference earth paste, the earth paste prepared with 0.3% NaSil and the same earth paste containing $Ca(OH)_2$ after setting.

When $Ca(OH)_2$ is added to the system, the earth paste presents the same behavior as the reference earth paste: calcium addition therefore allows a return to the initial state and reflocculates the particles. We now focus on the critical strain. When the critical strain is reached, the initial structure of the material has been sufficiently modified to produce the rupture of the particle network of interactions [13,51,52]. The order of magnitude of the critical strain highlights the ability of the structural network to be deformed under stress and defines the nature of interactions. Whereas a small critical strain is associated with short range links between the particles, a large critical strain involves non-contact interactions between the particles [13,53]. In cement suspensions, small critical strain (on the order of 10^{-2}%) are characteristic of short range links between C–S–H particles and associated with interaction between crystallised hydrated structures. Larger critical strain (of the order of 1%) can be associated with the breakage of the network of colloidal interactions (*i.e.*, van der Waals attractive forces) between cement particles [13,53].

In our study, the critical strain for the earth paste prepared with 0.3% NaSil is of the order of 10%, whereas the critical strain for the two other earth pastes is of the order of 1%. Our results are consistent with reports in literature. First because NaHMP and NaSil are efficient to disperse clays [15,16,17,18,19,20,21,22,23,24,25,26,27]. Secondly, by comparing these values with those found for cement suspensions [53], it is then possible to suggest that strong but short ranged colloidal interactions are at the origin of the cohesion of the earth paste and not C–S–H links. When 0.3% NaSil is added to the earth material, the critical strain increases, and the material can be even more deformed under stress than the reference paste, indicating that the interaction network between particles is long ranged. This observation suggests that NaSil modifies the van der Waals attractive force network at the origin of this critical strain. As mentioned above, NaSil is able to strongly deflocculate particles in suspension through its adsorption at the surface of the clay particles. We, therefore, suggest that adsorbed NaSil is at the origin of steric repulsive forces between particles, leading to a dispersion of particles into suspension. When $Ca(OH)_2$ is added, the critical strain of the earth paste is similar to the critical strain of the reference earth paste: the initial state is recovered. Adding calcium allows then a movement from a material displaying a critical strain of the order of 10% to a material with a critical strain of the order of 1%. The internal structure and the type of interactions of the material are not modified: after deflocculating the particles with the use of a dispersant, they reflocculate under their initial form under the effect of calcium. The fact that we observe an equivalent critical strain leads to the assumption that calcium addition annihilates the plasticizing effect of the dispersant by removing it from the clay surface. The amount of calcium introduced into the mixture corresponds to the Ca/Si ratios that are frequently used to produce synthetic C–S–H. It is therefore most likely that the dispersant has been consumed by the precipitation of C–S–H products into the interstitial solution. However, the amount of C–S–H produced is not sufficient to control the critical strain of the suspension as no inflexion of the curve is observed at 10^{-2}. The nature of the precipitation products needs to be identified to support this hypothesis.

Identification of Reaction Products

The reaction products have been identified through XRD and SEM on samples where only the dispersant and the calcium source have been mixed. When the NaSil is used as a dispersant, a fast disappearance of $Ca(OH)_2$ peaks (still present after three minutes but absent after three hours) and the precipitation of two phases are observed: formation of plombierite, a gel-like structure member of the calcium silicate hydrate (C–S–H) family, and sodium carbonate (Figure 9) [54]. Plombierite, named Tobermorite 14 Å, with the chemical composition $Ca_5Si_6O_{16}(OH)_2{\cdot}8H_2O$, is the most hydrated phase of the C–S–H group [55]. The SEM images show large plates of calcium hydroxide crystals after three minutes (Figure 10a) and a typical needles network of C–S–H after three hours (Figure 10b). These results confirm the previous hypothesis concerning the mechanism of coagulation of the soil by the addition of a calcium-containing mineral: the formation of C–S–H as an anti-plasticizer. Moreover, the process involves a coupling of portlandite dissolution and C–S–H precipitation, which delays the coagulation process providing time for processing the earth pastes in the fluid state.

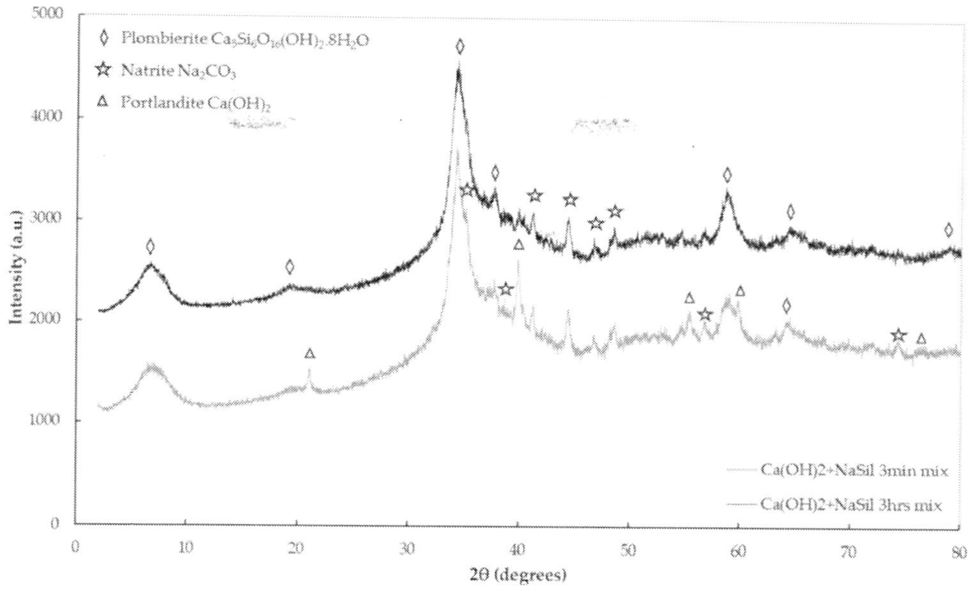

Figure 9: X-ray diffraction patterns of NaSil solution and $Ca(OH)_2$ at different mixing times, dried at 60 °C and 50 mbar.

Figure 10: SEM images of NaSil and $Ca(OH)_2$ precipitation products dried at 60 °C and 50 mbar after (**a**) 3 min; and (**b**) 3 h of mixing.

When NaHMP is used as a dispersant, we observe the rapid formation of hydroxyapatite [56,57]. This phosphate mineral precipitates in the suspension and, thus, removes the phosphate from the solution (Figure 11).

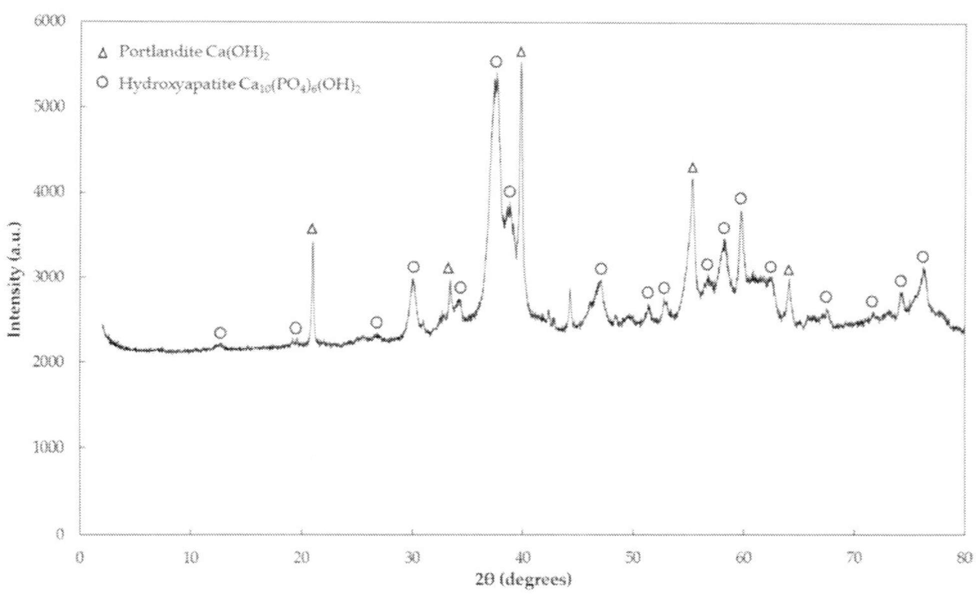

Figure 11: X-ray diffraction pattern of NaHMP solution and $Ca(OH)_2$ at 3 h mixing dried at 60 °C and 50 mbar.

CONCLUSIONS

The strategies that have been tested are promising and allow for the development of a process to cast a clay-based concrete as easily as a cement bound concrete. By modifying the clay properties with inorganic additives and a material that contains calcium, it is possible to deflocculate and flocculate clays during the casting process.

The use of sodium silicate or sodium hexametaphosphate as dispersant led to a strong deflocculation of clay particles by creating repulsive forces between particles. The yield stress of the clay paste is then considerably reduced, and good workability is obtained, which is required to pour the material in the formwork.

Furthermore, we showed that the addition of calcium hydroxide can reflocculate clay particles through a slow dissolution that releases calcium ions into the interstitial solution, which macroscopically brings the earth material back to its initial behavior. The released calcium ions lead to C–S–H precipitating with the silicate dispersant or hydroxyapatite precipitating with the phosphate based dispersant. This precipitation cancels the plasticizing effect. In this process, in contrast to most earth stabilized products, C–S–H precipitation is used not as a binding agent but as an anti-plasticizer that removes the inorganic dispersant additives. This is confirmed by the fact that the critical strain of the reflocculated clay suspension does not exhibit a behavior characteristic of a suspension controlled by C–S–H interaction.

This study presents a breakthrough compared to previous studies on castable earth. Instead of adding 5 to 10% of lime or cement, 0.2 to 0.5 wt % of lime are added as the amount of calcium introduced is there to counteract the action of the plasticizer used in very small proportion compared to earth material. This has fundamental consequences on the carbon footprint that is reduced by a factor 10 compared to the current practice.

However, further work on the development of a performant self-compacting clay concrete must be performed. This process preserves the initial material but does not improve its internal cohesion and resistance. As a consequence, additional removal of the water is still required to achieve higher strength.

ACKNOWLEDGMENTS

Financial supports by the Swiss National Fund (SNF) (project No. 205121-153368, SCCC–Self Compacted Clay based Concrete) and by an ETH SEED grant from the ETH Foundation are gratefully acknowledged. Plötze is warmly acknowledged for his help on XRD measurements and interpretation.

AUTHOR CONTRIBUTIONS

Gnanli Landrou conducted the study, which is a part of his PhD thesis focusing on the development of a Self Compacted Clay based Concrete. Coralie Brumaud was involved in all the phases of the project and wrote the article. Frank Winnefeld developed the methodology for the zeta potential measurements with clays. Robert J. Flatt identified the dispersants and helped in the understanding of their mechanisms of action. Guillaume Habert supervised and directed all the work. All authors discussed the results and implications and approved the final manuscript.

REFERENCES

1. Fontaine, L.; Anger, R. *Building with Earth—From Sand Grain to Architecture*; Belin/Cité des Sciences et de l'Industrie: Paris, France, 2009.
2. Morel, J.C.; Pkla, A.; Walker, P. Compressive strength testing of compressed earth blocks.*Constr. Build. Mater.* 2007, *21*, 303–309.
3. Shukla, A.; Tiwari, G.N.; Sodha, M.S. Embodied energy analysis of adobe house. *Renew. Energy*2009, *34*, 755–761.
4. Parra-Saldivar, M.L.; Batty, W. Thermal behaviour of adobe constructions. *Build. Environ.* 2006,*41*, 1892–1904.
5. Schneider, M.; Romer, M.; Tschudin, M.; Bolio, H. Sustainable cement production-present and future. *Cem. Concr. Res.* 2011, *41*, 642–650.
6. Moevus, M.; Fontaine, L.; Anger, R.; Doat, P. *Environmental Clay Concrete (B.A.E)*; Scientific Report, Programme C2D2. Ministère de l'Ecologie, du Développement Durable et de l'Energie: Paris, France, 2013.
7. Kang, N.; Hwang, H. Study on high strengthening of an earth wall using earth and high performance lime. In Proceedings of the International Conference on Earth Architecture in Asia, Mokpo, Korea, 11–14 October 2011.
8. Pellenq, R.J.M.; Lequeux, N.; van Damme, H. Engineering the bonding scheme in C–S–H: The iono-covalent framework. *Cem. Concr. Res.* 2008, *38*, 159–174.
9. Kovler, K.; Roussel, N. Properties of fresh and hardened concrete. *Cem. Concr. Res.* 2011, *41*, 775–792.
10. Horn, R.G. Surface forces and their action in ceramic materials. *J. Am. Ceram. Soc.* 1990, *73*, 1117–1135.
11. Van Damme, H.; Review of Horn, R.G. Surface forces and their action in ceramic materials. *J. Am. Ceram. Soc.* 1990, *73*, 1117–1135.
12. Yammine, J.; Chaouche, M.; Guerinet, M.; Moranville, M.; Roussel, N. From ordinary rheology concrete to self compacting concrete: A transition between frictional and hydrodynamic interactions. *Cem. Concr. Res.* 2008, *38*, 890–896.
13. Roussel, N.; Lemaître, A.; Flatt, R.J.; Coussot, P. Steady state flow of cement suspensions: A micromechanical state of the art. *Cem. Concr. Res.* 2010, *40*, 77–84.
14. Bergaya, F.; Theng, B.K.G.; Lagaly, G. *Handbook of Clay Science*; Elsevier: Amsterdam, The Netherlands, 2006; ISBN: 978-0-08-044183-2.
15. Andreola, F.; Castellini, E.; Ferreira, J.M.F.; Olhero, S.; Romagnoli, M. Effect of sodium hexametaphosphate and ageing on the rheological behaviour of kaolin dispersions. *Appl. Clay Sci.* 2005, *31*, 56–64.
16. Andreola, F.; Castellini, E.; Manfredinia, T.; Romagnoli, M. The role of sodium hexametaphosphate in the dissolution process of kaolinite and kaolin. *J. Eur. Ceram. Soc.* 2004,*24*, 2113–2124.

17. Adebowale, K.O.; Unuabonah, I.E.; Olu-Owolabi, B.I. Adsorption of some heavy metal ions on sulphate- and phosphate-modified kaolin. *Appl. Clay Sci.* 2005, *29*, 145–148.
18. Andreola, F.; Castellini, E.; Lusvardi, G.; Menabue, L.; Romagnoli, M. Release of ions from kaolinite dispersed in deflocculant solutions. *Appl. Clay Sci.* 2007, *36*, 271–278.
19. Rossington, K.R.; Senapati, U.; Carty, W.M. A critical evaluation of dispersants for clay-based systems. *Ceram. Eng. Sci. Proc.* 1998, *19*, 77–85.
20. Papo, A.; Piani, L.; Ricceri, R. Sodium tripolyphosphate and polyphosphate as dispersing agents for kaolin suspensions: Rheological characterization. *Colloids Surfaces A* 2002, *201*, 219–230.
21. Castellini, E.; Berthold, C.; Malferrari, D.; Bernini, F. Sodium hexametaphosphate interaction with 2:1 clay minerals illite and montmorillonite. *Appl. Clay Sci.* 2013, *83–84*, 162–170.
22. Worrall, W.E.; Worrall, M.D. *Clays and Ceramic Raw Materials*; Springer Science & Business Media: London, UK, 1986.
23. Manfredini, T.; Pellacani, G.C.; Pozzi, P. Sodium silicates as deflocculating agents for clays. *Ind. Ceram.* 1987, *7*, 85–87.
24. Rossington, K.R.; Senapati, U.; Carty, W.M. A critical evaluation of dispersants: Part II. Effects on rheology, pH, and specific adsorption. *Ceram. Eng. Sci. Proc.* 1999, *20*, 119–131.
25. Penner, D.; Lagaly, G. Influence of anions on the rheological properties of clay mineral dispersions. *Appl. Clay Sci.* 2001, *19*, 131–142.
26. Diz, H.M.M.; Rand, B. The mechanism of deflocculation of kaolinite by polyanions. *Br. Ceram. Trans. J.* 1990, *89*, 77–82.
27. Andreola, F.; Romagnoli, M. Role of the surface treatment in the deflocculation of kaolinite. *J. Am. Ceram. Soc.* 2006, *89*, 1107–1109.
28. Amoros, J.L.; Beltran, V.; Sanz, V.; Jarque, J.C. Electrokinetic and rheological properties of highly concentrated kaolin dipersions: Influence of particle fraction and dispersant concentration.*Appl. Clay Sci.* 2010, *49*, 33–43.
29. Zaman, A.A.; Demir, F.; Finch, E. Effects of process variables and their interactions on solubility of metal ions from crude kaolin particles: Results of a statistical design of experiments. *Appl. Clay Sci.* 2003, *22*, 237–250.
30. Zaman, A.A.; Mathur, S. Influence of dispersing agents and solution conditions on the solubility of crude kaolin. *J. Colloid Interface Sci.* 2004, *271*, 124–130.
31. Yuan, J.; Garforth, W.L.; Pruett, R.J. Influence of dispersant on the solubility of calcined kaolin.*Appl. Clay Sci.* 1998, *13*, 137–147.
32. Bergmann, J.; Kleeberg, R. Rietveld analysis of disordered layer silicates. In Proceedings of the 5th European Conference on Powder Diffraction (EPDIC 5), Materials Science Forum, Parma, Italy, 24–28 May 1997; pp. 300–305.
33. Bish, D.L.; Plötze, M. X-ray powder diffraction with emphasis on qualitative and quantitative analysis in industrial mineralogy. In *EMU Notes in Mineralogy Vol. 9, Advances in the Characterization of Industrial Minerals*; Christidis, G.E., Ed.; EMU and Mineralogical Society: Jena, Germany, 2011; pp. 35–76.
34. Gelardi, G.; Mantellato, S.; Marchon, D.; Palacios, M.; Eberhardt, A.B.; Flatt, R.J. Chemistry of chemical admixtures. In *Science and Technology of Concrete Admixtures*; Aïtcin, P.C., Flatt, R.J., Eds.; Elsevier: Cambridge, UK, 2016; pp. 149–218.
35. Nguyen, Q.D.; Boger, D.V. Yield stress measurement for concentrated suspensions. *J. Rheol.*1983, *27*, 321–349.
36. Nachbaur, L.; Mutin, J.C.; Nonat, A.; Choplin, L. Dynamic mode rheology of cement and tricalcium silicate pastes from mixing to setting. *Cem. Concr. Res.* 2001, *31*, 183–192.
37. Zingg, A.; Winnefeld, F.; Holzer, L.; Pakusch, J.; Becker, S.; Gauckler, L. Adsorption of polyelectrolytes and its influence on the rheology, zeta potential, and microstructure of various cement and hydrate phases. *J. Colloid Interface Sci.* 2008, *323*, 301–312.
38. Greenwood, R. Review of the measurement of zeta potentials in concentrated aqueous suspensions using electroacoustics. *Adv. Colloid Interface Sci.* 2003, *106*, 55–81.
39. Au, P.; Leong, Y.K. Rheological and zeta potential behaviour of kaolin and bentonite composite slurries. *Colloids Surfaces A* 2013, *436*, 530–541.
40. Thiery, M.; Dangla, P.; Belin, P.; Habert, G.; Roussel, N. Carbonation kinetics of a bed of recycled concrete aggregates: A laboratory stud on model materials. *Cem. Concr. Res.* 2013, *46*, 50–65.
41. Saak, A.W.; Jennings, H.M.; Shah, S.P. The influence of wall slip on yield stress and visco-elastic measurements of cement pastes. *Cem. Concr. Res.* 2001, *31*, 205–212.
42. Roussel, N. A theoretical frame to study stability of fresh concrete. *Mater. Struct.* 2006, *39*, 81–91.
43. Gelardi, G.; Flatt, R.J. Working mechanisms of water reducers and superplasticizers. In *Science and Technology of Concrete Admixtures*; Aïtcin, P.C., Flatt, R.J., Eds.; Elsevier: Cambridge, UK, 2016; pp. 257–278.
44. Flatt, R.J. Superplasticizers and the rheology of concrete. In *Understanding Rheology of Concrete*; Roussel, N., Ed.; Woodhead Publishing: Cambridge, UK, 2012; pp. 144–208.
45. Jeknavorian, A.A.; Jardine, L.A.; Koyata, H.; Ou, C.C.; Folliard, K. Interaction of superplasticizers with clay-bearing aggregates. In Proceedings of the 7th Canmet/ACI International Conference Superplasticizers and Other Chemical Admixtures in Concrete, Berlin, Germany, October 2003; pp. 1293–1316.
46. Lei, L.; Plank, J. A study on the impact of different clay minerals on the dispersing force of conventional and modified vinyl ether based polycarboxylate superplasticizers. *Cem. Concr. Res.*2014, *60*, 1–10.

47. Ng, S.; Plank, J. Interaction mechanisms between Na montmorillonite clay and MPEG-based polycarboxylate superplasticizers. *Cem. Concr. Res.* 2012, *42*, 847–854.

48. Zhang, L.; Lu, Q.; Xu, Z.; Liu, Q.; Zeng, H. Effect of polycarboxylate ether comb-type polymer on viscosity and interfacial properties of kaolinite clay suspensions. *J. Colloid Interface Sci.*2012, *378*, 222–231.

49. Marchon, D.; Mantellato, S.; Eberhardt, A.B.; Flatt, R.J. Adsorption of chemical admixtures. In*Science and Technology of Concrete Admixtures*; Aïtcin, P.C., Flatt, R.J., Eds.; Elsevier: Cambridge, UK, 2016; pp. 219–256.

50. Kura, G.; Ohashi, S. Complex formation of cyclic phosphate anions with bivalent cations. *J. Inorg. Nucl. Chem.* 1974, *36*, 1605–1609.

51. Maloney, C.E.; Lemaitre, A. Amorphous systems in athermal, quasistatic shear. *Phys. Rev.* E2006, *74*, 016118.

52. Zukoski, C.F. Particles and suspensions in chemical engineering: Accomplishments and prospects. *Chem. Eng. Sci.* 1995, *50*, 4073–4079.

53. Roussel, N.; Ovarlez, G.; Garrault, S.; Brumaud, C. The origins of thixotropy of fresh cement pastes. *Cem. Concr. Res.* 2012, *42*, 148–157.

54. Richardson, I.G. The calcium silicates hydrates. *Cem. Concr. Res.* 2008, *38*, 137–158.

55. Nonat, A. The structure and stoichiometry of C–S–H. *Cem. Concr. Res.* 2004, *34*, 1521–1528.

56. Fabbri, M.; Celotti, G.C.; Ravaglioli, A. Hydroxyapatite-based porous aggregates: Physico-chemical nature, structure, texture and architecture. *Biomaterials* 1995, *16*, 225–228.

57. Kay, M.I.; Young, R.A.; Posner, A.S. Crystal structure of hydroxyapatite. *Nature* 1964, *204*, 1050–1052.

CHAPTER 14

THERMAL PERFORMANCE OF HOLLOW CLAY BRICK WITH LOW EMISSIVITY TREATMENT IN SURFACE ENCLOSURES

Roberto Fioretti and Paolo Principi

Department of Industrial Engineering and Mathematical Sciences, Università Politecnica delle Marche, Via Brecce Bianche 12, 60131 Ancona, Italy

ABSTRACT

External walls made with hollow clay brick or block are widely used for their thermal, acoustic and structural properties. However, the performance of the bricks frequently does not conform with the minimum legal requirements or the values required for high efficiency buildings, and for this reason, they need to be integrated with layers of thermal insulation. In this paper, the thermal behavior of hollow clay block with low emissivity treatment on the internal cavity surfaces has been investigated. The purpose of this application is to obtain a reduction in the thermal conductivity of the block by lowering the radiative heat exchange in the enclosures. The aims of this paper are to indicate a methodology for evaluating the thermal performance of the brick and to provide information about the benefits that should be obtained. Theoretical evaluations are carried out on several bricks (12 geometries simulated with two different thermal conductivities of the clay), using a finite elements model. The heat exchange procedure is implemented in accordance with the standard, so as to obtain standardized values of the thermal characteristics of the block. Several values of emissivity are hypothesized, related to different kinds of coating. Finally, the values of the thermal transmittance of walls built with the evaluated blocks have been calculated and compared. The results show how coating the internal surface of the cavity provides a reduction in the thermal conductivity of the block, of between 26% and 45%, for a surface emissivity of 0.1.

INTRODUCTION

Energy consumption in the building sector has increased in recent years, representing approximately 40% of the overall energy required and generating 36% of greenhouse gases in Europe [1]. In European countries, the reduction in energy consumption for buildings is addressed by legislation, with a view to increasing energy efficiency, reducing greenhouse gas emissions and promoting the use of renewable energy. The target for 2020 is to have new buildings that do not require energy in the use stage (heating, cooling, illumination, hot water and other electric uses) [2].

One of the most widely used technologies for the building envelope involves walls made with hollow clay bricks or blocks, used for their good characteristics, such as high durability and excellent fire, thermal and sound insulation.

On the market there are several kinds of bricks, with different dimensions, with or without cavities, made with clay, concrete or wood and characterized by different thermal properties. The thermal properties that influence energy consumption for maintaining indoor microclimate comfort are the thermal transmittance and the thermal inertia of the envelopes. The parameter that mainly influences energy loss

Citation: Roberto Fioretti and Paolo Principi, Thermal Performance of Hollow Clay Brick with Low Emissivity Treatment in Surface Enclosures, doi:10.3390/coatings4040715.

during the cold seasons is the thermal transmittance of the envelopes, and some countries, including Italy, have imposed a limit value [3]. Thermal inertia mainly influences energy consumption during the warm seasons and thermal comfort, and walls built with block normally have acceptable values. Nowadays, clay blocks are a widespread solution for building external walls. The thermal insulation provided by these products is a result of the geometry of the brick and of the small pores present in the clay. Some studies evaluate the possibility of increasing the thermal resistance of the block through a change in the configuration of the enclosures [4,5,6,7,8,9] or by filling the enclosures with insulation material, such as perlite, mineral wood, polystyrene and other substances with low thermal conductivity [10].

Low emissivity coating, treatment or film is currently used to reduce radiative energy transmission in several fields comprising the building sector. For transparent surfaces, low emissivity coating is often used to reduce the *U*-value of the glass, with an infrared coating on the internal surface of the double or triple glazing [11,12,13]. Another application is on external walls or the roof, using a low emissivity and absorption treatment in order to reduce both the radiative heat exchange with the external environment and the solar absorption. One innovative application that uses low emissivity surfaces, in this case aluminum film, is thin multi-foil insulation that consists of a series of low emissivity films separated by wadding foams [14,15,16,17].

In a numerical analysis on the thermal behavior of fired clay hollow bricks [18], the reduction in the emissivities of cavities was investigated as a solution for improving the thermal resistance of the brick. The analysis was carried out on small and large-sized bricks, for different values of surface emissivity (between 0.3 and 0.7). The study demonstrates that the improvement depends on the configuration and the dimensions of the brick. For cavity surfaces with an emissivity of 0.3 applied on hollow bricks, this variation ranges between approximately 10% and 50%, depending on the kind of brick and the dimensions of the holes.

One theoretical and experimental analysis [19] investigated the increase in the thermal performance of a thermal block. In that study, an improvement in performance through the use of low emissivity coating and phase change material (PCM) was tested, by means of a theoretical analysis (conducted using FEM software) and an experimental analysis carried out with a thermo flux meter device. The results demonstrate the validity of these technical solutions for increasing thermal resistance (low emissivity coating) and thermal inertia (PCM) and the accuracy of the theoretical method used for the evaluation.

The present paper proposes the improvement in the energy performance of hollow clay block through the use of a low emissivity coating technique applied on the block by covering the enclosures with low emission and absorption paint, thereby reducing the overall heat transfer coefficient value of the block. For the study, several blocks and emissivities of the coating are hypothesized in order to obtain a large number of values of improvement. This technology is compared with that of filling the enclosures with an insulation material. A comparison between the results obtained and the previous evaluations is also discussed.

THEORETICAL EVALUATION

The evaluation of the thermal performance of the blocks with low emissivity treatment on the surface of the cavities was carried out using the COMSOL software [20]. To verify the thermal performance and the benefits provided by the low emissivity coating, a two-dimensional steady-state simulation was performed. The calculation method was previously validated [19] in accordance with ISO 10211-1 (Annex A) [21] and EN 1745 (Annex D) [22], and the results are reported in Table 1.

Table 1: Results of the numerical method and software validation, reprinted with permission from [19], copyright Elsevier 2012

Standard	Number of elements of the mesh	Degrees of freedom	Value comparison	Value standard	Value program	Accuracy/ Error	Max accuracy/Error
ANNEX D.4 case 1—EN 1745	21,024	42,313	conductance	0.6258 W/m²·K	0.6274 W/m²·K	0.25%	2%
ANNEX A case 1—UNI 10211-1	8,192	16,577	temperatures	all 28 values in the grid		<0.1 K	0.1 K
ANNEX A case 2—UNI 10211-1	16,572	33,815	temperatures	all 9 values in the grid		<0.1 K	0.1 K
			thermal flux	9.5 W/m²	9.49 W/m²	0.01 W/m²	0.1 W/m²

To simulate the heat exchange inside the cavities, a simplified and standardized method using equivalent conductivity was used. The algorithm used was developed according to the calculation procedure suggested in EN 1745 (Annex D) and EN ISO 6946 (Annex B-C) [23], which combines the effects of heat conduction, convection and radiation through the use of equivalent conductivity, thus approaching the problem exclusively from the conduction heat transfer perspective. The equivalent conductivity of the core is obtained by the following equation:

$$\lambda_{eq} = \frac{d}{R_g} \quad (1)$$

where the thermal resistance of the cavity may be calculated using the following equation:

$$R_g = \frac{1}{h_a + 1/2 \cdot E \cdot h_{ro}(1 + \sqrt{1 + d^2/b^2} - d/b)} \quad (2)$$

where E is the emittance of the surface:

$$E = \frac{1}{1/\varepsilon_1 + 1/\varepsilon_2 - 1} \tag{3}$$

which depends on the emissivity of the surfaces of the cavity and where ε_1 and ε_2 are, respectively, the emissivities of the emitting and receiving surfaces.

The convective heat exchange coefficient h_a is the maximum value between 1.25 W/m^2·K and 0.025/d, where d is the width of the void in the direction of the thermal flux. The lowest heat exchange coefficient h_a is for a value of d equal to or less than 0.02 m.

The coefficient h_{ro} is the radiative heat exchange coefficient of the black body, and the value depends on the mean temperature of the cavity. For this evaluation the temperature hypothesized was 10 °C [the mean value between the standard internal temperature (20 °C) and the standard external temperature during the winter (0 °C)], and the corresponding value is equal to 5.1 W/m^2·K (Table A.1 UNI EN ISO 6946) [23].

The main assumptions considered for the simulation process are:

- two-dimensional model;
- steady-state condition;
- all thermophysical properties kept constant;
- isotropic conductivity;
- equivalent conductivity (convective and radiative part);
- thermal resistance of the voids calculated according to B.3 of EN ISO 6946:1996;
- no mass transfer;
- emissivity values of the emitting and receiving surfaces are equal ($\varepsilon_1 = \varepsilon_2$);
- thermal resistance of the holes with non-rectangular shape is assumed equal to a rectangular hole with the same area and the same dimensional ratio (d/b) [23].

Steady-state analysis to determine the benefits of applying low emissivity coating was performed on several kinds of blocks characterized by different dimensions, geometry and void fractions. To have a sufficient sample in order to be able to discuss the benefits provided by the application, 12 different block geometries (Figure 1) were evaluated, with several values of total thickness and the number and dimensions of holes. The simulation was also performed with two different values of thermal conductivity of the material, 0.90 W/m·K and 0.45 W/m·K, the first related to a high density clay of 2400 kg/m^3 and the second to a clay with pore density of 1500 kg/m^3. In order to verify the effect of surface emissivity on the overall thermal resistance of the block, the emissivity was varied from 0.9 to 0.1. An emissivity of 0.9 corresponds to the thermal performance of the standard block without coating.

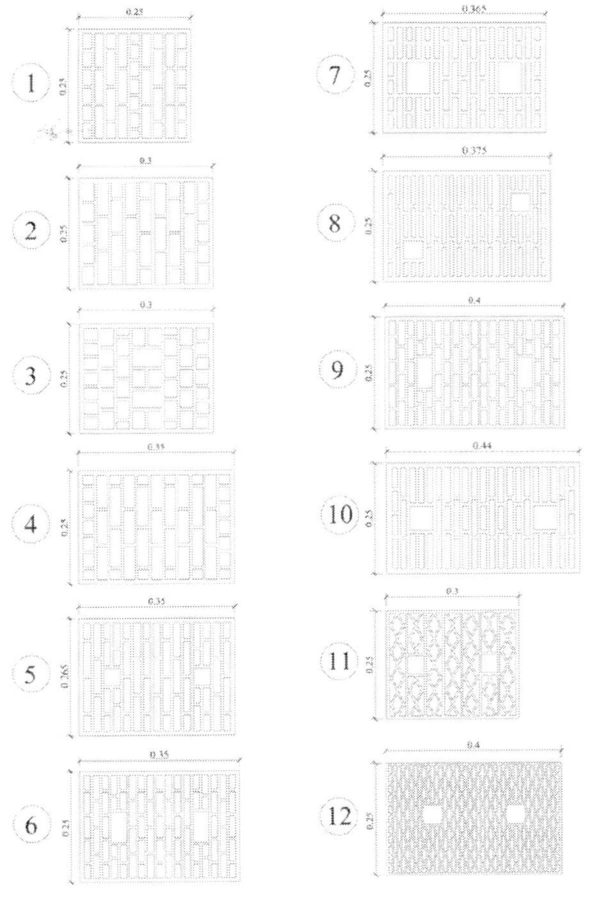

Figure 1: Images of the evaluated blocks (plan view).

In order to set a reachable limit value, an evaluation without radiative heat exchange was carried out. This case is not real, because it corresponds to an emissivity equal to zero, but it is the lowest value of heat exchange in the cavities, in the case of only a convective heat exchange. The geometric and thermal characteristics of the blocks are reported in Table 2, which also lists the number of the elements of the meshes and the number of degrees of freedom of the model built in the software.Figure 1 shows the geometry and the dimensions of the investigated blocks.

Alternative simulations were carried out on blocks with EPS filling (extruded polystyrene; thermal conductivity of 0.04 W/m·K) in the voids in order to obtain a comparison with the B.A.T. (best available technology) of a high thermal performance block.

Finally, the thermal transmittances (U-value) of a wall built with the evaluated block units have been calculated using the procedure set down in UNI EN 1745 [22] and UNI EN ISO 6946 [23]. The U-value is the most important parameter used to verify the conformity of the wall to national legislation [24,25,26] and to evaluate the thermal heat loss and energy consumption. To evaluate the U-value, a sample wall has been hypothesized, built with an internal layer of plaster (thickness = 0.02 m; thermal conductivity = 0.4 W/m·K), masonry made with block and a horizontal mortar joint (height = 0.015 m; thermal conductivity of 1.0 W/m·K) and a layer of external plaster (thickness = 0.02 m; thermal conductivity = 0.5 W/m·K). The boundary conditions to evaluate the U-value were R_{se} = 0.04 m^2·K/W and R_{si} = 0.13 m^2·K/W in accordance with the Standard UNI EN ISO 6946 [23].

Table 2: Characteristics and thermal proprieties of the investigated blocks

Sample	Length (m)	Width (m)	Height (m)	Number of voids	Rows of voids	Void fraction	Mesh	Degrees of freedom	Conductivity of the material (W/m·K)
1A 1B	0.25	0.25	0.25	39	9	61.97%	14,144	28,469	0.9 0.45
2A 2B	0.25	0.3	0.25	34	9	65.19%	13,532	27,251	0.9 0.45
3A 3B	0.25	0.3	0.25	44	8	62.91%	13,640	27,461	0.9 0.45
4A 4B	0.25	0.35	0.25	45	11	65.58%	17,152	34,517	0.9 0.45
5A 5B	0.265	0.35	0.25	57	15	57.25%	17,888	35,977	0.9 0.45
6A 6B	0.25	0.35	0.25	77	17	59.27%	20,872	41,977	0.9 0.45
7A 7B	0.25	0.365	0.25	64	17	50.55%	27,800	55,949	0.9 0.45
8A 8B	0.25	0.375	0.25	67	21	46.03%	28,152	56,641	0.9 0.45
9A 9B	0.25	0.4	0.25	86	19	60.48%	26,100	52,529	0.9 0.45
10A 10B	0.25	0.45	0.25	61	21	55.95%	21,856	44,027	0.9 0.45
11A 11B	0.25	0.30	0.25	96	21	60.63%	32,516	65,235	0.9 0.45
12A 12B	0.25	0.40	0.25	252	41	57.38%	44,876	90,037	0.9 0.45

RESULTS AND DISCUSSION

In order to verify the impact of the low emissivity coating on the brick thermal conductivity, the numerical simulations were conducted for different values of emissivity ranging from 0.1 to 0.9. The evaluation shows that the equivalent thermal resistances of the simulated blocks, in comparison with their initial resistance, increase with the decrease in cavity surface emissivity. This decrease in heat exchange is consequent to a reduction in the radiative heat exchange in the voids, due to the low emissivity coating. The reduction in the radiative part causes a decrease in the equivalent thermal conductivity of the cavity, while the convective heat coefficient remains the same in the different evaluations.

As shown in Table 3, the change in the emissivity of the cavity surfaces leads to a reduction in the equivalent conductivity of the cavities, which is dependent on the emissivities and the dimensions (*b*) and (*d*) of the cavities. The equivalent conductivity of all three types of cavities drops by between 55% and 60%, when the emissivity is reduced to 0.1, assuming a value between 0.028 and 0.077 W/m·K (while the value in the uncoated block was between 0.062 and 0.232 W/m·K). The best value attainable without radiative heat exchange was 0.025–0.066 W/m·K, corresponding to a reduction of 59%–75%.

Table 3: Dimensions and equivalent thermal conductivity of the cavities

Cavity	*d* (m)	*b* (m)	h_a (W/m²·K)	λ_{eq} (W/m·K)									
				ε = 0.9	ε = 0.8	ε = 0.7	ε = 0.6	ε = 0.5	ε = 0.4	ε = 0.3	ε = 0.2	ε = 0.1	No radiation
1-1	0.02	0.033	1.250	0.090	0.078	0.068	0.059	0.052	0.045	0.039	0.034	0.029	0.025
1-2	0.02	0.073	1.250	0.099	0.085	0.073	0.064	0.055	0.047	0.041	0.035	0.030	0.025
1-3	0.02	0.017	1.250	0.082	0.071	0.063	0.055	0.048	0.042	0.037	0.033	0.029	0.025
1-4	0.02	0.031	1.250	0.089	0.078	0.067	0.059	0.051	0.045	0.039	0.034	0.029	0.025
2-1	0.026	0.041	1.250	0.117	0.101	0.088	0.077	0.067	0.058	0.051	0.044	0.038	0.033
2-2	0.026	0.072	1.250	0.125	0.108	0.093	0.081	0.070	0.061	0.052	0.045	0.038	0.033
2-3	0.026	0.033	1.250	0.113	0.098	0.086	0.075	0.065	0.057	0.050	0.043	0.038	0.033
3-1	0.031	0.027	1.250	0.128	0.111	0.097	0.085	0.075	0.066	0.058	0.051	0.0e44	0.039
3-2	0.029	0.017	1.250	0.113	0.099	0.087	0.077	0.068	0.060	0.053	0.047	0.041	0.036
3-3	0.029	0.04	1.250	0.128	0.111	0.096	0.084	0.073	0.064	0.056	0.049	0.042	0.036
3-4	0.04	0.064	1.250	0.180	0.156	0.135	0.118	0.103	0.090	0.078	0.068	0.058	0.050
4-1	0.024	0.021	1.250	0.099	0.086	0.075	0.066	0.058	0.051	0.045	0.039	0.034	0.030
4-2	0.024	0.031	1.250	0.105	0.091	0.079	0.069	0.060	0.053	0.046	0.040	0.035	0.030
4-3	0.024	0.072	1.250	0.116	0.100	0.087	0.075	0.065	0.056	0.049	0.042	0.036	0.030
5-1	0.016	0.036	1.563	0.080	0.070	0.061	0.054	0.047	0.042	0.037	0.032	0.029	0.025
5-2	0.016	0.077	1.563	0.086	0.074	0.065	0.057	0.050	0.043	0.038	0.033	0.029	0.025
5-3	0.016	0.055	1.563	0.083	0.073	0.063	0.056	0.049	0.043	0.038	0.033	0.029	0.025
5-4	0.038	0.038	1.250	0.160	0.139	0.121	0.106	0.093	0.082	0.072	0.063	0.055	0.048
6-1	0.015	0.033	1.667	0.076	0.067	0.059	0.052	0.046	0.041	0.036	0.032	0.028	0.025
6-2	0.015	0.055	1.667	0.080	0.070	0.061	0.054	0.047	0.042	0.037	0.032	0.029	0.025
6-3	0.015	0.045	1.667	0.079	0.069	0.060	0.053	0.047	0.041	0.037	0.032	0.028	0.025
6-4	0.035	0.068	1.250	0.161	0.140	0.121	0.105	0.092	0.080	0.069	0.060	0.051	0.044
7-1	0.012	0.035	2.083	0.068	0.060	0.053	0.047	0.042	0.038	0.034	0.031	0.028	0.025
7-2	0.012	0.07	2.083	0.071	0.063	0.055	0.049	0.044	0.039	0.035	0.031	0.028	0.025
7-3	0.053	0.07	1.250	0.232	0.201	0.175	0.153	0.134	0.117	0.102	0.089	0.077	0.066
8-1	0.01	0.031	2.500	0.061	0.054	0.049	0.044	0.040	0.036	0.033	0.030	0.027	0.025
8-2	0.01	0.065	2.500	0.064	0.057	0.051	0.045	0.041	0.037	0.033	0.030	0.027	0.025
8-3	0.01	0.081	2.500	0.064	0.057	0.051	0.046	0.041	0.037	0.033	0.030	0.028	0.025
8-4	0.042	0.042	1.250	0.176	0.153	0.134	0.117	0.103	0.090	0.079	0.069	0.060	0.053
9-1	0.015	0.033	1.667	0.076	0.067	0.059	0.052	0.046	0.041	0.036	0.032	0.028	0.025
9-2	0.015	0.054	1.667	0.080	0.070	0.061	0.054	0.047	0.042	0.037	0.032	0.029	0.025
9-3	0.015	0.045	1.667	0.079	0.069	0.060	0.053	0.047	0.041	0.037	0.032	0.028	0.025
9-4	0.034	0.069	1.250	0.158	0.136	0.118	0.103	0.089	0.078	0.067	0.058	0.050	0.043
10-1	0.013	0.045	1.923	0.073	0.064	0.056	0.050	0.044	0.040	0.035	0.031	0.028	0.025
10-2	0.013	0.065	1.923	0.074	0.065	0.057	0.051	0.045	0.040	0.036	0.032	0.028	0.025
10-3	0.013	0.079	1.923	0.075	0.066	0.058	0.051	0.045	0.040	0.036	0.032	0.028	0.025
10-4	0.052	0.052	1.250	0.218	0.190	0.166	0.145	0.128	0.112	0.098	0.086	0.075	0.065
11-1	0.017	0.031	1.47	0.081	0.071	0.062	0.055	0.048	0.042	0.037	0.033	0.029	0.025
11-2	0,009	0.025	2.78	0.057	0.051	0.046	0.042	0.038	0.035	0.032	0.029	0.027	0.025
11-3	0.010	0.054	2.50	0.063	0.056	0.050	0.045	0.041	0.037	0.033	0.030	0.027	0.025
11-4	0.036	0.041	1.25	0.154	0.134	0.117	0.102	0.089	0.078	0.069	0.060	0.052	0.045
11-5	0.010	0.022	2.50	0.059	0.053	0.048	0.043	0.039	0.035	0.032	0.030	0.027	0.025
12-1	0.008	0.018	3.13	0.053	0.047	0.043	0.039	0.036	0.033	0.031	0.029	0.027	0.025
12-2	0.006	0.019	4.17	0.047	0.043	0.039	0.036	0.034	0.032	0.030	0.028	0.026	0.025
12-3	0.006	0.039	4.17	0.048	0.044	0.040	0.037	0.034	0.032	0.030	0.028	0.026	0.025
12-4	0.044	0.042	1.25	0.184	0.160	0.140	0.122	0.107	0.094	0.083	0.072	0.063	0.055

Figure 2 shows the value corresponding to the thermal conductivity of EPS (red line). For most of the enclosures, the value of equivalent thermal conductivity obtained with an emissivity equal to 0.01 is less than the thermal conductivity of the EPS. In particular, if the thickness (*d*) of the cavity is less than 0.029 m, the equivalent thermal conductivity is lower than the value of EPS.

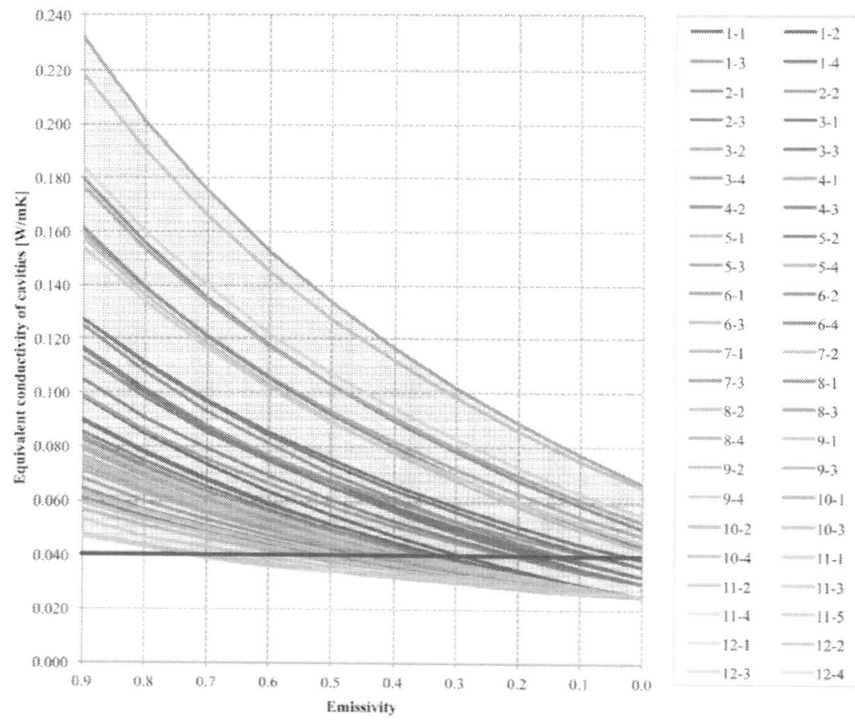

Figure 2: Impact of surface emissivity on the variation in equivalent thermal conductivity of the cavities [red line = thermal conductivity of EPS (extruded polystyrene)].

Values of thermal resistance of the evaluated block are shown in Table 4. A reduction in the equivalent thermal conductivity of the enclosures produces an increase in the thermal resistance value of the block of +32%–82% (Table 5), from values of 2.65–0.96 $m^2 \cdot K/W$ ($\varepsilon = 0.9$) to values of 4.20–1.34 $m^2 \cdot K/W$ ($\varepsilon = 0.1$) (Table 4).

Table 4: Thermal resistance of the evaluated block ($m^2 \cdot K/W$)

Sample	$\varepsilon = 0.90$	$\varepsilon = 0.80$	$\varepsilon = 0.70$	$\varepsilon = 0.60$	$\varepsilon = 0.50$	$\varepsilon = 0.40$	$\varepsilon = 0.30$	$\varepsilon = 0.20$	$\varepsilon = 0.10$	No radiation	EPS
1A	1.04	1.11	1.18	1.25	1.33	1.38	1.45	1.51	1.57	1.64	1.45
1B	1.46	1.58	1.71	1.84	1.97	2.12	2.25	2.39	2.54	2.70	2.26
2A	1.14	1.22	1.31	1.39	1.47	1.55	1.63	1.70	1.79	1.82	1.76
2B	1.55	1.70	1.84	1.99	2.15	2.30	2.47	2.64	2.83	2.92	2.77
3A	0.96	1.02	1.07	1.12	1.17	1.21	1.25	1.29	1.34	1.35	1.34
3B	1.39	1.49	1.59	1.70	1.80	1.90	2.00	2.09	2.20	2.30	2.27
4A	1.39	1.50	1.59	1.70	1.79	1.89	1.98	2.08	2.17	2.25	2.11
4B	1.90	2.08	2.25	2.44	2.62	2.82	3.01	3.22	3.43	3.68	3.28
5A	1.55	1.65	1.75	1.84	1.93	2.03	2.12	2.21	2.29	2.37	2.09
5B	2.15	2.33	2.51	2.68	2.87	3.07	3.25	3.45	3.63	3.84	3.20
6A	1.57	1.66	1.76	1.84	1.93	2.00	2.08	2.17	2.24	2.30	2.06
6B	2.22	2.39	2.57	2.73	2.92	3.08	3.25	3.45	3.58	3.78	3.21
7A	1.39	1.45	1.53	1.59	1.65	1.71	1.76	1.82	1.86	1.91	1.75
7B	2.03	2.16	2.30	2.43	2.55	2.68	2.80	2.93	3.04	3.17	2.79
8A	1.57	1.64	1.71	1.77	1.84	1.91	1.97	2.02	2.07	1.88	1.88
8B	2.28	2.41	2.53	2.69	2.82	2.95	3.09	3.21	3.33	3.45	2.90
9A	1.76	1.87	1.97	2.07	2.17	2.25	2.34	2.43	2.49	2.59	2.34
9B	2.48	2.67	2.87	3.06	3.27	3.45	3.65	3.86	4.03	4.24	3.58
10A	1.87	1.97	2.08	2.17	2.27	2.36	2.44	2.52	2.61	2.68	2.40
10B	2.65	2.84	3.04	3.23	3.43	3.62	3.80	3.99	4.20	4.38	3.73
11A	1.70	1.78	1.85	1.93	1.99	2.05	2.12	2.17	2.23	2.28	2.03
11B	2.33	2.49	2.63	2.79	2.91	3.05	3.20	3.33	3.47	3.58	3.02
12A	2.77	2.89	3.01	3.11	3.21	3.29	3.37	3.46	3.55	3.60	3.07
12B	3.73	3.93	4.16	4.34	4.55	4.70	4.86	5.05	5.24	5.35	4.28

Table 5: Thermal resistance of the evaluated block: percentage of improvement

Sample	ε = 0.90	ε = 0.80	ε = 0.70	ε = 0.60	ε = 0.50	ε = 0.40	ε = 0.30	ε = 0.20	ε = 0.10	No radiation	EPS
1A	100%	107%	114%	120%	128%	133%	139%	145%	151%	157%	140%
1B	100%	109%	118%	126%	135%	145%	154%	164%	174%	185%	155%
2A	100%	107%	115%	122%	129%	135%	143%	149%	156%	160%	154%
2B	100%	109%	119%	128%	138%	148%	159%	170%	182%	188%	178%
3A	100%	106%	111%	116%	121%	126%	130%	134%	139%	140%	139%
3B	100%	107%	115%	122%	130%	137%	144%	151%	159%	166%	164%
4A	100%	107%	114%	122%	129%	136%	142%	149%	156%	162%	151%
4B	100%	109%	118%	128%	138%	148%	158%	169%	180%	193%	172%
5A	100%	107%	113%	119%	125%	132%	137%	143%	148%	154%	135%
5B	100%	109%	117%	125%	134%	143%	151%	161%	169%	179%	149%
6A	100%	106%	112%	117%	123%	128%	133%	138%	143%	147%	131%
6B	100%	108%	116%	123%	131%	139%	147%	155%	162%	171%	145%
7A	100%	105%	110%	115%	119%	123%	127%	131%	134%	138%	126%
7B	100%	106%	113%	119%	125%	132%	138%	144%	150%	156%	137%
8A	100%	105%	109%	113%	118%	122%	126%	129%	132%	120%	120%
8B	100%	106%	111%	118%	124%	130%	136%	141%	147%	151%	127%
9A	100%	106%	112%	117%	123%	128%	133%	138%	141%	147%	132%
9B	100%	108%	116%	123%	132%	139%	147%	156%	162%	171%	144%
10A	100%	106%	111%	116%	122%	126%	130%	135%	140%	143%	129%
10B	100%	107%	115%	122%	130%	137%	143%	151%	158%	165%	141%
11A	100%	105%	109%	114%	117%	121%	125%	128%	131%	134%	120%
11B	100%	107%	113%	119%	125%	131%	137%	143%	149%	154%	129%
12A	100%	104%	108%	112%	116%	119%	122%	125%	128%	130%	111%
12B	100%	105%	111%	116%	122%	126%	130%	135%	140%	143%	115%

Table 6 shows the values of the equivalent thermal conductivity of the analyzed block, which is derived by dividing the thickness of a block by its thermal resistance. While for the standard block ($\varepsilon = 0.9$), the equivalent thermal conductivity obtained is between 0.311 and 0.157 W/m·K, for the improved block ($\varepsilon = 0.1$), the values are 0.224–0.096 W/m·K, corresponding to a reduction of 24%–45% (Table 7). The equivalent conductivity of the block without radiative heat exchange was 0.222–0.091 W/m·K, corresponding to a reduction of 17%–48%. For the block with polystyrene, the value of equivalent conductivity calculated is 0.224–0.150 W/m·K. All of the blocks evaluated with a low emissivity coating of 0.1 have a thermal behavior equal to or higher than the block simulated with the polystyrene inside the void. The block with EPS has the same behavior as the block with low emissivity coating for a surface emissivity between 0.4 and 0.1. The mean value of the reduction in the equivalent thermal conductivity obtained for an emissivity equal to 0.1 is 35%, compared with 30% for the block with EPS inside the cavity.

Table 6: Equivalent thermal conductivity of the evaluated block (W/m·K)

Sample	ε = 0.90	ε = 0.80	ε = 0.70	ε = 0.60	ε = 0.50	ε = 0.40	ε = 0.30	ε = 0.20	ε = 0.10	No radiation	EPS
1A	0.240	0.225	0.211	0.201	0.188	0.181	0.173	0.166	0.159	0.153	0.172
1B	0.171	0.158	0.146	0.136	0.127	0.118	0.111	0.104	0.098	0.093	0.111
2A	0.263	0.245	0.229	0.216	0.204	0.194	0.184	0.176	0.168	0.164	0.170
2B	0.193	0.177	0.163	0.151	0.140	0.130	0.121	0.114	0.106	0.103	0.108
3A	0.311	0.295	0.280	0.268	0.257	0.248	0.239	0.232	0.224	0.222	0.224
3B	0.217	0.201	0.188	0.177	0.167	0.158	0.150	0.143	0.136	0.130	0.132
4A	0.251	0.234	0.220	0.206	0.195	0.185	0.177	0.168	0.161	0.155	0.166
4B	0.184	0.169	0.156	0.144	0.133	0.124	0.116	0.109	0.102	0.095	0.107
5A	0.227	0.212	0.200	0.190	0.181	0.172	0.165	0.158	0.153	0.147	0.167
5B	0.163	0.150	0.140	0.131	0.122	0.114	0.108	0.101	0.096	0.091	0.109
6A	0.223	0.210	0.199	0.190	0.181	0.175	0.168	0.162	0.156	0.152	0.170
6B	0.158	0.147	0.136	0.128	0.120	0.114	0.108	0.102	0.098	0.092	0.109
7A	0.263	0.251	0.239	0.230	0.222	0.214	0.207	0.201	0.196	0.191	0.209
7B	0.180	0.169	0.159	0.150	0.143	0.136	0.130	0.125	0.120	0.115	0.131
8A	0.239	0.229	0.219	0.212	0.203	0.197	0.190	0.185	0.181	0.200	0.200
8B	0.165	0.155	0.148	0.139	0.133	0.127	0.121	0.117	0.112	0.109	0.129
9A	0.227	0.214	0.203	0.194	0.185	0.178	0.171	0.164	0.161	0.155	0.171
9B	0.161	0.150	0.139	0.131	0.122	0.116	0.110	0.104	0.099	0.094	0.112
10A	0.241	0.228	0.217	0.207	0.198	0.191	0.185	0.179	0.173	0.168	0.187
10B	0.170	0.158	0.148	0.139	0.131	0.124	0.119	0.113	0.107	0.103	0.121
11A	0.177	0.169	0.162	0.155	0.151	0.146	0.142	0.138	0.134	0.132	0.148
11B	0.129	0.121	0.114	0.108	0.103	0.098	0.094	0.090	0.087	0.084	0.099
12A	0.144	0.139	0.133	0.129	0.124	0.122	0.119	0.116	0.113	0.111	0.130
12B	0.107	0.102	0.096	0.092	0.088	0.085	0.082	0.079	0.076	0.075	0.093

Table 7: Equivalent thermal conductivity of the evaluated block: percentage of reduction

Sample	ε = 0.90	ε = 0.80	ε = 0.70	ε = 0.60	ε = 0.50	ε = 0.40	ε = 0.30	ε = 0.20	ε = 0.10	No radiation	EPS
1A	0%	−6%	−12%	−17%	−22%	−25%	−28%	−31%	−34%	−36%	−28%
1B	0%	−8%	−15%	−21%	−26%	−31%	−35%	−39%	−43%	−46%	−35%
2A	0%	−7%	−13%	−18%	−22%	−26%	−30%	−33%	−36%	−37%	−35%
2B	0%	−8%	−16%	−22%	−28%	−32%	−37%	−41%	−45%	−47%	−44%
3A	0%	−5%	−10%	−14%	−17%	−20%	−23%	−26%	−28%	−29%	−28%
3B	0%	−7%	−13%	−18%	−23%	−27%	−31%	−34%	−37%	−40%	−39%
4A	0%	−7%	−13%	−18%	−22%	−26%	−30%	−33%	−36%	−38%	−34%
4B	0%	−8%	−15%	−22%	−27%	−33%	−37%	−41%	−45%	−48%	−42%
5A	0%	−6%	−12%	−16%	−20%	−24%	−27%	−30%	−32%	−35%	−26%
5B	0%	−8%	−14%	−20%	−25%	−30%	−34%	−38%	−41%	−44%	−33%
6A	0%	−6%	−11%	−15%	−19%	−22%	−25%	−27%	−30%	−32%	−24%
6B	0%	−7%	−14%	−19%	−24%	−28%	−32%	−36%	−38%	−41%	−31%
7A	0%	−5%	−9%	−13%	−16%	−19%	−21%	−24%	−26%	−27%	−21%
7B	0%	−6%	−12%	−16%	−20%	−24%	−27%	−31%	−33%	−36%	−27%
8A	0%	−4%	−8%	−12%	−15%	−18%	−20%	−23%	−24%	−17%	−17%
8B	0%	−6%	−10%	−15%	−19%	−23%	−26%	−29%	−32%	−34%	−22%
9A	0%	−6%	−11%	−15%	−19%	−22%	−25%	−28%	−29%	−32%	−24%
9B	0%	−7%	−14%	−19%	−24%	−28%	−32%	−36%	−38%	−41%	−31%
10A	0%	−5%	−10%	−14%	−18%	−21%	−23%	−26%	−28%	−30%	−22%
10B	0%	−7%	−13%	−18%	−23%	−27%	−30%	−34%	−37%	−39%	−29%
11A	0%	−4%	−8%	−12%	−14%	−17%	−20%	−22%	−24%	−25%	−16%
11B	0%	−6%	−11%	−16%	−20%	−23%	−27%	−30%	−33%	−35%	−23%
12A	0%	−4%	−8%	−11%	−14%	−16%	−18%	−20%	−22%	−23%	−10%
12B	0%	−5%	−10%	−14%	−18%	−21%	−23%	−26%	−29%	−30%	−13%

Table 8 shows the *U*-value of the walls built with the evaluated blocks, calculated considering a hypothetical configuration of the layers (external plaster, masonry with horizontal mortar and internal plaster). While for the standard block ($\varepsilon = 0.9$), the value of thermal transmittance is between 0.32–0.87 W/m²·K, for the improved block ($\varepsilon = 0.1$), the values are 0.26–0.69 W/m²·K, corresponding to a reduction of 18%–37% (Table 9). For the block with polystyrene, the *U*-value calculated is 0.29–0.69 W/m²·K.

The comparison with the two previous evaluations [18,19] offers insight into whether the percentages of improvement found in this paper could be extended to other kinds of brick or block, with different geometric and material properties. Figure 3 shows the values of reduction obtained in this work compared with the values obtained in a previous analysis, using the same methodology, but carried out on a different block, and an analysis carried out on walls made with traditional bricks. The results obtained in the previous study [19] do not differ greatly from the values obtained in the present work, with a variation of less than 3% with respect to the mean values.

On the contrary, the comparison made with a previous analysis [18] carried out on walls with traditional brick indicates a substantial difference. In fact, only four of the eight walls analyzed have a behavior that is comparable with the mean values obtained in the present work (difference < 6%), while for the remaining walls, the values obtained are clearly different.

The possible causes of these differences are:

- evaluation made considering not only the brick but also the mortar;
- dimension of the cavity and the block;
- percentage of voids;
- number of voids;
- thermal conductivity of the clay.

Further analysis has been suggested in order to evaluate the performance of a large number of samples that could represent most of the blocks or bricks available on the market. The evaluation of the impact of the surface emissivity on the *U*-value of the wall could also be useful in order to investigate energy consumption and economic benefits, under real conditions.

Table 8: Overall thermal transmittance of the wall built with the evaluated block (W/m²·K)

Sample	ε = 0.90	ε = 0.80	ε = 0.70	ε = 0.60	ε = 0.50	ε = 0.40	ε = 0.30	ε = 0.20	ε = 0.10	No radiation	EPS
1A	0.84	0.80	0.76	0.74	0.70	0.68	0.66	0.64	0.63	0.61	0.66
1B	0.66	0.62	0.59	0.56	0.53	0.51	0.49	0.47	0.45	0.43	0.49
2A	0.77	0.74	0.70	0.67	0.65	0.62	0.60	0.58	0.56	0.55	0.57
2B	0.62	0.58	0.55	0.52	0.49	0.47	0.45	0.43	0.41	0.40	0.41
3A	0.87	0.84	0.81	0.78	0.76	0.74	0.72	0.71	0.69	0.69	0.69
3B	0.67	0.64	0.61	0.58	0.56	0.54	0.52	0.50	0.48	0.47	0.47
4A	0.66	0.63	0.60	0.58	0.55	0.53	0.51	0.50	0.48	0.47	0.49
4B	0.53	0.50	0.47	0.44	0.42	0.40	0.38	0.36	0.35	0.33	0.36
5A	0.62	0.59	0.56	0.54	0.52	0.50	0.49	0.47	0.46	0.45	0.49
5B	0.48	0.46	0.43	0.41	0.39	0.38	0.36	0.35	0.34	0.32	0.37
6A	0.61	0.58	0.56	0.54	0.52	0.51	0.50	0.48	0.47	0.46	0.50
6B	0.47	0.45	0.43	0.41	0.39	0.38	0.36	0.35	0.34	0.33	0.36
7A	0.66	0.64	0.62	0.60	0.59	0.57	0.56	0.55	0.54	0.52	0.56
7B	0.50	0.48	0.46	0.44	0.43	0.41	0.40	0.39	0.38	0.37	0.40
8A	0.61	0.59	0.57	0.55	0.54	0.52	0.51	0.50	0.49	0.48	0.53
8B	0.46	0.44	0.43	0.41	0.40	0.38	0.37	0.36	0.35	0.34	0.39
9A	0.55	0.53	0.51	0.49	0.47	0.46	0.45	0.44	0.43	0.42	0.45
9B	0.43	0.41	0.39	0.37	0.35	0.34	0.33	0.31	0.31	0.30	0.33
10A	0.52	0.50	0.48	0.47	0.45	0.44	0.43	0.42	0.41	0.40	0.43
10B	0.40	0.38	0.37	0.35	0.34	0.32	0.31	0.30	0.29	0.28	0.32
11A	0.58	0.56	0.55	0.53	0.52	0.51	0.50	0.49	0.48	0.47	0.51
11B	0.46	0.44	0.43	0.41	0.40	0.39	0.37	0.36	0.35	0.35	0.39
12A	0.40	0.39	0.37	0.37	0.36	0.35	0.35	0.34	0.33	0.33	0.37
12B	0.32	0.31	0.30	0.29	0.28	0.28	0.27	0.26	0.26	0.25	0.29

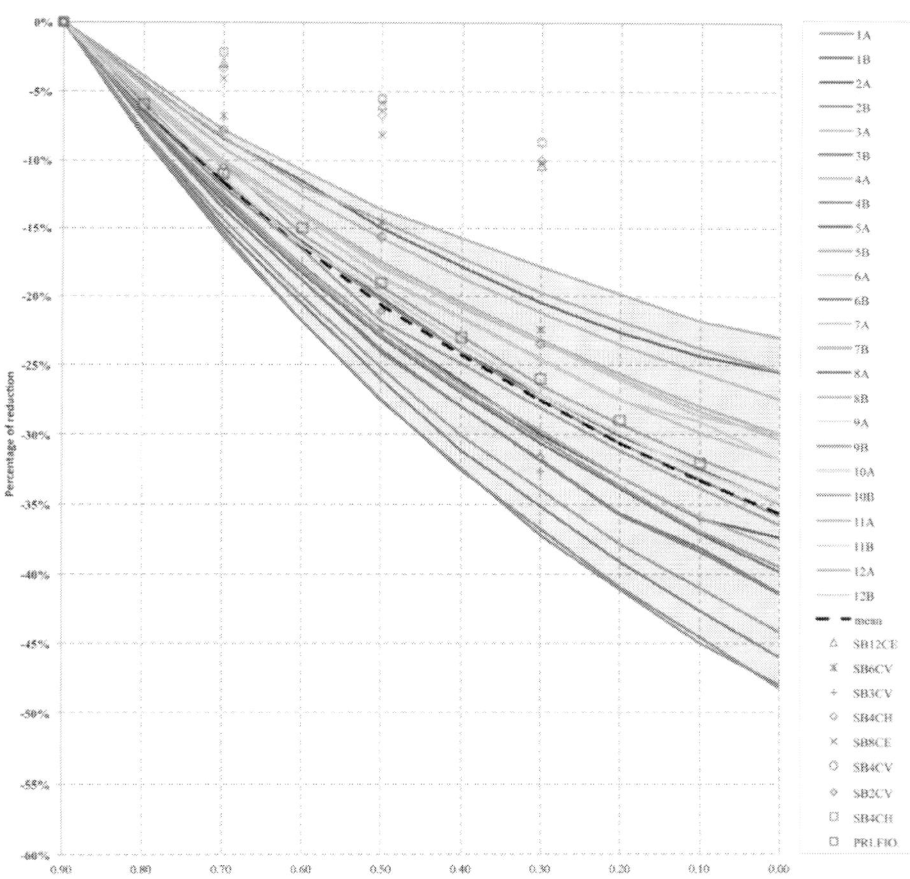

Figure 3: Comparison between the impacts of surface emissivity on the variation in equivalent thermal conductivity of the blocks obtained in the present work (grey lines) and in a previous investigation (red point [18], blue point [19]).

Table 9: Overall thermal transmittance of the wall built with the evaluated block: percentage of reduction

Sample	ε = 0.90	ε = 0.80	ε = 0.70	ε = 0.60	ε = 0.50	ε = 0.40	ε = 0.30	ε = 0.20	ε = 0.10	No radiation	EPS
1A	0%	−5%	−9%	−12%	−16%	−18%	−21%	−23%	−25%	−27%	−21%
1B	0%	−6%	−11%	−15%	−19%	−23%	−26%	−29%	−32%	−35%	−26%
2A	0%	−5%	−9%	−13%	−16%	−19%	−22%	−25%	−27%	−28%	−27%
2B	0%	−6%	−11%	−16%	−21%	−24%	−28%	−31%	−34%	−36%	−34%
3A	0%	−4%	−7%	−10%	−13%	−15%	−17%	−19%	−21%	−21%	−21%
3B	0%	−5%	−10%	−13%	−17%	−20%	−23%	−26%	−28%	−30%	−30%
4A	0%	−5%	−9%	−13%	−17%	−20%	−23%	−25%	−28%	−29%	−26%
4B	0%	−6%	−11%	−16%	−21%	−25%	−28%	−31%	−34%	−37%	−32%
5A	0%	−5%	−9%	−12%	−15%	−18%	−20%	−23%	−25%	−27%	−20%
5B	0%	−6%	−11%	−15%	−19%	−22%	−25%	−28%	−31%	−33%	−25%
6A	0%	−4%	−8%	−11%	−14%	−16%	−18%	−21%	−23%	−24%	−18%
6B	0%	−5%	−10%	−14%	−18%	−21%	−24%	−27%	−29%	−31%	−23%
7A	0%	−3%	−7%	−9%	−12%	−14%	−16%	−18%	−19%	−21%	−16%
7B	0%	−4%	−9%	−12%	−15%	−18%	−21%	−23%	−25%	−27%	−20%
8A	0%	−3%	−6%	−9%	−11%	−13%	−15%	−17%	−18%	−19%	−12%
8B	0%	−4%	−7%	−11%	−14%	−17%	−20%	−22%	−24%	−25%	−16%
9A	0%	−4%	−8%	−11%	−14%	−16%	−19%	−21%	−22%	−24%	−19%
9B	0%	−5%	−10%	−14%	−18%	−21%	−24%	−27%	−29%	−31%	−23%
10A	0%	−4%	−8%	−11%	−13%	−16%	−18%	−20%	−22%	−23%	−17%
10B	0%	−5%	−10%	−13%	−17%	−20%	−23%	−25%	−28%	−30%	−22%
11A	0%	−3%	−6%	−9%	−11%	−13%	−15%	−16%	−18%	−19%	−12%
11B	0%	−4%	−8%	−12%	−14%	−17%	−20%	−22%	−24%	−25%	−16%
12A	0%	−3%	−6%	−8%	−10%	−11%	−13%	−14%	−16%	−17%	−7%
12B	0%	−4%	−7%	−10%	−12%	−14%	−16%	−18%	−20%	−21%	−9%

CONCLUSIONS

This paper presents the evaluation of the thermal performance of hollow clay bricks and blocks with low emissivity treatment of the internal cavity surfaces.

The numerical evaluation was carried out in order to determine the increase in the thermal resistance provided by low emissivity coating applied on the void surfaces of the brick. For the evaluation, a set of 20 samples was investigated, characterized by different shapes and the thermal conductivity of the materials. Each sample was simulated for surface emissivities of 0.9–0.1. The values obtained with the numerical simulation were compared with the standard block and with the block insulated with EPS in the voids.

The results showed that a low emissivity coating leads to an increase in the thermal resistance provided by the block, through the reduction in the radiative heat exchange in the voids. Using the low emissivity coating, it is possible to reduce the equivalent thermal conductivity of the cavities by between 55%–70% (ε = 0.1) to a value of 0.027 W/m·K, which is lower than the thermal conductivity of common thermal insulation.

The effect of the reduction in the thermal heat exchange in the void has a significant impact on the thermal resistance of the block. In fact, the results show that the application of low emissivity coating could reduce the equivalent thermal conductivity of the block by at least 24% (for an emissivity of 0.1). The consequent reduction in the *U*-value of a wall built with the improved blocks is 18%–37%.

In all cases, the thermal behavior of the treated block is improved with respect to the standard block, and with an emissivity of 0.1, it is better than the block insulated with polystyrene.

The theoretical procedure to calculate the thermal performance of the block is easily repeatable and is employed to verify conformity with the current standard.

Further studies are suggested to include a wide range of blocks, to test the real emissivity of the coated surface with different kinds of coating and to evaluate performance decay in real conditions.

AUTHOR CONTRIBUTIONS

Numerical evaluations have been conducted by Roberto Fioretti. Analysis and interpretation of the results as well as conclusions have been conducted by Roberto Fioretti and Paolo Principi. The manuscript has been written by Roberto Fioretti and Paolo Principi.

REFERENCES

1. *Directive 2010/31/EU of the European Parliament and of the Council of 19 May 2010 on the Energy Performance of Buildings*; European Commission: Brussels, Belgium, 2010.
2. *Directive 2012/27/EU of the European Parliament and of the Council of 25 October 2012 on Energy Efficiency,*

Amending Directives 2009/125/EC and 2010/30/EU and Repealing Directives 2004/8/EC and 2006/32/EC Text with EEA Relevance; European Commission: Brussels, Belgium, 2012.

3. *Decreto Legislativo 29 dicembre 2006, n. 311. Disposizioni correttive ed integrative al Decreto Legislativo 19 agosto 2005, n. 192, recante attuazione della direttiva 2002/91/CE, relativa al rendimento energetico nell'edilizia*; Gazzetta Ufficiale della Repubblica Italiana: Roma, Italy, 2006. (In Italian)
4. Del Coz Díaz, J.J.; García Nieto, P.J.; Betegón Biempica, C.; Prendes Gero, M.B. Analysis and optimization of the heat-insulating light concrete hollow brick walls design by the finite element method. *Appl. Therm. Eng.* 2007, *27*, 1445–1456.
5. Morales, M.P.; Juárez, M.C.; López-Ochoa, L.M.; Doménech, J. Study of the geometry of a voided clay brick using non-rectangular perforations to optimize its thermal properties. *Energy Build.* 2011, *43*, 2494–2498.
6. Morales, M.P.; Juárez, M.C.; Muñoz, P.; Gómez, J.A. Study of the geometry of a voided clay brick using rectangular perforations to optimize its thermal properties. *Appl. Therm. Eng.* 2011, *31*, 2063–2065.
7. Sun, J.; Fang, L.; Han, J. Optimization of concrete hollow brick using hybrid genetic algorithm combining with artificial neural networks. *Int. J. Heat Mass Transf.* 2012, *53*, 5509–5518.
8. Li, L.P.; Wu, Z.G.; He, Y.L.; Lauriat, G.; Tao, W.Q. Optimization of the configuration of 290 × 140 × 90 hollow clay bricks with 3-D numerical simulation by finite volume method. *Energy Build.* 2008, *40*, 1790–1798.
9. Del Coz Dìaz, J.J.; Garcìa Nieto, P.J.; Suàrez Sierra, J.L.; Betegòn Biempica, C. Nonlinear thermal optimization of external light concrete multi-holed brick walls by the finite element method. *Int. J. Heat Mass Transf.* 2008, *51*, 1530–1541.
10. Zukowski, M.; Haese, G. Experimental and numerical investigation of a hollow brick filled with perlite insulation. *Energy Build.* 2010, *42*, 1402–1408.
11. Schaefer, C.; Bräuer, G.; Szczyrbowski, J. Low emissivity coatings on architectural glass. *Surf. Coat. Technol.* 1997, *1*, 37–45.
12. Martin-Palma, R.J.; Vazquez, L.; Martinez-Duart, J.M.; Malats-Riera. Silver-based low-emissivity coatings for architectural windows: Optical and structural properties. *Sol. Energy Mater. Sol. Cells* 1998, *1–2*, 55–66.
13. Miyazaki, M.; Ando, E. Durability improvement of Ag-based low-emissivity coatings. *J. Non-Cryst. Solids* 1994, *3*, 245–249.
14. Pasztory, Z.; Peralta, P.N.; Peszlen, I. Multi-layer heat insulation system for frame construction buildings. *Energy Build.* 2011, *43*, 713–717.
15. Saber, H.H.; Maref, W.; Swinton, M.C.; St-Onge, C. Thermal analysis of above-grade wall assembly with low emissivity materials and furred airspace. *Build. Environ.* 2011, *7*, 1403–1414.
16. Saber, H.H.; Maref, W.; Swinton, M.C. Thermal response of basement wall systems with low-emissivity material and furred airspace. *J. Build. Phys.* 2012, *35*, 353–371.
17. Saber, H.H.; Maref, W.; Sherrer, G.; Swinton, M.C. Numerical modeling and experimental investigations of thermal performance of reflective insulations. *J. Build. Phys.* 2012, *36*, 163–177.
18. Bouchair, A. Steady state theoretical model of fired clay hollow bricks for enhanced external wall thermal insulation. *Build. Environ.* 2008, *43*, 1603–1618.
19. Principi, P.; Fioretti, R. Thermal analysis of the application of pcm and low emissivity coating in hollow bricks. *Energy Build.* 2012, *51*, 131–142.
20. COMSOL Multiphysics. Modeling and simulation software. Available online: http://www.comsol.it/ (accessed on 28 July 2014).
21. *EN ISO 10211-1:1995 Thermal Bridges in Building Construction—Heat Flows and Surface Temperatures—Part 1: General Calculation Methods*; European Committee for Standardization: Brussels, Belgium, 1995.
22. *BS EN 1745:2002 Masonry and Masonry Products. Methods for Determining Design Thermal Values*; European Committee for Standardization: Brussels, Belgium, 2002.
23. *EN ISO 6946:1996 Building Components and Building Elements—Thermal Resistance and Thermal Transmittance—Calculation Method*; European Committee for Standardization: Brussels, Belgium, 1996.
24. *Attuazione Della Direttiva 2002/91/CE Relativa al Rendimento Energetico Nell'edilizia*; Italian Legislative Decree No. 192/2005; Gazzetta Ufficiale della Repubblica italiana: Roma, Italy, 2005. (In Italian)
25. *Disposizioni Correttive ed Integrative al Decreto Legislativo 19 Agosto 2005, n. 192, Rec Ante Attuazione Della Direttiva 2002/91/CE Relativa al Rendimento Energetico Nell'edilizia*; Italian Legislative Decree No. 311/2006; Gazzetta Ufficiale della Repubblica italiana: Roma, Italy, 2006. (In Italian)
26. *Regolamento di Attuazione Dell'articolo 4, Comma1, Lettere a) e b),del decreto19 agosto 2005, n.192, Concernente Attuazione Della Direttiva 2002/91/CE sul Rendimento Energetico in Edilizia*; Italian Decree No. 59/2009; Gazzetta Ufficiale della Repubblica italiana: Roma, Italy, 2009. (In Italian)

CHAPTER 15

RESIDUAL MECHANICAL PROPERTIES OF CONCRETE MADE WITH CRUSHED CLAY BRICKS AND ROOF TILES AGGREGATE AFTER EXPOSURE TO HIGH TEMPERATURES

Ivana Miličević [1], Nina Štirmer [2]and Ivana Banjad Pečur [2]

[1]Department for Materials and Structures, Faculty of Civil Engineering, University of Osijek, Osijek 31000, Croatia

[2]Department for Materials, Faculty of Civil Engineering, University of Zagreb, Zagreb 10000, Croatia

ABSTRACT

This paper presents the residual mechanical properties of concrete made with crushed bricks and clay roof tile aggregates after exposure to high temperatures. One referent mixture and eight mixtures with different percentages of replacement of natural aggregate by crushed bricks and roof tiles are experimentally tested. The properties of the concrete were measured before and after exposure to 200, 400, 600 and 800 °C. In order to evaluate the basic residual mechanical properties of concrete with crushed bricks and roof tiles after exposure to high temperatures, ultrasonic pulse velocity is used as a non-destructive test method and the results are compared with those of a destructive method for validation. The mixture with the highest percentage of replacement of natural aggregate by crushed brick and roof tile aggregate has the best physical, mechanical, and thermal properties for application of such concrete in precast concrete elements exposed to high temperatures.

INTRODUCTION

The production of mineral raw materials in the European Union member states is very dynamic because it is fairly dependent on the realization of and investment in projects approved for a defined period of time [1]. Aggregate production in the 15 European Union countries comprising the members of the European Aggregates Association (UEPG) amounted to 2620 million tonnes of sand, gravel and crushed stone in 2000, which represents a production of 6.9 tonnes per capita [1]. It is obvious that this level of annual aggregate production has a significant impact on the environment. It is our responsibility to reduce consumption of natural mineral resources and to increase the use of secondary raw materials, *i.e.*, recycled materials. Aggregate production is economically and socially important as well: the value of natural raw materials produced annually amounts to approximately 35 billion euros, and the aggregate production industry employs around 250,000 people. The estimated demolition waste material in the EU member states amounts to approximately 180 million tons per annum, which is 1.3 kg a day per capita. There are significant differences in produced construction waste quantities within the European Union; for instance, Germany and the Netherlands produce about 1.9 kg per capita daily, whereas Sweden, Greece and Ireland produce less than 0.5 kg [1].

More than 80% of the total construction obtained by waste in certain European countries, such as The Netherlands, Belgium and Denmark, is recycled [2]. In this way, these highly developed countries simultaneously decrease the need

Citation: Ivana Miličević, Nina Štirmer and Ivana Banjad Pečur, Residual Mechanical Properties of Concrete Made with Crushed Clay Bricks and Roof Tiles Aggregate after Exposure to High Temperatures, doi:10.3390/ma9040295.

for new exploitation of pristine natural areas and resolve the problems of construction waste, which is otherwise often disposed of in landfills. Since the usage of recycled materials in all areas of human activity is a global trend, which is true for the construction sector as well, it is now necessary to use recycled construction waste in the production process of concrete within Croatian borders.

Crushed clay bricks and roof tiles are among the best aggregates, for use in concrete are known have to resistance at high temperatures, and concrete with such aggregates performs much better than similar concretes containing granite aggregate [3]. Clay bricks and roof tile aggregates are thermally stable. This is probably why they perform well in concrete subjected to high temperatures. The resistance of clay bricks and roof tiles to high temperatures is an important characteristic: it has been recognized that brickwork masonry and clay roofing are very effective in resisting and preventing the spread of fire. When brick and roof tile materials are used as aggregates in concrete, they are not expected to negatively affect the resistance to high temperatures. To ensure the resistance to high temperatures, crushed brick and roof tile aggregate concrete should be kept dry. When they are wet, the internal steam pressure that is created at high temperatures can cause spalling [4]. The low thermal conductivity of crushed clay brick and roof tile aggregate concrete also improves its resistance to high temperatures. Such concrete is much better protected against early heating, as it keeps its structural integrity under high temperatures for a longer period than ordinary concrete.

EXPERIMENTAL INVESTIGATION

Materials and Methods

Crushed brick (CB), crushed roof tile (CT) and dolomite/limestone (natural aggregate—NA) have been used in this study as aggregates for concrete mixtures. There was no need for additional processing of dolomite because it already was separated into fractions. The bricks and roof tiles were crushed at the brick and roof tile factories and delivered to the concrete-mixing location. Since both CB and CT represent waste products in the process of making new bricks and roof tiles, no additional processing of these materials was required. If CB and CT are to be used as aggregates for concrete, it is necessary to prove that they are adequate replacements for conventional NA. The CB and CT were sieved into 0–4 mm, 4–8 mm and 8–16 mm fractions. Based on the granulometric composition, particle size groups were determined as shown in Table 1. The particle shape of the coarse aggregates was determined as a Shape Index in accordance with [5], and it was selected in accordance with HRN EN 933-4 [6]. The Concrete Constructions Technical Regulations (TPBK) indicate the highest allowed particle size for concrete constructions, which is SI_{20} [7]. Table 1 shows that all three aggregate types comply with the mandatory regulations: the fine particle content is determined in accordance with HRN EN 933-8 [8] and the TPBK. Particle size in relation to fine particle percentage is determined in accordance with [5]. The largest acceptable values of the fine particle size prescribed by the TPBK are:f_3 for natural fine aggregate, f_{10} for crushed and mixed fine aggregates, and $f_{1.5}$ for the coarse fraction. Based on the listed limits, as shown in Table 1, all three aggregates comply with the mandatory regulations. The bulk density was tested in accordance with standard HRN EN 1097-3 [9]. The TPBK dictate that the aggregate bulk density must comply with the project terms or customer demands. Specific density and water absorption were tested in accordance with standard HRN EN 1097-6 [10]. Like for the bulk density, the TPBK dictates that the specific density of aggregate particles must comply with the project terms or the customer demands. The data in Table 1 indicate that CB and CT have a lower density than NA. It was expected, then, that concrete mixtures with CB and CT will have lower density values depending on the replacement percentage. Table 1 shows that the water absorption of CB and CT is significantly higher than for dolomite. It is necessary to take into account the water absorption when designing the concrete mixture composition, since the values are high: 16%–19% for bricks and 9%–11% for roof tiles, depending on the fraction.

Table 1: Properties of CB, CT and NA determined according to [5]

Aggregate Type	CB		CT		NA	
	(Fine)	(Coarse *)	(Fine)	(Coarse *)	(Fine)	(Coarse *)
Particle shape (%)	-	7	-	11	-	6
Shape index	-	(SI15)	-	(SI15)	-	(SI15)
Particle size group	GF85	GC90/15	GF85	GC90/15	GF85	GC90/15
Specific density SSD (mg/m^3)	2.13	2.16	2.29	2.25	2.86	2.88
Cleanliness (%)	4.4	0.4	3.6	0.8	3.2	0.8
Fines quality class	(f10)	(f1.5)	(f10)	(f1.5)	(f10)	(f1.5)
Los Angeles number/class	-	40 (LA30)	-	37 (LA30)	-	24 (LA25)
Water absorption (%)	19.05	16.71	11.50	9.25	2.29	0.50
Freezing/thaw resistance (%)	-	3.0	-	4.0	-	10.7
Magnesium sulphate class	-	(MS18)		(MS18)		(MS18)

* Coarse aggregate comprises fractions 4–8 mm and 8–16 mm.

The coarse aggregate abrasion test (Los Angeles–LA) was conducted in accordance with the standard HRN EN 1097-2/A1 [11] and the TPBK. It is evident from the results in Table 2 that all three aggregate types comply with the mandatory regulations and that the brick and roof tile waste show satisfactory results, despite the expectations of a reduced abrasion resistance.

Table 2: Concrete composition

		Mix Proportions (kg/m³)								
Code		RM	BM42	B10	BM19	BM26	BM34	BM35	BM36	BM50
Cement		300	300	360	400	300	400	400	400	400
Water		150	150	165.6	160	120	200	200	200	160
Water/Cement		0.50	0.50	0.46	0.40	0.40	0.50	0.50	0.50	0.40
Superplasticizer		-	-	1.08	2.00	1.50	-	-	-	-
Filler		208.17	208.17	197.68	-	-	-	-	-	-
Fine aggregate	CB	-	113.1	214.9	132.9	147.3	125.5	125.5	125.5	266.6
	CT	-	118.9	-	142.9	316.8	-	269.8	269.8	143.3
	NA	608.1	304.1	288.7	356.9	197.8	505.5	168.5	168.5	178.9
Coarse aggregate	CB	-	205.9	273.6	-	257.7	438.2	-	438.9	233.1
	CT	-	438.3	83.3	484.4	268.4	228.6	228.6	228.6	242.9
	NA	1170.9	292.8	611.6	620.0	687.2	292.6	877.9	292.6	621.7
Target strength (N/mm²)		40	25	30	20	20	25	30	40	40

Aggregate freezing and thaw resistance were tested in accordance with standard HRN EN 1367-2 [12], by which a magnesium sulphate test was also conducted. The particle size distribution that aggregates need to comply with depends on the end use of the concrete and was determined by standard [5] and the TPBK. Based on the results of the conducted tests shown in Table 1, it can be concluded that all three aggregate types comply with the mandatory regulations and are all in the same magnesium sulfate class.

The grading curves of CB, CT and NA in further tests are shown in Figure 1, Figure 2 and Figure 3. The curves represent the mean values of three siftings of an individual aggregate type. Based on the grading composition, the aggregate classes were determined as shown in Table 1. The table shows that the CB and CT satisfy the mandatory conditions of standard HRN EN 12620 [5] and the Concrete Constructions Technical Regulations (TPBK) [7], and can therefore be used to replace conventional NA.

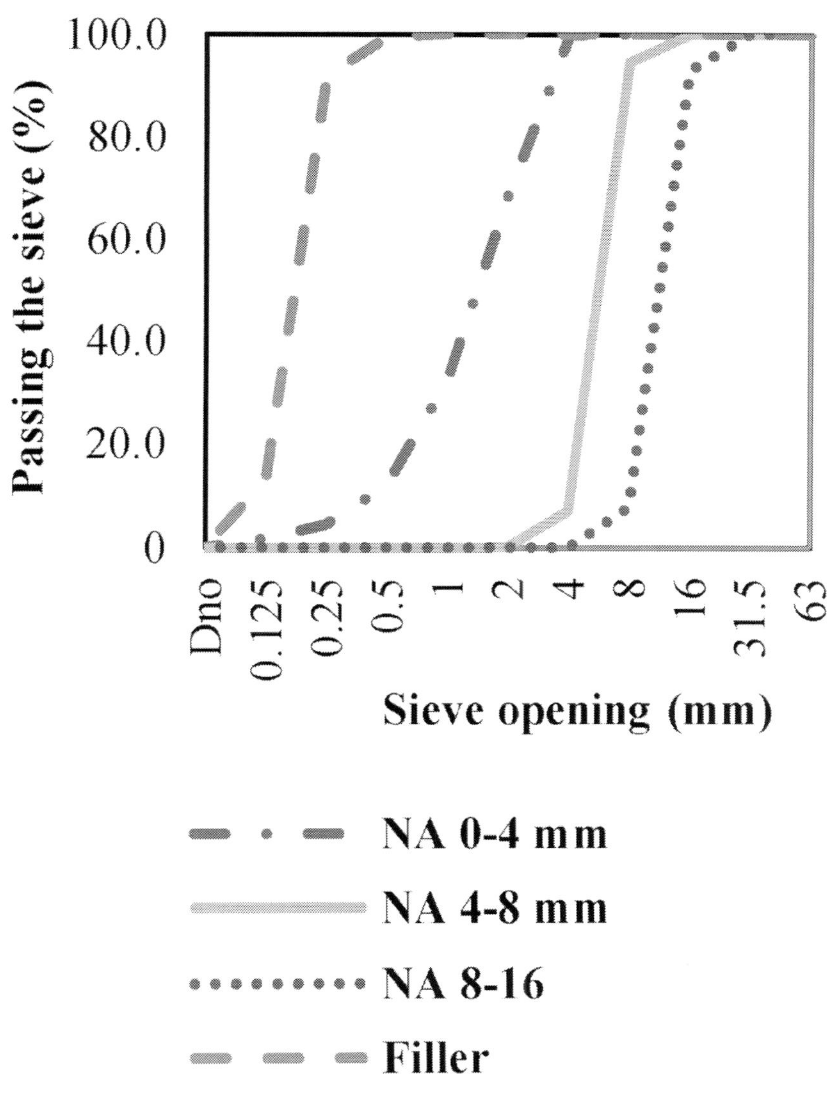

Figure 1: Grading curves for dolomite.

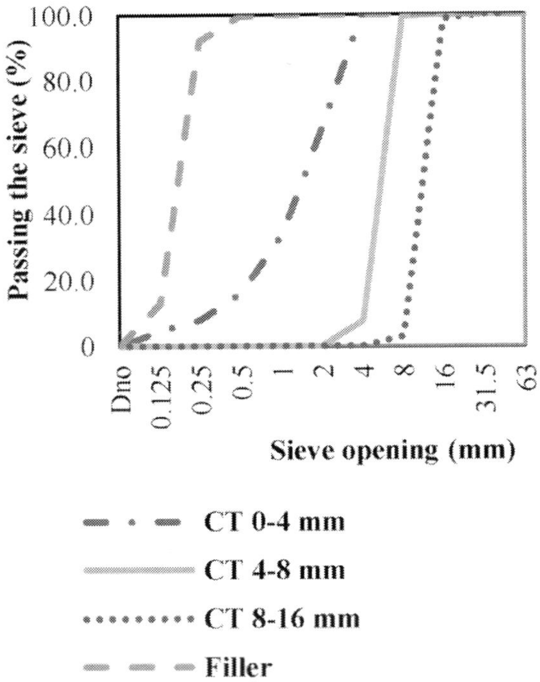

Figure 2: Grading curves of crushed roof tiles.

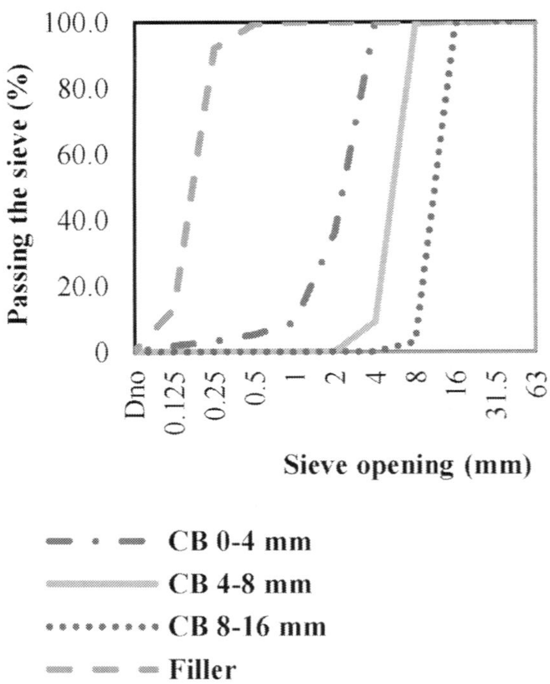

Figure 3: Grading curves of crushed bricks.

The mixture compositions are shown in Table 2. Crushed aggregate has been mixed with the water before placing it in the container. After the aggregate has reached the saturated dry-surface state, it was stored in the container to stay in this condition until the usage. The container was closed hermetically in order to prevent the loss of humidity. The aggregates were stored for 24 h under these conditions before the beginning of concrete mixing in accordance with previous research [3,4], in which it was determined that 24 h are sufficient to reach saturated dry-surface state. When mixing the components, the aggregates were in a saturated, dry-surface state. The total amount of water added to the mixture is shown in Table 2. NA powder was used as a filler. By analyzing the test results for CB and CT aggregates, it was possible to conclude the following:

CB and CT can be used as full or partial replacement for NA (in this study dolomite aggregates).

Special attention must be paid to the concrete composition with regard to the water quantity required and the percentage of replacement of NA by brick and roof tile waste, depending on concrete requirements.

Portland cement CEM I 42.5 N was used throughout the preliminary experimental test phase. Its characteristic mechanical, physical and chemical properties are in compliance with standard HRN EN 197-1 [13]. Polycarboxylate based superplasticizer Sika ViscoCrete (20 Gold) was added to obtain a high workability and early strength development; it is a brown, water-soluble fluid. The characteristics of the superplasticizer were tested by the manufacturer.

Concrete was mixed in a laboratory mixer of 70 L maximum volume. Each of the nine compositions was mixed three times in order to determine the repeatability of the test results. Concrete was consolidated on a vibrating table with 150 Hz frequency.

RESULTS AND DISCUSSION

Properties of Fresh Concrete

After the concretes (Table 3) were mixed, the fresh-state properties were tested, as reported inTable 3. Tests were conducted in accordance with the HRN EN 12350 [14] group of standards for fresh concrete. Each test was repeated three times for each individual mixture, and the mean values are shown in Table 3.

Table 3: Properties of fresh concrete mixtures

Code	RM	BM42	B10	BM19	BM26	BM34	BM35	BM36	BM50
Density (kg/m^3)	2658	2296	2353	2149	2048	2088	2152	2185	2150
Slump (mm)	31.7	11.7	31.7	55.0	30.0	30.0	60.0	30	20
Slump class	S1	S1	S1	S2	S1	S1	S2	S1	S1
Air content (%)	2.5	4.4	4.3	5.0	4.6	5.4	4.8	4.4	5.5

By comparing the mixtures in Table 3, the following can be concluded:

As expected, the densities of mixtures containing CB and CB aggregates are lower than the density of the reference mixture with NA.

The consistency of all nine mixtures is in accordance with the recommendations of the standard HRN EN 206-1 [15].

The air content is lower in the RM mixture (2.53%) than in the mixtures with CB and CT (4% to 5.5%), which was expected due to the more irregular particle shape of coarse aggregates.

Properties of Hardened Concrete at a Temperature of 20 ± 5 °C

Concrete mixtures were cast in molds and the samples were covered with a plastic sheet; they were then left in molds for 24 h. After they were demolded, the concrete samples were placed in a water tank in a digitally controlled temperature environment of 20 °C until they were tested. Table 4 andTable 5 show the results of the hardened concrete properties at 28 and 56 days. The tests were conducted with three samples of an individual mixture and the mean values are given in Table 4 andTable 5.

Table 4: Properties of 28-day-old hardened concrete

Code	RM	BM42	B10	BM19	BM26	BM34	BM35	BM36	BM50
Compressive strength (N/mm^2)	39.36	25.53	31.95	22.24	20.53	24.02	34.2	41.72	42.67
Flexural strength (N/mm^2)	5.71	4.34	5.31	5.68	4.72	5.07	5.32	5.79	5.88
Density of hardened concrete (kg/m^3)	2479	2165	2250	2042	2005	2069	2147	2193	2125
Ultrasonic speed velocity (km/s)	5.03	3.89	4.14	3.52	3.46	3.77	4.79	5.12	5.43

Table 5: Properties of 56-day-old hardened concrete

Code	RM	BM42	B10	BM19	BM26	BM34	BM35	BM36	BM50
Compressive strength (N/mm^2)	45.33	33.42	37.33	23.32	21.33	24.43	34.80	42.24	44.80
Flexural strength (N/mm^2)	6.03	4.96	5.38	5.73	4.91	5.22	5.48	5.81	5.92
Density of hardened concrete (kg/m^3)	2476	2100	2184	1842	1839	1945	1978	1986	1973
Modulus of elasticity (MPa)	41.85	19.26	22.61	20.8	17.15	17.83	21.17	19.78	21.74
Ultrasonic speed velocity (km/s)	507	3.96	4.23	3.74	3,52	3.81	4.95	5.33	5.61
Gas permeability $\times 10^{-16}$ (m^2)	3.49	4.10	5.25	3.84	3.72	3.96	3.57	3.61	3.56

The density of CB and CT aggregate concrete is lower than that of the reference mixture, as was expected. Any further increase of the CB and CT percentage may lead to the production of lightweight concretes.

Table 4 and Table 5 show that the compressive strengths of all nine concrete mixtures comply with the requirements of standards HRN EN 15037-1 [16] and HRN EN 15037-2 [17], which define the minimum concrete classes C12/15 for load-bearing floor blocks and C20/25 for precast-concrete load-bearing beams. The compressive strengths of the 28-day-old mixtures with CB and CT aggregates are on average 28% lower than that of the RM, as expected. The compressive strengths are comparable with conventional concrete containing NA and were achieved for concrete with CB and CT as aggregates, taking into account the 45% limit for coarse and 50% limit for fine particle fractions in substitution of NA with CB and CT.

The flexural strength of mixtures with CB and CT aggregates is on average 9% lower than that of the NA concrete mixture.

The gas permeability of concrete was measured on cylinder-shaped samples, 50 mm in diameter and 50 mm in length. All tests were conducted on three samples and Table 5 shows the mean values. The test results show that each of the nine concrete mixtures falls into the category of poor-quality concrete with regard to gas permeability. This can be explained by the preparation of the samples in plastic tubes, 50 mm in diameter, wherein it was not possible to properly consolidate the concrete that caused the presence of cavities. The tests should be repeated on 10 cm samples in order to validate the results. The results also indicate that concretes with crushed bricks and roof tiles have a higher gas permeability than traditional concrete. It is presumed that the reason for the higher gas permeability is the particle shape of CB and CT, which is sharper and the particles are more porous than dolomite grains.

According to Table 5, the statistical modulus of elasticity of concrete mixtures with CB and CT aggregates is up to 50% lower than that of concrete with NA. When designing load-bearing construction elements using concrete with CB and CT as aggregate, for which shrinkage is relevant, it is important to take into consideration the lower modulus of elasticity. A lower modulus of elasticity reduces the stiffness of the construction. Since the stiffness is lower, the use of concrete with CB and CT is recommended for example in vertical elements as filling material (for instance, in an infilled frame). It is also suggested that concrete blocks with such CB- and CT-infilled frames could provide a better earthquake resistance. The use of such concrete in horizontal construction elements (floor beams, floor slabs, *etc.*) would result in elements with a higher drying shrinkage contribution at identical strain when compared to concrete with NA. However, in short elements such as floor blocks, the modulus of elasticity plays no significant role, and the use of CB and CT in resisting and semi-resisting floor blocks may therefore be considered possible.

By comparison with the ultrasonic speed velocity results listed in Table 4 and Table 5, it can be concluded that CB and CT samples have a lower ultrasonic speed velocity, which may indicate the appearance of defects and cracks, a higher porosity and other non-homogeneities. Since all mixtures have ultrasonic speed velocity above 3.60 km/s, it may be concluded that they are in the high-quality concrete class [18].

The drying shrinkage of concrete, as shown in Figure 4, was measured with samples constantly exposed to humidity conditions of 95% ± 5% relative humidity. Like for all other hardened-concrete properties, drying shrinkage was tested with three samples of the same mixture. Drying shrinkage of each sample was measured in two different measuring lines facing each other on days three, four, seven, and afterwards every seven days until the 91st day, which is 15 times in total.

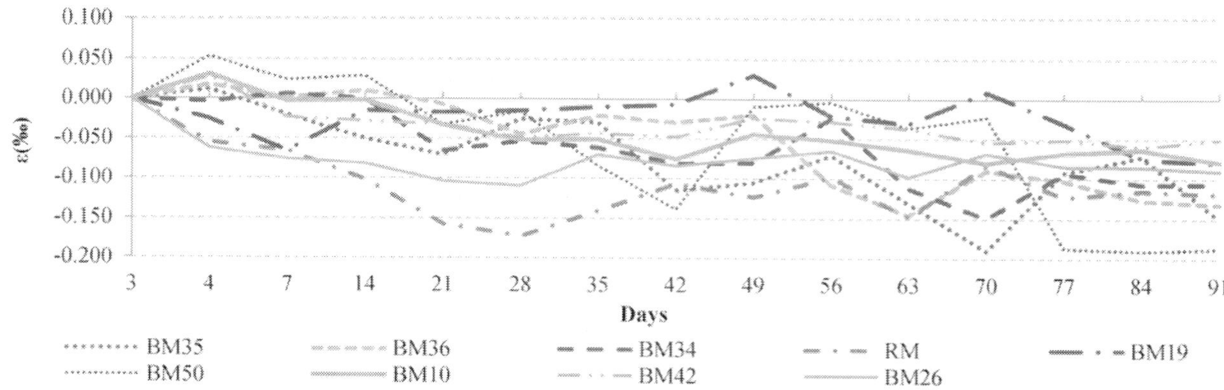

Figure 4: Comparison of drying shrinkage of concrete mixtures

Figure 4 shows a comparison of drying shrinkage measured for 91-day-old mixtures. Reason for the inconsistent trend could be the relatively high percentage range of relative humidity, and this issue will be addressed in future research. Despite inconsistent trends for particular concrete mixtures, the results indicate that the lowest level of drying shrinkage was observed for samples with CB and CT. This implies a positive effect for the final use of these kinds of aggregates in precast concrete elements.

Hardened Properties of Concrete after Exposure to High Temperatures

Samples for testing mixture properties after exposure to high temperatures were made using identical mixture compositions, and the samples were cured in an identical manner up until 28 days of age compared to samples tested at room temperature. After 28 days, before their placement in an electric furnace at 105 °C, the samples were kept at room temperature for seven days in order to avoid the emergence of cracks due to the sudden drying at 105 °C. The drying procedure was carried out until the samples were dried completely. In order to be able to design concrete structures as well as to assess the safety with regard to thermal events or after a real accident according to [19], it is necessary to have detailed knowledge of the concrete properties for a temperature range from 20 °C to at least 750 °C. In some cases, the data of concrete properties are needed up to 1000 °C. The samples were then exposed to high temperatures in the electric furnace. Prior to exposure in the electric furnace, the weight of samples was determined to check humidity in the period between the drying and high temperature exposure. If the weight increase was higher than 1% in comparison to dry weight, the samples were returned to drying procedure as described above. The temperature rate in the electric furnace was 1 °C/min and was measured via a thermo probe located on the furnace and a control probe placed in the furnace. The heating regime was in accordance with the recommendations of RILEM TC 200-HTC [19]. Thermocouples integrated into the three samples of each mixture enabled the register temperature rise in the individual samples (Figure 5, Figure 6 andFigure 7). Based on previous research of the authors and based on recomendations RILEM TC 200-HTC, [19] to obtain reproductible results, specimens were at identical curing conditions after exposure to high temperatures. Specimens were cured at a temperature of 20 ± 5 °C and in relative humidity 50% ± 5%, under conditions without any moisture exchange, for seven days after which testing of mechanical properties took place.

Figure 5: Electric furnace.

Figure 6: Samples in furnace.

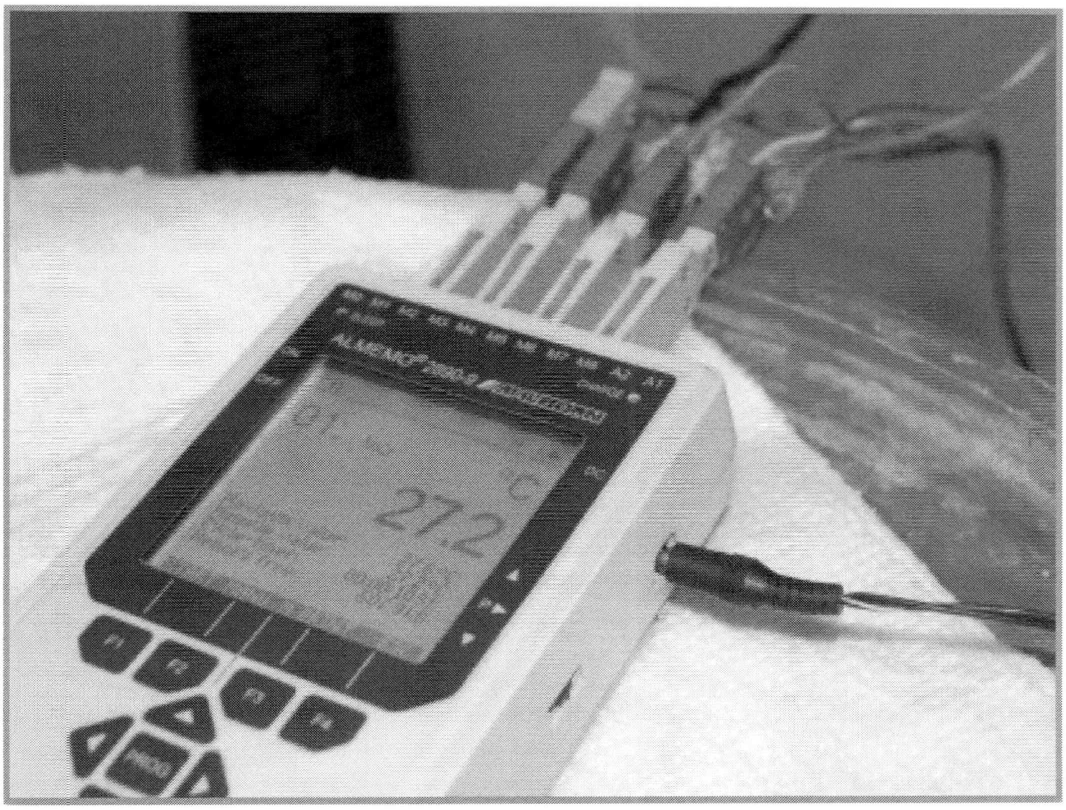

Figure 7: Measurement of the temperature in samples.

The heating and cooling regimes of samples exposed to temperatures of 200 °C and 800 °C are shown in Figure 8 and Figure 9. Figure 8 indicates that, during the heating of samples at 200 °C, the temperature rise in all three mixtures was approximately the same. During exposure to a temperature of 800 °C, a higher temperature rise is observed in the RM samples, as shown in Figure 9. The reason for this elevated temperature rise in the RM samples may be found in the higher thermal conductivity of RM concrete when compared to concretes with CB and CT, as was confirmed by a thermal conductivity coefficient test carried out for concrete in [20]. Figure 8 also shows a lower temperature rise in samples with CB and CT as aggregates, which may imply that these concretes have a higher heat conductivity when compared to concrete with NA.

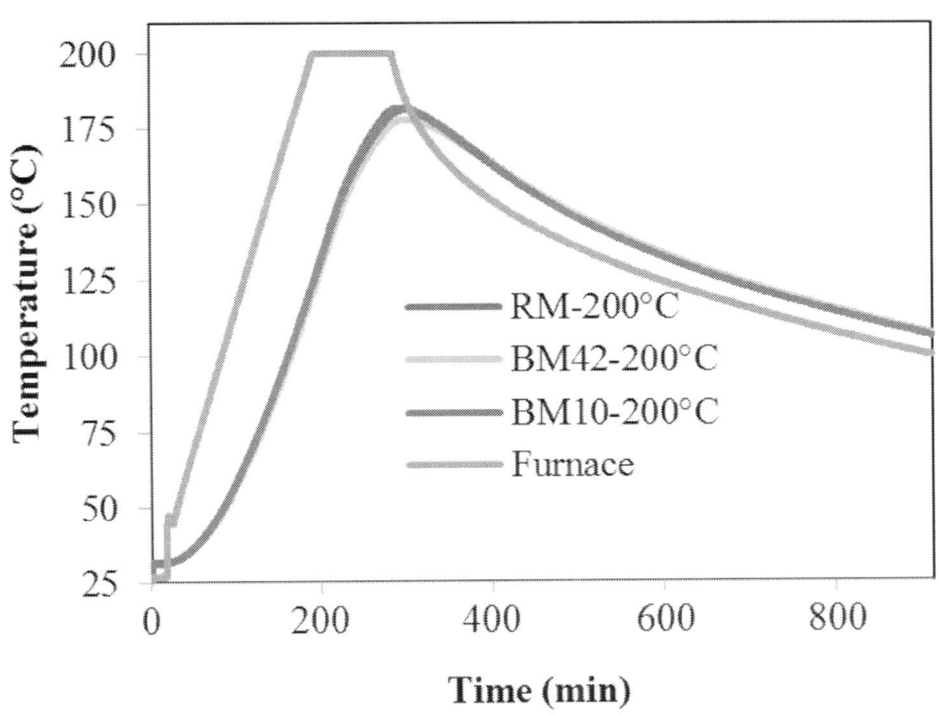

Figure 8: Heating regime of mixtures RM, BM10 and BM42 for 200 °C.

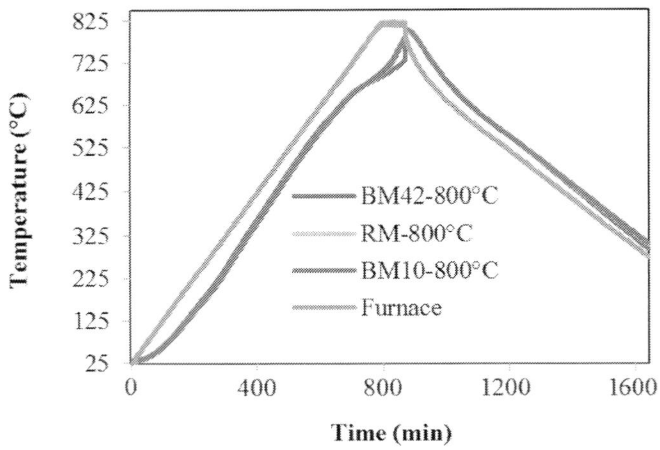

Figure 9: Heating regime of mixtures RM, BM10 and BM42 for 800 °C.

Concrete, like the majority of materials, changes mass under the influence of temperature. The weight was measured before and after high temperature exposure on digital scales with a precision of ±0.1 g.

Mass losses after heating are shown in Figure 10. The residual mass, the percentage of the remaining mass in comparison to the initial mass before heating, was approximately 98% for all samples exposed to 200 °C. After the exposure of samples to temperatures of 600, 800 and 1000 °C, the RM lost significant mass as opposed to its initial mass, whereas the mixtures with CB and CT aggregates displayed a better resistance to mass loss due to heating.

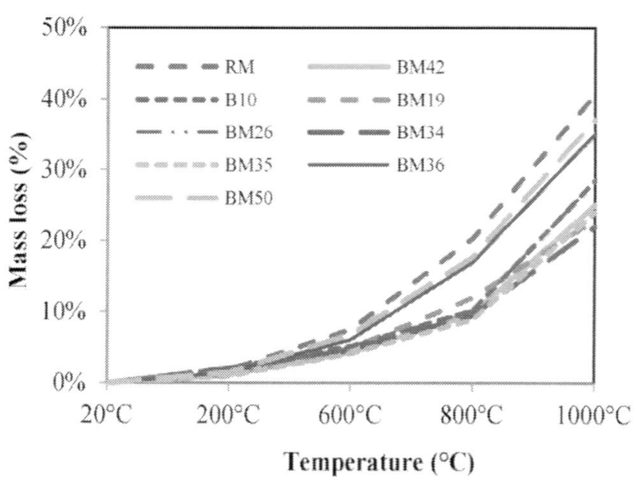

Figure 10: Mass loss after exposure to high temperatures.

The concrete mass loss after heating was calculated using Equation (1) (illustrated in Figure 10):

$$M_{loss} = \left(M_{before} - M_{after}\right) / M_{before} \quad (1)$$

It is also observable that the lowest mass loss was obtained for the mixtures with the highest percentage of CB and CT aggregates. The mass loss in concrete with CB and CT aggregates is about 25% after exposure to 1000 °C, while the referent mixture mass loss is 40.6%. This indicates that it is possible to achieve a significantly lower concrete mass loss during high temperature exposure when replacing 50% of fine particles and 75% of coarse particles of NA with CB and CT. Since all samples had a weight increase of up to 1% before heating, the mass loss presumably originates from water evaporation. It is presumed that this results in a lower pressure, causing less concrete deterioration, but it may also result in better integrative properties of concrete components, *i.e.*, higher residual mechanical properties.

Concrete changes its volume when exposed to different temperatures. Therefore, the volume of each sample was measured before and after exposure to high temperatures. Figure 11 shows the percentage of sample volume increase after exposure to high temperatures. The results indicate that the least observable change in volume was noted for BM36 samples. For example, the sample volume increase after exposure to 800 °C for RM was 36%, whereas, for BM36, it was 9.5% and for B10 12.3%. These results indicate that concrete with CB and CT as aggregates have an approximately 25% less volume change due to high temperature exposure.

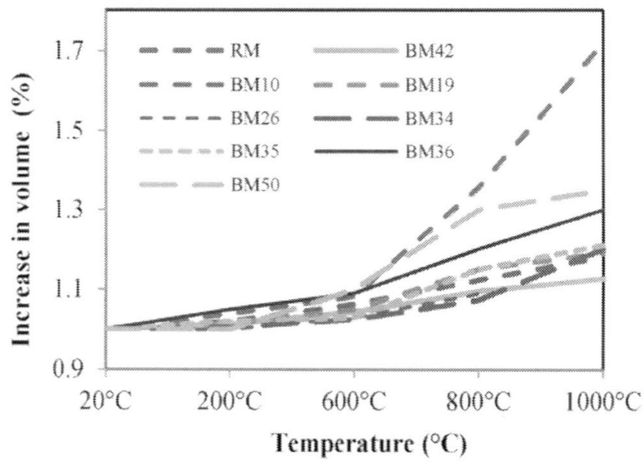

Figure 11: Relative volume as a function of high temperatures.

Since a fire causes stress in a construction, the resulting stresses are higher for mixture RM. If the space for the change of length is inadequate, secondary moments may appear, causing the construction to collapse.

Figure 12 shows the residual compressive strengths of concrete samples after they were exposed to high temperatures. Residual compressive strengths were calculated as a percentage of the remaining compressive strength after high temperature exposure, as compared to the initial compressive strength. The results in Figure 12 indicate that, after exposure to a temperature of 200 °C, the compressive strengths of all samples increased compared to the compressive strengths of samples cured at room temperature of 20 °C. The compressive strength increase compared to the initial values for the RM samples is 16%, and, for concrete containing CB and CT aggregates, the increase is between 2%–6%. According to [21,22,23], the compressive strength increase is caused by chemical changes in the cement paste. A loss of humidity occurs, due to which interparticle forces between hydration products increase, which subsequently leads to an increase in compressive strength. Another increase in compressive strength is presumably a shorter high temperature exposure for 200 °C compared to the duration of exposure at 600 °C, which also affects reason for the hydration reaction [22].

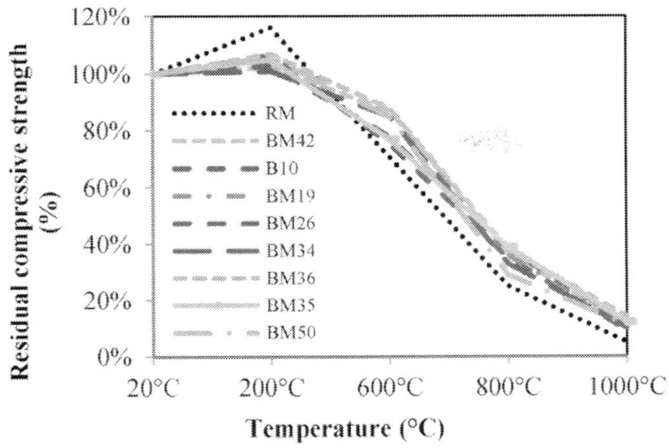

Figure 12: Residual compressive strengths.

Concrete with CB and CT as aggregate displays a similar strength development profile at high temperatures as concrete with natural aggregate. The chemical changes of minerals for hardened cement paste and aggregates are dependent on the inner temperature of concrete.

Exposure of samples to 400 °C and higher temperatures cause a dehydration of calcium-hydroxide (CSH) in hardened cement paste. According to the results shown in Figure 12, it is possible to conclude that, up to approximately 400 °C, the decrease of the initial compressive strength of samples is small, and that the compressive strength loss is within a 10% limit. Presumably, the 10% decrease is the result of the elimination of free water and CSH dehydration from concrete [24,25].

After exposure of concrete to a temperature of 600 °C, the loss of compressive strength becomes more significant: for RM, it is 29%, and for CB and CT concrete, it is 13% to 25%. The samples with the highest percentages of CB and CT aggregate (BM36) show the lowest compressive strength loss at 600 °C, as expected. Crushed bricks and roof tile aggregates concrete enable a better passage of gas through concrete.

At temperatures higher than 800 °C, decarbonization of limestone occurs and the aggregate containing $CaCO_3$ deteriorates, releasing CO_2 as gas from the concrete. If the gas exit route is blocked, the resulting pressure causes microcracking and deterioration of the concrete structure. Since the brick and roof tile production process includes a basic limitation of the clay chemical composition, which is a maximum percentage of calcium-carbonate of 4%. CB and CT aggregates of concrete will presumably have a significantly better integration and higher remaining strength after high temperature exposure when compared to dolomite aggregate concrete, which contains around 30% calcium-carbonate. After exposure to high temperatures, the decarbonization of dolomite results in porous dolomite lime with a large specific surface area. The dolomite lime increases its volume under the influence of the earlier described reaction and gas leaves the concrete.

All samples exposed to 1000 °C show a low residual compressive strengths, e.g., RM 5%, BM36 11% and BM50 7%. The samples were tested immediately after cooling to room temperature, a day after they were heated. All samples that remained for seven days at room temperature, and humidity conditions deteriorated and fractures occurred as a result of dolomite expansion, as described in the previous section.

CB and CT aggregates are sharper and coarse particles, which may result in a better bond between the aggregate particles and cement paste, subsequently resulting in better mechanical properties of concrete. The presented results indicate that CB and CT are better at tolerating high temperatures when compared to dolomite, due to the lower percentage of CaO in the chemical composition and the geometrical properties of its particles, as is evident in the percentage of the remaining compressive strengths illustrated in Figure 12. The residual flexural strengths are shown in Figure 13. All conclusions listed for residual compressive strengths of concretes RM to BM50 can be drawn also for residual flexural strengths. The samples of the BM36 mixture show the best properties with regard to residual flexural strengths.

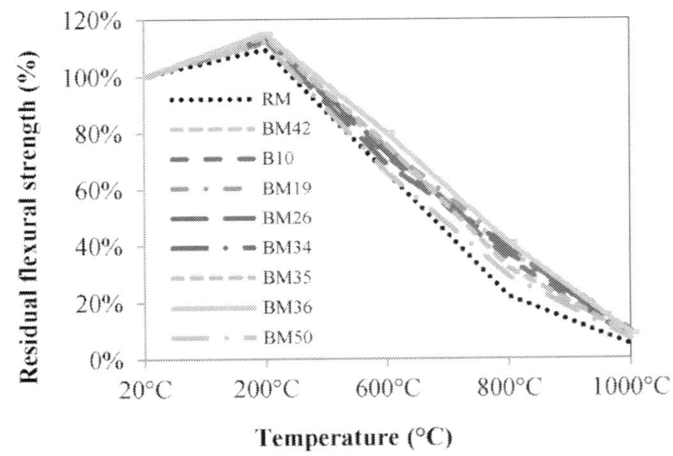

Figure 13: Residual flexural strengths.

The modulus of elasticity of concrete decreases after exposure to high temperatures, as is shown inFigure 14. Samples exposed to 1000 °C could not be tested because their outer surface peeled, making it impossible to properly position the specimen in the testing device and to attain proper results. Therefore, the tests were conducted only on samples exposed to 200, 600 and 800 °C. The results show that the residual modulus of elasticity of concretes with CB and CT as aggregates during exposure to high temperatures are continuously higher, about 10% to 13%. The highest residual modulus of elasticity after exposure to all temperatures was obtained for mixture BM36. By comparingFigure 12 and Figure 14, it can be seen that the reduction in modulus of elasticity obtained at increasing temperature is higher than the compressive strength reduction due to the deformation increase for the same strain value. Presumably, the decrease in deformation after exposure to high temperatures is due to the superior thermal properties of CB and CT compared to dolomite, which could imply the possibility of the application of such concrete in resisting or semi-resisting floor blocks. The ultrasonic pulse velocity through samples was determined before and after exposure to high temperatures, and the results are shown in Figure 15. Since the pore structure changes during the exposure of concrete to high temperature, this also affects mechanical properties. Measurement of the ultrasonic pulse velocity through samples after exposure to high temperatures was used to determine whether the appearance of cracks in concrete causes a reduction in mechanical characteristics. FromFigure 15, it is evident that RM has an approximately 5%

lower residual ultrasonic pulse velocity for all temperatures compared to mixtures with CB and CT aggregates, indicating that more cracks develop in concrete with NA.

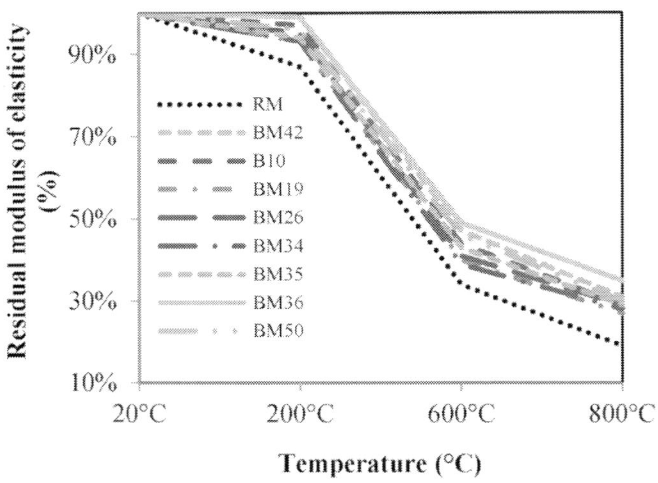

Figure 14: Residual modulus of elasticity.

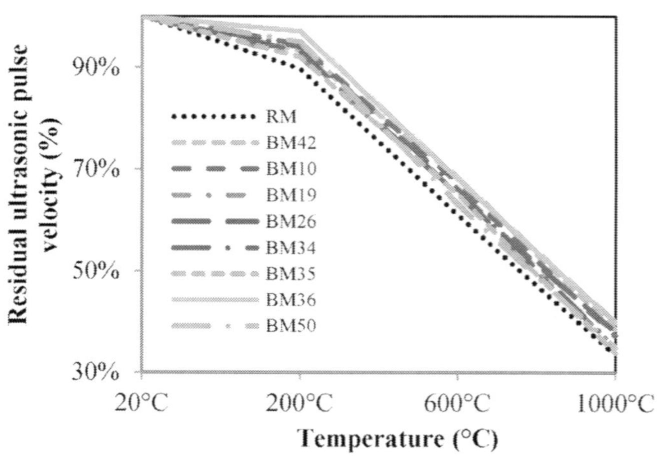

Figure 15: Residual ultrasonic pulse velocity.

Considering that floor blocks will not be exposed to outside environmental conditions (exposure class X0), it is presumed that the permeability is the property with most pronounced with regard to durability. Since methods of laboratory testing of gas permeability have not been standardized yet but have been described and recommended by Technical Committee RILEM 116-PCD within the European organization RILEM [26], the tests were conducted in line with recommendations of Cembureau (RILEM TC 116–PCD). Figure 16 indicates that RM has a lower gas permeability coefficient than mixtures with CB and CT aggregates, making it the most suitable concrete with regard to gas permeability after high temperature exposure. It is also evident that all concrete mixtures fall into the category of poor-quality concretes with regard to gas permeability, and all earlier conclusions with regard to gas permeability tests at room temperature also apply for gas permeability in samples exposed to high temperatures.Figure 16 shows that the gas permeability is up to 47 times higher after exposure to temperatures above 600 °C than after exposure to a temperature of 200 °C. The reason for this is the appearance of cracks in the concrete. Surface cracks on samples of RM appear after the exposure to a temperature of 600 °C, and the number increases continuously at increasing temperature when they are exposed to 800 and 1000 °C. With all the other mixtures containing CB and CT aggregates, surface cracks appear only after the samples have been exposed to a temperature of 800 °C. Presumably, the cracks are smaller in size and fewer in number in samples with CB and CT due to the lower content of calcium-carbonate in these aggregates compared to dolomite.

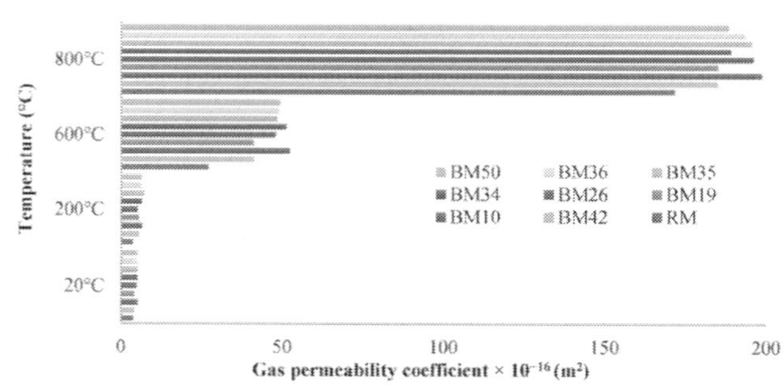

Figure 16: Gas permeability after exposure to high temperatures.

Figure 17a,b show samples seven days after their exposure to a temperature of 1000 °C. In Figure 17a,b shows spalling of the concrete surface above the aggregate particles in RM samples, whereas, in samples with CB and CT aggregates, the surface spalling occurs crossing the concrete matrix and dolomite particles. Brick and roof tile particles kept their integrity, underlining that CB and CT are appropriate for the production of fire-resistant concretes, as previously stated. Likewise, it may be claimed that the concrete matrix properties need to be improved with regard to its thermal properties. The use of highly-fire-resistant cement (for example, aluminium cement) or brick dust and roof tile dust as a partial cement substitute is recommended, which needs further research.

(a) (b)

Figure 17: Examples of fire spalling of concrete surface after exposure to a temperature of 1000 °C for different concrete mixtures. (**a**) Concrete with CB and CT aggregates; (**b**) Mixture RM.

CONCLUSIONS

The review of possible applications for crushed clay bricks (CB) and roof tiles (CT) waste as partial substitution for natural aggregates (NA) in concrete identified the research area of this paper. The identification of research needs leads to the in depth research and proof of the possibility for using concrete with CB and CT as aggregates for the production of precast concrete elements.

The first step in the process of determining the properties of concrete with CB and CT as aggregates comprised experimental tests of CB and CT, which proved that such material may be used as a partial or full substitute for NA, depending on the required concrete properties.

Basic concrete properties were determined in order to optimize concrete with CB and CT as aggregates.

The densities of concrete mixtures with crushed brick and roof tiles (mixtures BM) are 11% to 14% lower than those of concrete with natural aggregate (mixture RM).

Experimental testing on BM and RM samples in the hardened state at room temperature showed that samples with CB and CT as aggregates compared to the reference samples RM had on average: a 27% lower compressive strength, a 16% lower flexural strength, a 19% lower hardened-state density, a 50% lower modulus of elasticity, a 20% lower ultrasonic pulse velocity, a 18% higher air permeability, and 28% less drying shrinkage.

It was concluded that special emphasis must be placed on using such concrete for horizontal and vertical structural elements (beams and columns) due to the lower modulus of elasticity and the possibility of larger deformations occurring.

The conducted laboratory experimental tests on mixtures BM with CB and CT aggregates after exposure to temperatures of 20, 200, 600, 800 and 1000 °C, and the comparison of the results with RM, lead to the conclusions:

The BM36 mixture with the highest percentage of replacement of NA with CB and CT aggregates has the best physical, mechanical, and thermal properties with regard to the application of precast concrete elements exposed to high temperatures.

In general, mixtures with CB and CT as aggregates have better physical and mechanical properties than regular concrete after exposure to high temperatures.

It is important to mention that this research is not applicable for crushed bricks and tiles in general (e.g., obtained from demolition of masonry structures), but for waste products of the process of making new bricks and roof tiles only. Furthermore, in order to make conclusions about the applicability of concrete with CB and CT aggregates in precast concrete elements in general (regardless of temperature to which the concrete is exposed), further research will be carried out to address the durability properties of the concrete.

ACKNOWLEDGMENTS

This research was performed within project "ECO-SANDWICH" funded in the frame of CIP ECO Innovation programme (ECO/11/304438/SI2.626301).

AUTHOR CONTRIBUTIONS

I.M. conceived and designed the experiments; N.Š. and I.B.P. performed the experiments; I.M. and N.Š. analysed the data; I.M. contributed materials/analysis tools; I.M. wrote the paper.

REFERENCES

1. Cullum-Kenyon, S. *Market Development Study for Recycled Aggregate Products*; Report to Waste Reduction Advisory Committee; Thurber Engineering Ltd.: Calgary, AB, Canada, 2001; pp. 1–32.
2. Živković, S.A.; Krasić, D. Mineral Resources Management Strategy of Croatia. 2008. Available online: http://www.zavod.pgz.hr/docs/zzpuHR/documents/301/Original.pdf (accessed on 15 April 2016).
3. Khalaf, F.M.; DeVenny, A.S. Performance of Brick Aggregate Concrete at High Temperatures. *J. Mater. Civ. Eng.* 2004, *16*, 556–565.
4. Hansen, T.C. *Recycling of Demolished Concrete and Masonry*; RILEM Rep. 6; E & FN Spon: London, UK, 1992.
5. HZN Standards Publication. *Aggregates for Concrete*; HRN EN 12620; Croatian Standards Institute: Zagreb, Croatia, 2008.
6. HZN Standards Publication. *Tests for Geometrical Properties of Aggregates—Part 4: Determination of Particle Shape—Shape Index*; HRN EN 933-4; Croatian Standards Institute: Zagreb, Croatia, 2008.
7. Ministry of Environmental and Nature Protection, Concrete Constructions Technical Regulations (TPBK). Available online: http://narodne-novine.nn.hr/clanci/sluzbeni/2009_11_139_3381.html(accessed on 13 April 2016).
8. HZN Standards Publication. *Tests for Geometrical Properties of Aggregates—Part 8: Assessment of Fines—Sand Equivalent Test*; HRN EN 933-8; Croatian Standards Institute: Zagreb, Croatia, 2015.
9. HZN Standards Publication. *Tests for Mechanical and Physical Properties of Aggregates—Part 3: Determination of Loose Bulk Density and Voids*; HRN EN 1097-3; Croatian Standards Institute: Zagreb, Croatia, 2004.
10. HZN Standards Publication. *Tests for Mechanical and Physical Properties of Aggregates—Part 6: Determination of Particle Density and Water Absorption*; HRN EN 1097-6; Croatian Standards Institute: Zagreb, Croatia, 2013.
11. HZN Standards Publication. *Tests for Mechanical and Physical Properties of Aggregates—Part 2: Methods for the Determination of Resistance to Fragmentation*; HRN EN 1097-2:2004/A1:2007; Croatian Standards Institute: Zagreb, Croatia, 2007.
12. HZN Standards Publication. *Tests for Thermal and Weathering Properties of Aggregates—Part 2: Magnesium Sulfate Test*; HRN EN 1367-2:2010; Croatian Standards Institute: Zagreb, Croatia, 2010.
13. HZN Standards Publication. *Cement—Part 1: Composition, Specifications and Conformity Criteria*

for Common Cements; HRN EN 197-1:2005; Croatian Standards Institute: Zagreb, Croatia, 2005.

14. HZN Standards Publication. *Testing Fresh Concrete*; HRN EN 12350; Croatian Standards Institute: Zagreb, Croatia, 2009.
15. HZN Standards Publication. *Concrete—Part 1: Specification, Performance, Production and Conformity*; HRN EN 206-1; Croatian Standards Institute: Zagreb, Croatia, 2006.
16. HZN Standards Publication. *Precast Concrete Products. Beam-and-Block Floor Systems. Beams*; HRN EN 15037-1; Croatian Standards Institute: Zagreb, Croatia, 2008.
17. HZN Standards Publication. *Precast Concrete Products. Beam-and-Block Floor Systems. Concrete Blocks*; HRN EN 15037-2; Croatian Standards Institute: Zagreb, Croatia, 2011.
18. Bjegović, D.; Serdar, M.; Stipanović Oslaković, I. *Assessment of Concrete Properties in Structures, Handbook of Concrete Durability*; Kim, S.-H., Ann, K.Y., Eds.; Middleton Publishing Inc.: Seoul, Korea, 2010; pp. 131–179.
19. RILEM TC 200-HTC. Mechanical Concrete Properties at High Temperature—Modelling and Applications. *Mater. Struct.* 2005, *38*, 913–919.
20. Miličević, I.; Bjegović, D.; Siddique, R. Experimental research of concrete floor blocks with crushed bricks and tiles aggregate. *Constr. Build. Mater.* 2015, *94*, 775–783.
21. Zaina, M.F.M.; Safiuddina, M.; Mahmud, H. Development of high performance concrete using silica fume at relatively high water-binder ratios. *Cem. Concr. Res.* 2000, *30*, 1501–1505.
22. Chen, B.; Li, C.; Chen, L. Experimental study of mechanical properties of normal-strength concrete exposed to high temperatures at an early age. *Fire Saf. J.* 2009, *44*, 997–1002.
23. Behnood, A.; Ghandehari, M. Comparison of compressive and splitting tensile strength of high-strength concrete with and without polypropylene fibers heated to high temperatures. *Fire Saf. J.*2009, *44*, 1015–1022.
24. Savva, A.; Manita, P.; Sideris, K.K. Influence of elevated temperatures on the mechanical properties of blended cement concretes prepared with limestone and siliceous aggregates. *Cem. Concr. Compos.* 2005, *27*, 239–248.
25. Xing, Z.; Beaucour, A.-L.; Hebert, R.; Noumowe, A.; Ledesert, B. Influence of the nature of aggregates on the behaviour of concrete subjected to elevated temperature. *Cem. Concr. Res.*2011, *41*, 392–402.
26. RILEM TC 116-PCD. Permeability of concrete as a criterion of its durability. Recommendations 561 Tests for gas permeability of concrete. *Mater. Struct.* 1999, *32*, 174–179.

CHAPTER 16

USE OF NATURAL-FIBER BIO-COMPOSITES IN CONSTRUCTION VERSUSTRADITIONAL SOLUTIONS: OPERATIONAL AND EMBODIED ENERGY ASSESSMENT

Carmen Galan-Marin , Carlos Rivera-Gomez and Antonio Garcia-Martinez

Institute of Architecture and Construction Science, University of Seville, Seville 41012, Spain

ABSTRACT

During the last decades natural polymers have become more and more frequent to replace traditional inorganic stabilizers in building materials. The purpose of this research is to establish a comparison between the most conventional building material solutions for load-bearing walls and a type of biomaterial. This comparison will focus on load-bearing walls as used in a widespread type of twentieth century dwelling construction in Europe and still used in developing countries nowadays. To carry out this analysis, the structural and thermal insulation characteristics of different construction solutions are balanced. The tool used for this evaluation is the life cycle assessment throughout the whole lifespan of these buildings. This research aims to examine the environmental performance of each material assessed: fired clay brick masonry walls (BW), concrete block masonry walls (CW), and stabilized soil block masonry walls (SW) stabilized with natural fibers and alginates. These conventional and new materials are evaluated from the point of view of both operational and embodied energy.

INTRODUCTION

Spain is, among other countries, a signatory of the United Nations Commission on Sustainable Development (CSD). As stated by the organizing committee of the 2015 United Nations Climate Change Conference, COP 21 or CMP 11 that was held in Paris, from the outset of the talks, the expected key result was an agreement to set a goal of limiting global warming to less than two degrees Celsius (°C) compared to pre-industrial levels [1]. The construction sector has a significant impact on the environment [2]. According to some studies, around 20% of the total impacts are related to manufacturing, construction, demolition processes, and final disposal of building materials, elements and systems [3], but the operational energy is, in many cases, considered the most significant aspect.

Energy in buildings can be categorized into two types: firstly by energy for the maintenance/servicing of a building during its useful life, namely operational energy (O.E.) and, secondly, by energy capital that goes into production of a building using various building materials, named embodied energy (E.E.). Study of both types of energy consumption is required for a complete understanding of building energy needs. Embodied energy of buildings can vary over wide limits depending upon the choice of building materials and building techniques. Reinforced concrete walls, fired clay brick masonry, concrete block masonry, bream and block slabs represent common conventional systems forming the main structure of buildings in Spain. Similar building systems can be found in many other developed and developing countries.

Citation: Carmen Galan-Marin, Carlos Rivera-Gomez and Antonio Garcia-Martinez, Use of Natural-Fiber Bio-Composites in Construction versus Traditional Solutions: Operational and Embodied Energy Assessment, doi:10.3390/ma9060465.

Alternative building technologies, such as stabilized soil blocks, can be used for minimizing the embodied energy of buildings [4,5,6,7,8,9,10]. Generally, the materials used for the structure of buildings represent more than 50% of the embodied energy in the building [11]. In this sense, the use of alternative materials, such as mortar/concrete blocks, stabilized soil blocks, or fly-ashes, instead of materials with a high embodied energy, such as reinforced concrete, could save 20% of the cumulative energy over a 50-year life cycle [12]. In addition, recycling building materials [13,14] is essential to reduce the embodied energy in the building. For instance, the use of recycled steel and aluminum saves more than 50% in embodied energy [15].

There are many studies in the literature dealing with energy analysis in buildings [16], some of them have been published thus far on the evaluation of environmental impacts using a life cycle assessment (LCA) tool on the operational energy in the life cycle of residential dwellings. These studies have appeared during the last 10 years [17,18]. Furthermore, in different countries some other LCA case studies for operational energy of residential dwellings were performed, as in the following examples:

Sartori and Hestnes [19] performed a literature survey on buildings' life cycle energy use of 60 cases from nine countries. This study concluded that operating energy represents by far the largest part of energy demand in a building during its life cycle.

Ortiz *et al.* [20] have applied LCA methodology to evaluate environmental impacts on a Spanish Mediterranean house located in Barcelona with a total area of 160 m^2 and a projected 50-year life span. The same authors studied [21] the importance of operational energy in life cycle assessment (LCA). The comparison was applied within the residential building sector for two buildings, one in a developed country (Spain), and one in a developing country (Colombia).

Chung *et al.* [22] conducted an energy input–output (E-IO) analysis in Korea. The results showed that accounting for energy intensities and greenhouse gas (GHG) emission intensities is becoming an essential step in proper understanding of the energy usage structure.

Thormark [23], in Sweden, analyzed, within its CEPHEUS project (cost efficient passive houses as European standard) in the European Thermie program, on how far the design phase of housing was relevant with regard to reducing operational energy and how the choice of building materials may affect both embodied energy and recycling potential during the 50-year life span of the building.

Stephan and Stephan [24] quantized the life cycle energy and cost requirements associated with 22 different energy-reduction measures targeting embodied, operational, and user-transport requirements. It evaluates a case study apartment building in Sehaileh, Lebanon.

Koesling *et al.* [25] established a model valid for various types of building sector and applied it to estimate the amount of embodied energy in the building envelopes of 20 dairy farms in Norway.

Guan *et al.* [26] studied the energy used in all three phases of construction, operation, and demolition of eight residential buildings in and around Brisbane, Queensland, Australia. It was found that the main contribution to the operational energy in residential buildings comes from the use of general appliances in homes.

Brown *et al.* [27] analyzed how to mitigate climate change through operational energy reduction in existing Swedish residential buildings.

Zhu *et al.* [28] developed a new optimization method for building envelope design in order to get the lowest carbon emissions of building operational energy consumption using an orthogonal experimental design.

Pinky and Sivakumar [29] presented a case study of life cycle energy analysis of a residential development consisting of 96 identical apartment-type homes located in Southern India. They considered that the life cycle energy of the building includes the construction energy, operational energy and demolition energy. Construction refers to initial construction as well as recurring maintenance and repair work.

Praseeda *et al.* [30] discussed the embodied energy and operational energy assessment of a few residential buildings in different climatic locations in India. The study shows that the balance of O.E. and E.E. in LCA greatly depends upon the types of materials used in construction and extent of space conditioning adopted. In some cases E.E. can exceed O.E. in the whole life cycle. Buildings with reinforced concrete frames and monolithic reinforced concrete walls have very high E.E.

Stephan *et al.* [31] presented a framework which takes into account energy requirements at the building scale, *i.e.*, the embodied and operational energy of the building and its refurbishment, and at the city scale, *i.e.*, the embodied energy of nearby infrastructures and the transport energy (direct and indirect) of its users. This framework has been implemented through the development of a software tool which allows the rapid analysis of the life cycle energy demand of buildings at different scales.

Islam *et al.* [32] described the life cycle assessment and life cycle cost analysis of a typical Australian house under the design phase in their paper. The implications of life cycle environmental impacts and life cycle costs were evaluated and the optimum assemblage design is reported using an optimization algorithm. A set of best solutions is found depending on different factors: the model assumptions, range of environmental and economic indicators considered, and the chosen quantitative criteria.

Iddon and Firth [33] developed a building information model (BIM) tool to simultaneously estimate embodied and operational carbon over a 60 year life span for a typical four bedroom detached house. Using the tool, four different construction scenarios are evaluated, representing a range of current construction methods used in present day house buildings in the UK.

Finally, Ibn-Mohammed *et al.* [34] took a retrospective approach to critically review the relationship between embodied and operational emissions over the lifecycle of buildings in their paper. This is done to highlight and demonstrate the increasing proportion of embodied emissions, which is a consequence of the efforts to decrease operational emissions.

The current study takes an environmental perspective when comparing various conventional technologies for building walls to others that use new low-impact materials.

By identifying and quantifying the materials used in the manufacturing and construction processes and the consequent operational energy, by applying LCA methodology, we identify the environmental impact of each alternative building material studied. Therefore, the aim of this research is to compare the environmental aspects and potential impact associated with the construction, maintenance, use, and disposal of walls in three-storey buildings, determining the option with the lowest negative impact in relation to insulation and material characteristics. A life cycle assessment was made of three models of housing blocks erected with load-bearing walls that varied according to their material structure. The options compared involved conventional and unconventional building materials.

There are several previous studies above mentioned comparing different structural system in terms of LCA, but tend to focus just in one of the aspects. This study implements three different kinds of parameters in a single case study, the structural comparison, the material comparison, and the environmental comparison. The last variable included is to compare the results in two real climate conditions and real scenarios.

The aim of this research is to determinate which material type produces the least environmental impact during the different stages of its life cycle. The purpose in this article is to demonstrate the relationship between materiality, architecture, and design, and the environmental impact produced by different construction systems. In order to establish this relationship, life cycle assessment is the tool used to analyze the environmental impact produced by the construction materials studied in its life cycle.

The main limitations of the study are related to the variables used. Only two alternatives to the stabilized earth block have been selected, being the most widely used materials in this kind of load-bearing wall. A specific building type has been chosen as a case study, whereas this decision comes from the abundance of its use in Europe in the twentieth century. Two climate scenarios have been selected, the most different ones among the Spanish climate zones. Just one type of insulating material has been introduced, being the most used in this type of construction solutions. Previous studies, comparing three different insulation materials, provided no significant differences in results, increasing the complexity of the data presentation and understanding.

CASE STUDY

Building Description

The case study is a typical apartment block, very common in the European scenario and a scheme widely used for social housing in the last century. The building is a three-storey case (maximum height allowed in social housing in Spain that does not require the introduction of an elevator) and a typical distribution of two apartments per floor (see Figure 1), balancing the use of the common space *versus*housing spaces. Surface area per floor is 147.83 m^2, so the total built area of the building is 443.49 m^2, with a total of six houses.

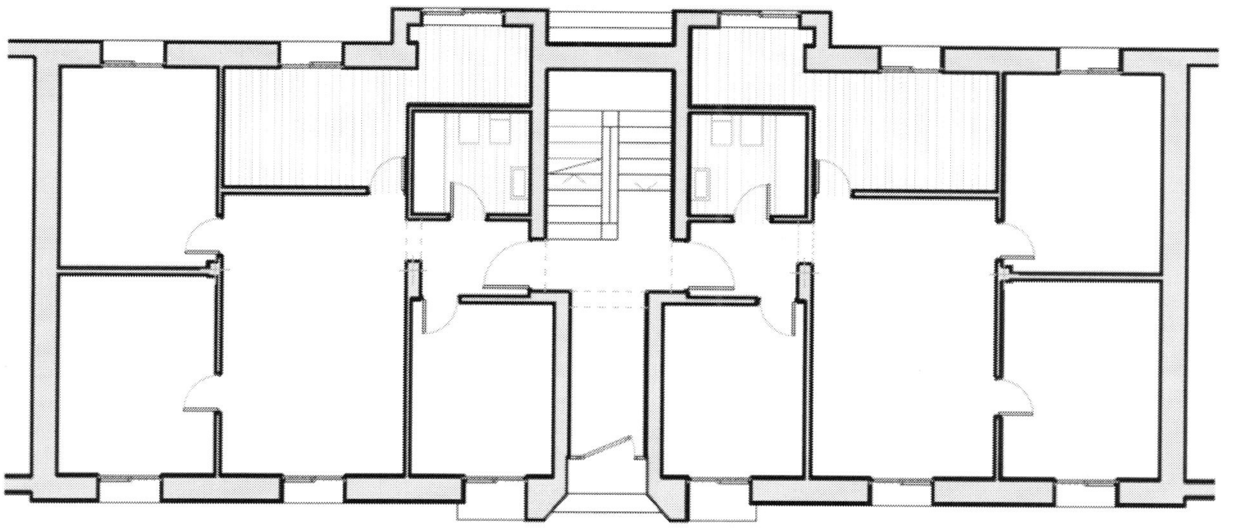

Figure 1: Building plan of the analyzed case study.

It is a mixed-structure building: façade load-bearing walls and inner concrete pillars. The floor slabs are a beam and block system and reinforced concrete beams, thus optimizing the structural function of the materials used in the envelope walls, while the pillars inside allow a free design of the interior plant.

In order to analyze the influence of the construction materials, several material options for the façade load-bearing walls have been studied. The different building construction systems used are BW (fired brick walls); CW (concrete block walls); and SW (stabilized soil walls). These systems are detailed in the next section. Inner concrete pillars are considered for all three cases. All of the ACV calculations are done according to ISO 14040 [35] and ISO 14044 [36], which implies a 50 year lifespan of the building. For the purpose of our study, the building is located in Spain. In order to analyze the influence of the operational energy demand, the building is located in two different Spanish climates (see Figure 2). These climates are named Location 1, corresponding to the Mediterranean climate and Location 2, corresponding to the inner continental areas of the peninsula.

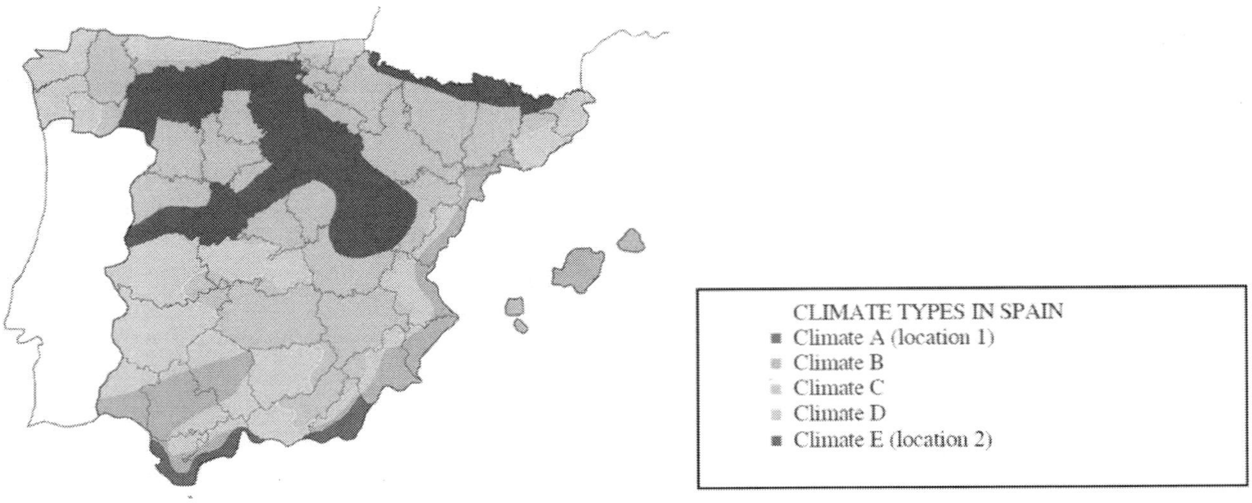

Figure 2: Climate types in Spain according to CTE-DB-HE [38].

The thicknesses of the load bearing walls have been calculated in order to mechanically respond to the loads of the different levels. The insulation thickness has been calculated for all three types of materials and both climates considered in order to meet the standards for the building envelope U values. The software used to establish the structure of the models proposed is CYPECAD. Madrid (Spain) (2015) [37], which is used for the analysis and design of building structures in homes, buildings, and civil engineering projects that have horizontal and vertical loads. This program is adapted to international regulations and automatically generates hypotheses for any user-defined combination according to the stated premises. Users can also define their own project situations to personalize the combinations to be taken into account in the calculations for the structural elements of the project. Entering data, such as the physical parameters of the different materials and the building characteristics, in the software gives us the wall dimensions. The results are displayed in Table 1.

Table 1: Thickness of the load-bearing walls and insulation thickness for the different levels

External Wall Type	Building Level	Wall Thickness (m)	Insulation Thickness (m) PUR	
			LOCATION I	LOCATION II
Ceramic Brick (BW)	Level 3	0.250	0.02	0.04
	Level 2	0.250	0.02	0.04
	Level 1	0.375	0.01	0.03
Concrete Block (CW)	Level 3	0.200	0.02	0.04
	Level 2	0.200	0.02	0.04
	Level 1	0.400	0.02	0.04
Stabilized Soil (SW)	Level 3	0.300	0.02	0.04
	Level 2	0.350	0.02	0.03
	Level 1	0.450	0.01	0.03

According to Spanish regulations, CTE_DB-HE [38] the limitation for the building envelope U values are, for the warm Mediterranean climate (Location 1) zone A3: 0.94 $W/m^2{\cdot}K$, and for the continental climate (Location 2) zone E1: 0.57 $W/m^2{\cdot}K$. Polyurethane has been used as an insulation material and the thickness required varies between 2 and 4 cm.

Materials Used

To establish a standard of comparison, we have chosen common, and not so common, building materials widely used for a specific building typology. Such conventional materials are: fired clay brick masonry walls (BW) and concrete block masonry walls (CW). The unconventional material is a natural fiber bio-composite stabilized soil block, also used for masonry walls (SW).

Fired Clay Brick Walls (BW)

Fired bricks are among the oldest and most enduring of mankind's building materials. The current fired bricks require a considerable amount of thermal energy during the burning process because they are fired at temperatures between 1000 °C and 1200 °C, depending on the clay. Light-colored clays usually require higher firing temperatures than dark-colored ones. This translates into a thermal energy of 3.75–4.75 MJ per brick. An average value of 4.25 MJ per brick (standard size in Spain: 240 × 115 × 70 mm^3) has been considered for the comparison and computation of energy content of buildings and masonry.

Concrete Block Walls (CW)

Concrete blocks are light-weight/low-density blocks very commonly used for the construction of envelope walls in multi-storey buildings in many countries. The basic composition of the blocks consists of cement, sand, and coarse aggregates (less than 4 mm in size). The energy content of the block will mainly depend on the cement percentage. Energy spent for crushing of coarse aggregate will also contribute to the block energy. The cement percentage generally varies between 7% and 10% by weight. The quality of the block, particularly its compressive strength, is the deciding factor for cement percentage. Energy content of the concrete block of size 400 × 200 × 200 mm^3 will be in the range of 12.3–15.0 MJ.

Stabilized Soil Block Walls (SW)

The stabilized soil blocks considered in this research are made by the combination of clay soil, water, a natural polymer as a stabilizer, and animal fiber reinforcement. The polymer used is calcium alginate, which is added to the mixture in the proportion of 1.2% by weight [39]. Calcium alginate production is a chemical synthesized from wet, chopped seaweed adding calcium chloride and sodium carbonate. The animal fiber is wool, used cut and raw, without washing or processing; the proportion used is 0.25% by weight. Blocks are cured at room temperature. The energy consumption is mainly by transport and extracting since they are not fired or steam cured. Barber and Pellow [40] reported results for the impact category of Energy Use for the on-farm phase of production for merino wool of 13.42 MJ per kg greasy wool or 22.55 MJ per kg wool fiber. The wool process gives the greatest uncertainty in energy use scouring and since this contributed almost 90% of the total processing. Processing was the single largest use of energy in the supply chain but the authors recognize the great variability in the results for the processing phase. For the block manufacturing, the wool used is unprocessed raw material (see Table 2).

Table 2: Energy use and greenhouse gas emissions (GWP) for 1 tone merino wool top as reported by Barber and Pellow [40]

Impact Category	On Farm	Processing	Transport	Total
Total energy (MJ/t wool top)	22,550	21,700	1490	45,730
GWP (T CO_2-eq/t wool top)	1655	471	103	2229

LIFE-CYCLE ASSESSMENT GOAL AND SCOPE

The particular focus of the application of the life-cycle assessment (LCA) in this study is to obtain the values of the embodied energy and global warming potential impacts (GWP) categories associated with the construction of three types of bearing walls: fired clay brick masonry walls (BW), concrete block masonry walls (CW), and stabilized soil block masonry walls (SW). A three storey construction is evaluated. To establish a suitable comparison framework, after calculating the necessary thickness of the walls, thermal conductivities have been unified in order to obtain equivalent operational energy parameters for all three materials.

According to the proposed framework, this study should answer the following question: what are the impacts produced by the processes related to the construction for each one of the combinations proposed?

According to the objective of this study the functional unit established is the total surface of walls in each case.

The assessed system is composed of every process that takes part in the production, construction, maintenance, deconstruction, and final disposal of every component of the building structure. It has excluded every process related to the operational phase of the dwelling. The system includes the following processes:

Manufacturing of the building products phase. For each building material involved in the building every good and service from cradle to grave are considered. The manufacturing of employed machinery and territorial infrastructure processes has been considered.

Assembly and construction phase. This covers every process aimed at integrating all products and services in the site in each studied dwelling. The transportation of building materials from the factory to the site, the placement of building products has been considered

Maintenance and repair phase. This includes all repair operations and maintenance of building components. The renewal of those materials which have a lower durability has been considered.

Dismantling and demolition phase. Every process carried out at the end of the life of the building to remove and demolish the dwelling has been taken into consideration: demolition, removal of building elements, and transportation of demolition materials to recycling or disposal have been included.

Disposal and recycling phase. This covers all processes suffered by demolition materials after dismantling *i.e.*, the deconstruction of building materials.

The environmental data of wool and algae have been extracted from the recent studies conducted by Barber and Pellow [40] and Resurreccion *et al.* [41], respectively. The environmental data of the rest of the building materials were obtained from the well-recognized ECOINVENT database [42]. The calculation procedure to obtain the life cycle inventory was the described by García-Martínez [43]:

1. Identification and quantification of the initial building products and auxiliary materials—including replacement materials that take part in the life cycle.
2. Identification and quantification of the basic processes associated with the construction and deconstruction. The determination of the energy consumed during the construction and demolition is obtained as a factor of the total building material volume, following the procedure as described by Kellenberger *et al.* [44]. The following procedure has been taken:

a. Basic materials have been grouped into unit processes (Table 3). Construction systems, structural elements, walls and roofs, windows, doors, and finishing materials (from floors, ceiling, and walls) has been considered. The dimensioning of these elements is according to the obtained structural and thermal values (Table 1).

Table 3: Inventory list of the materials used and corresponding name in the ECOINVENT database. Unitary values for Global Warming Potential and Embodied Energy

Component	ID	Name (ECOINVENT)	Unit	Global Warming Potential 100a CML 2001 kg CO_2-eq	Embodied Energy Cumulative Energy Demand MJ
Gravel	464	gravel, round, at mine	kg	0,0024	0,0576
Limestone	466	limestone, at mine	kg	0,0019	0,0283
Concrete	504	concrete, normal, at plant	m^3	265,2200	1447,2335
Bricks	512	ceramic tiles, at regional storage	kg	0,7651	14,9540
Plaster Board	517	gypsum plaster board, at plant	kg	0,3600	6,2652
Mortar	537	cement mortar, at plant	kg	0,1953	1,5182
Clay plaster	538	clay plaster, at plant	kg	0,0195	0,5181
Glass	805	flat glass, coated, at plant	kg	0,6855	15,0414
Aluminium	1059	aluminum, secondary, from new scrap, at plant	kg	0,4102	8,4313
Reinforcing steel	1141	reinforcing steel, at plant	kg	1,3442	20,9352
Steel	1154	steel, low-alloyed, at plant	kg	1,6294	26,1590
Paint	1670	alkyd paint, white, 60% in solvent, at plant	kg	2,5115	84,3073
Bitumen	1814	bitumen sealing, at plant	kg	1,0177	51,0742
Polyethylene	1829	polyethylene, HDPE, granulate, at plant	kg	1,8921	79,8534
Polyurethane	1838	polyurethane, flexible foam, at plant	kg	4,4208	101,2692
Polyvinylchloride	1840	polyvinylchloride, at regional storage	kg	2,1625	59,0158
Rubber	1847	synthetic rubber, at plant	kg	3,1972	101,3493
Water	2288	tap water, at user	kg	0,0003	0,0062
Wood	2507	sawn timber, softwood, planed, kiln dried, at plant	m^3	713,1300	12792,1890
Sand	464	gravel, round, at mine (Tierra)	kg	0,0024	0,0576
Algae *	A001	Algae, at regional storehouse	kg	0,0200	20,0000
Sheep Wool **	P001	Wool mat, at plant	kg	0,9850	13,4200
Electricity (Const-Dem)	698	electricity mix	kWh	0,5004	10,9038
Diesel (Const-Dem)	559	diesel, burned in building machine	MJ	0,0910	1,3799
Transport (Const-Dem)	1943	transport, lorry 32 t	tkm	0,1663	2,8149
Electricity (Operational)	698	electricity mix	kWh	0,5004	10,9038
Natural Gas (Operational)	1327	natural gas, high pressure, at consumer	MJ	0,0101	1,1359
Brick final disposal	2005	disposal, building, brick, to final disposal	kg	0,0141	0,3110
Concrete final disposal	2007	disposal, building, cement (in concrete) and mortar, to final disposal	kg	0,0148	0,3217
Glass final disposal	2019	disposal, building, glass sheet, to final disposal	kg	0,0108	0,2614
Polyurethane final disposal	2040	disposal, building, polyurethane foam, to final disposal	kg	2,4699	1,3770
Polyvinylchloride final disposal	2043	disposal, building, polyvinylchloride products, to final disposal	kg	2,2223	12,5207
Steel final disposal	2048	disposal, building, reinforcement steel, to final disposal	kg	0,0678	1,1252
Wood final disposal	2052	disposal, building, waste wood, untreated, to final disposal	kg	1,4743	0,2038
Aluminium final disposal	2090	disposal, aluminium, 0% water, to municipal incineration	kg	0,0369	0,7668
Bitumen final disposal	2217	disposal, bitumen, 1.4% water, to sanitary landfill	kg	0,1079	0,3407
Inert material final disposal	2221	disposal, inert material, 0% water, to sanitary landfill	kg	0,0128	0,3330

* Data provided by Resurreccion *et al.* [41]. ** Data provided by Barber and Pellow [40].

b. Division into groups and listing product specifications. Each case study has been divided into building elements according to the Building Cost Data Base of Andalusia (BCCA). The materials, the building machinery, and the labor has been related to each building element.

c. he building elements have been quantified using the construction management software Presto V.8 by RIB Software AG, Stuttgart (Germany).

d. The basic materials used in each case study has been obtained from the results given by Presto.

e. Once quantify each basic material and considering its physics properties, its mass and volume has been obtained.

3. Determination of input and output of each unit process. The ECOINVENT database and published LCA studies have been used to obtain environmental information of unit processes (see Table 3). Final disposal processes for the plastics, metals, bitumen-based and wood materials have been considered. Other materials have been considered inert from the point of view of their final disposal. The quantification of the final disposal processes has been obtained from the quantities the initial basic products (Table 4).

Table 4: Inventory list of the materials used for every type of building system and climate

Stage	Component	Unit	BW		CW		SW	
			LOCATION I	LOCATION II	LOCATION I	LOCATION II	LOCATION I	LOCATION II
Manufacture PH_1	Gravel	kg	199.316,16	199.316,16	199.316,16	199.316,16	199.316,16	199.316,16
	Limestone	kg	776,96	776,96	776,96	776,96	776,96	776,96
	Concrete	m^3	158,36	158,36	185,97	185,97	158,36	158,36
	Bricks	kg	196.924,82	196.924,82	83.044,34	83.044,34	83.044,34	83.044,34
	Plaster Board	kg	10.071,90	10.071,90	10.071,90	10.071,90	10.071,90	10.071,90
	Mortar	kg	86.316,30	86.316,30	20.461,47	20.461,47	10.446,01	10.446,01
	Clay plaster	kg	19.666,60	19.666,60	19.666,60	19.666,60	19.666,60	19.666,60
	Glass	kg	3.583,13	3.583,13	3.583,13	3.583,13	3.583,13	3.583,13
	Aluminium	kg	2.329,44	2.329,44	2.329,44	2.329,44	2.329,44	2.329,44
	Reinforcing steel	kg	6.550,74	6.550,74	6.550,74	6.550,74	6.550,74	6.550,74
	Steel	kg	314,62	314,62	314,62	314,62	314,62	314,62
	Paint	kg	8.263,95	8.263,95	8.263,95	8.263,95	8.263,95	8.263,95
	Bitumen	kg	258,39	258,39	258,39	258,39	258,39	258,39
	Polyethylene	kg	45,81	45,81	45,81	45,81	45,81	45,81
	Polyurethane	kg	155,93	27,46	187,11	467,78	155,93	311,85
	Polyvinylchloride	kg	27,46	27,46	27,46	27,46	27,46	27,46
	Rubber	kg	79,06	79,06	79,06	79,06	79,06	79,06
	Water	kg	34.664,05	34.664,05	26.845,78	26.845,78	101.112,30	101.112,30
	Wood	m^3	4,11	4,11	4,11	4,11	4,11	4,11
	Sand	kg	0,00	0,00	0,00	0,00	226.324,80	226.324,80
	Algae	kg	0,00	0,00	0,00	0,00	2.715,90	2.715,90
	Sheep Wool	kg	0,00	0,00	0,00	0,00	565,81	565,81

Construction and Demolition PH_2	Electricity (Const-Dem)	kWh	34.058,76	34.501,07	29.467,66	30.131,13	31.651,75	32.762,28
	Diesel (Const-Dem)	MJ	286.093,61	289.808,99	247.528,31	253.101,48	29.737,79	275.203,15
	Transport (Const-Dem)	tkm	111.348,48	111.370,00	97.721,92	97.754,20	94.982,97	97.186,61
Operational PH_5	Electricity (Operational)	kWh	767.477,57	814.227,25	767.477,57	814.227,25	767.477,57	814.227,25
	Natural Gas (Operational)	MJ	3.867.080,73	7.658.781,91	3.867.080,73	7.658.781,91	3.867.080,73	7.658.781,91
Final Disposal PH_4	Brick final disposal	kg	196.924,82	196.924,82	83.044,34	83.044,34	83.044,34	83.044,34
	Concrete final disposal	kg	395.893,90	395.893,90	464.924,50	464.924,50	395.893,90	395.893,90
	Glass final disposal	kg	3.583,13	3.583,13	3.583,13	3.583,13	3.583,13	3.583,13
	Polyurethane final disposal	kg	155,93	343,04	187,11	467,78	155,93	311,85
	Polyvinylchloride final disposal	kg	27,46	27,46	27,46	27,46	27,46	27,46
	Steel final disposal	kg	6.865,36	6.865,36	6.865,36	6.865,36	6.865,36	6.865,36
	Wood final disposal	kg	2.874,80	2.874,80	2.874,80	2.874,80	2.874,80	2.874,80
	Aluminum final disposal	kg	2.329,44	2.329,44	2.329,44	2.329,44	2.329.44	2.329,44
	Bitumen final disposal	kg	258,39	258,39	258,39	258,39	258,39	258,39
	Inert material final disposal	kg	324.671,08	324.671,08	258.816,25	258.816,25	478.407,30	478.407,30

4. Inventory and assessment. The impact assessment is carried out using the CML 2001 method in relation to the GWP impact category. The "cumulative energy demand" in relation to the embodied primary energy (Table 3).
5. Operational energy data are considered according to the benchmarks included in the Spanish ministry report SPAHOUSE [45] for houses located in Spain, considering two of the climates mentioned, location 1, corresponding to the warm Mediterranean climate and location 2, corresponding to the inner continental areas of the peninsula. Including insulation, all building envelope U values are considered equal for all three materials and subsequently, the operational energy will be considered to be the same for all three cases. Insulation is placed inside the building envelope so as to minimize the differences of thermal inertia among them.

RESULTS

The results of the inventory list of the construction materials used for every type of construction system and climate are shown in Table 4.

The following phases have been considered for the LCA: manufacturing, (PH1); construction/demolition (PH2); transport (PH3); final disposal (PH4); operational energy (PH5). Table 4 shows the inventory list of the construction materials used for every stage, type of building construction system, and climate.

Figure 3, Figure 4, Figure 5 and Figure 6 show a graphic comparison for embodied energy values and the global warming potential of all three types of materials, with both locations considered.

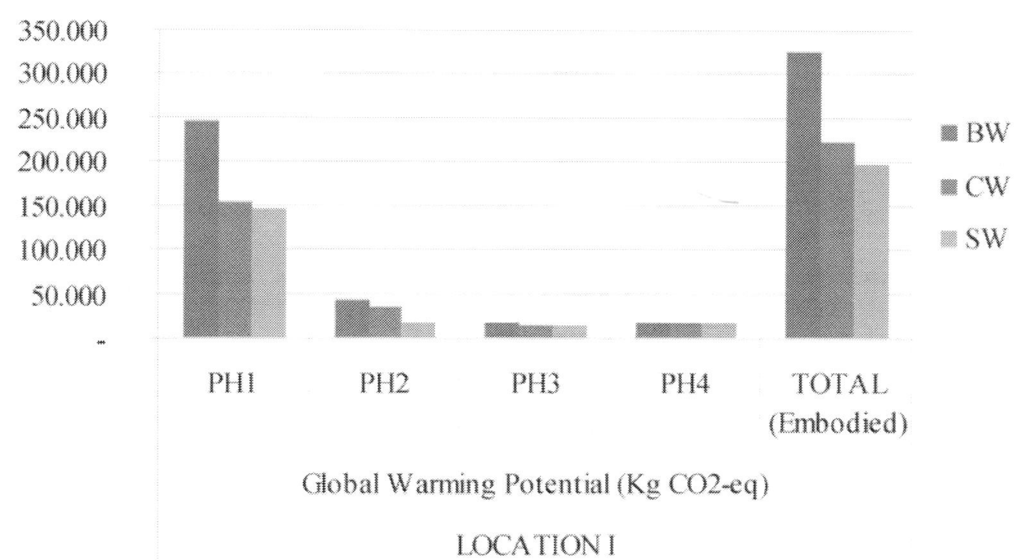

Figure 3: Global Warming Potential (GWP) 100a (Kg eq-CO_2) for each type of building construction system and Mediterranean climate (location 1). PH1: manufacturing; PH2: constructing–demolishing; PH3: transport; PH4: final disposal. BW: ceramic brick walls; CB: concrete block walls; SW: stabilized soil walls.

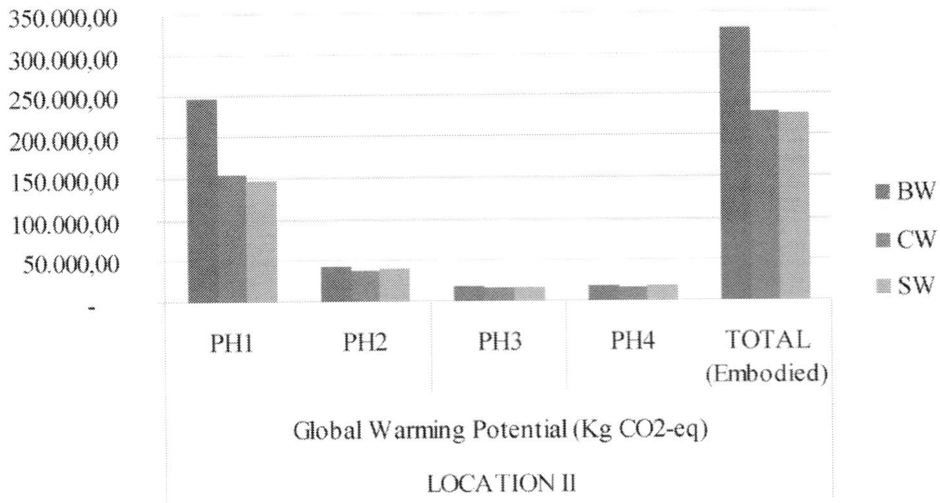

Figure 4: GWP 100a (Kg eq-CO_2) for each type of building construction system and continental climate (location 2). PH1: manufacturing; PH2: constructing–demolishing; PH3: transport; PH4: final disposal. BW: ceramic brick walls; CB: concrete block walls; SW: stabilized soil walls.

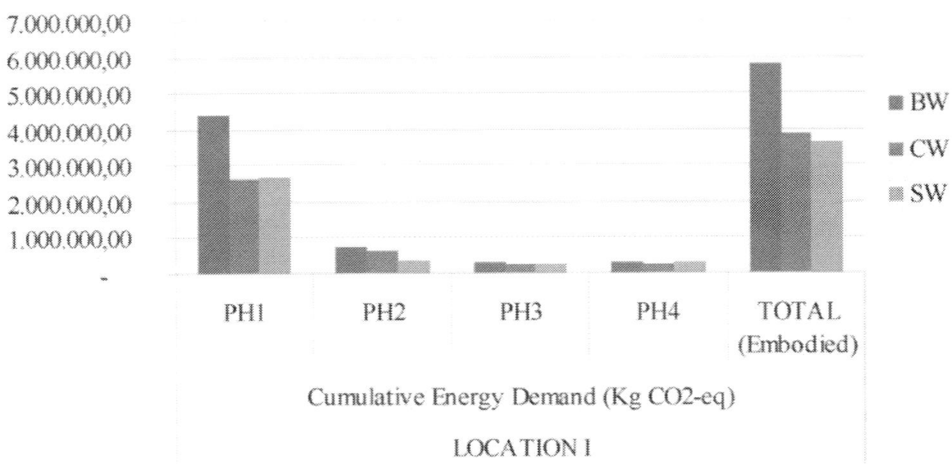

Figure 5: Cumulative energy demand (MJ) GWP 100a (Kg eq-CO_2) for each type of building construction system and continental climate (location 1). PH1: manufacturing; PH2: constructing–demolishing; PH3: transport; PH4: final disposal. BW: ceramic brick walls; CB: concrete block walls; SW: stabilized soil walls.

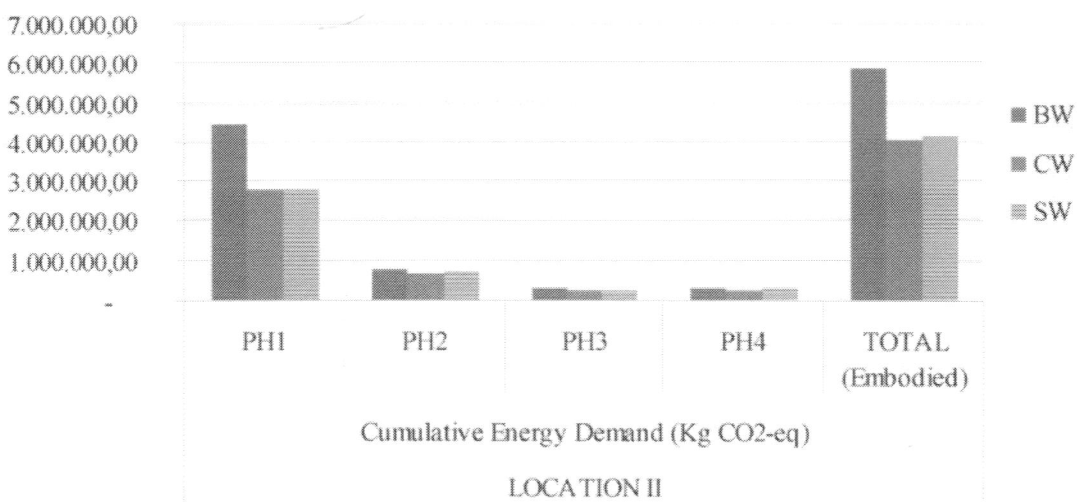

Figure 6: Cumulative energy demand (MJ) GWP 100a (Kg eq-CO_2) for each type of building construction system and continental climate (location 2). PH1: manufacturing; PH2: constructing–demolishing; PH3: transport; PH4: final disposal. BW: ceramic brick walls; CB: concrete blocks walls; SW: stabilized soil walls.

Table 5 shows an extensive list of environmental impact values shown by type of building construction system and climate location that allow a comparison of embodied *versus* operational energy in different materials and locations. Results are given globally for the entire building, per square meter and year.

Table 5: Mean environmental impact values showed by type of building construction system and climate location

Total Results		Global Warming Potential (kg CO_2-eq /m^2 year				Cumulative Energy Demand (MJ-eq/m^2 year)			
		PH1, PH4	PH2, PH3	Total Embodied	Total Operational PH5	PH1, PH4	PH2, PH3	Total Embodied	Total Operational PH 5
LOCATION I	Ceramic Bricks Wall (BW)	264.949,39	61.586,88	326.536,27	422.971,93	11.833,84	2.447,38	14.281,22	12.760.933,10
	Concrete Blocks Wall (CW)	171.385,34	53.515,04	224.900,38	422.971,93	8.014,74	2.104,76	10.119,50	12.760.933,10
	Stabilized Soil Wall (SW)	164.542,52	34.340,63	198.883,15	422.971,93	8.182,94	2.214,66	10.397,61	12.760.933,10
LOCATION II	Ceramic Bricks Wall (BW)	266.238,71	63.256,98	329.495,69	484.546,96	16.693,07	2.411,88	19.104,95	17.577.594,30
	Concrete Blocks Wall (CW)	174.206,50	54.115,19	228.321,69	484.546,96	8.106,53	2.139,21	10.245,74	17.577.594,30
	Stabilized Soil Wall (SW)	168.551,38	56.449,75	225.001,13	484.546,96	8.219,03	2.233,47	10.452,50	17.577.594,30
Results per m^2 and year		PH1, PH4	PH2, PH3	Total Embodied	Total Operational PH5	PH1, PH4	PH2, PH3	Total Embodied	Total Operational PH 5
LOCATION I	Ceramic Bricks Wall (BW)	11,95	2,78	14,73	19,07	0,53	0,11	0,64	575,48
	Concrete Blocks Wall (CW)	7,73	2,41	10,14	19,07	0,36	0,09	0,46	575,48
	Stabilized Soil Wall (SW)	7,42	1,55	8,97	19,07	0,37	0,10	0,47	575,48
LOCATION II	Ceramic Bricks Wall (BW)	12,01	2,85	14,86	21,85	0,75	0,11	0,86	792,69
	Concrete Blocks Wall (CW)	7,86	2,44	10,30	21,85	0,37	0,10	0,46	792,69
	Stabilized Soil Wall (SW)	7,60	2,55	10,15	21,85	0,37	0,10	0,47	792,69

DISCUSSION

Regarding cumulative energy demand and global warming potential (GWP), for the different LCA phases, a proportion can be seen between the consumption of the three different materials employed. Manufacturing phase PH1 is the most relevant phase for embodied energy for both CO_2-eq and MJ. The contribution of the manufacturing phase to these results is significant, representing mean percentages in relation to the total stages. In this phase there are significant differences that make the brick wall a more unfavorable material from energy consumption and emissions associated point of view, being more than 65% higher than the stabilized soil wall. The values of both soil and concrete block walls are quite similar. Nevertheless, in all cases the embodied energy is lower for the SW than for the other two materials except for the cumulative energy demand of location 2 due to greater energy employed in soil transportation.

Taking into account the different locations, there are no big differences between the values of embodied energy for both climates. The differences between materials are higher in global warming potential than in the cumulative energy demand.

If different phases are compared, both CO_2 emissions and cumulative energy demand, the greatest differences among both climates takes place in PH2. Analyzing Table 5, consumption data per m^2 and year, they show remarkable differences between SW and BW values of total embodied energy. CO_2emissions for BW rise up to 1.6 times higher in cold climates. For warm climates, BW exceeds 1.5 times the values of SW.

The operational energy emissions of the building are higher than the ones associated with the embodied energy for all three materials. Accordingly the operational energy emissions of the building represent more than 200% of the embodied energy for the SW case. However for BW operational energy represents only 130% of embodied energy.

Accordingly with these results it could be said that the energy consumed to build houses with brick walls in warm climates represents 165% of the energy invested to erect the same building with stabilized earth walls.

CONCLUSIONS

Analyzing the results obtained in the different phases of the lca, both for energy demand and emissions, it can be stated that the manufacturing phase of the construction materials is the most relevant chapter for both for cumulative energy demand and global warming potential of a building's embodied energy. Since this is the most relevant phase, determinations or modifications of it are fundamental, and so it is the material choice during the design phase of a building. According to the results obtained, approximately three SW homes could be erected with the same amount of CO_2emissions and cumulative energy demand for every two homes built with BW.

Future possible research directions that would improve or expand this work are related to the comparison with other types of construction solutions, not only for load-bearing walls, but for envelope blocks without structural function. Other building types, other climates, and geographical conditions could be evaluated. The potential recyclability of these materials could also be assessed, something that is not reflected in the limits of this LCA.

ACKNOWLEDGMENTS

The authors wish to acknowledge the IUACC "Instituto Universitario de Arquitectura y Ciencias de la Construcción" for the necessary support to develop this research. We would also like to express our gratitude to the Foundation INNOVARCILLA, also known as the Andalusian Ceramics Technology Centre.

This research project and the costs to publish in open access have been partially funded by the TEP-1988 project PANEL of the 2012 call of the Dirección General de Investigación,

Tecnología y Empresa. Consejería de Economía, Innovación, Ciencia y Empleo., Junta de Andalucía.

AUTHOR CONTRIBUTIONS

Carmen Galan-Marin, Carlos Rivera-Gomez and Antonio Garcia-Martinez conceived and designed the experiments; Carmen Galan-Marin performed the experiments; Carmen Galan-Marin and Carlos Rivera-Gomez analyzed the data and wrote the paper.

CONFLICTS OF INTEREST

The authors declare no conflict of interest. The founding sponsors had no role in the design of the study; in the collection, analyses, or interpretation of data; in the writing of the manuscript, and in the decision to publish the results.

REFERENCES

1. Adoption of the Paris Agreement—Proposal by the President—Draft Decision -/CP.21 2015. Available online: http://unfccc.int/resource/docs/2015/cop21/eng/l09r01.pdf (accessed on 19 April 2016).
2. Cabeza, L.F.; Rincón, L.; Vilariño, V.G.; Pérez, G.; Castell, A. Life cycle assessment (LCA) and life cycle energy analysis (LCEA) of buildings and the building sector: A review. *Renew. Sustain. Energy Rev.* 2014, *29*, 394–416.
3. Uihlein, A.; Eder, P. *Towards Additional Policies to Improve the Environmental Performance of Buildings*; EUR 23775 EN—2009; JRC Scientific and Technical Report: Seville, Spain, 2009.
4. Venkatarama Reddy, B.V.; Jagadish, K.S. Embodied energy of common and alternative building materials and technologies. *Energy Build.* 2003, *35*, 129–137.
5. Vázquez, M. Building and Impact on the environment: The case of the earth and other materials (Construcción e impacto sobre el ambiente: El caso de la tierra y otros materlales). *Inf. Constr.*2001, *52*, 29–43.
6. Janssen, G.M.T.; Hendriks, C.F. Sustainable use of recycled materials in building construction.*Adv. Build. Technol.* 2002, *II*, 1399–1406.
7. Walker, P. Strength, durability and shrinkage characteristics of cement stabilised soil blocks.*Cem. Concr. Compos.* 1995, *17*, 301–310.
8. Fitzmaurice, R. *Manual on Stabilised Soil Construction for Housing*; United Nations: New York, NY, USA, 1958.
9. Walker, P.; Venkatarama Reddy, B.V.; Mesbah, A.; Morel, J.C. The case for compressed Earth block construction. In Proceedings of the 6th International Seminar on Structural Masonry for Developing Countries, Bangalore, India, 11–13 October 2000; Allied Publishers Ltd.: New Delhi, India, 2000; pp. 27–35.
10. Shukla, A.; Tiwari, G.N.; Sodha, M.S. Embodied energy analysis of adobe house. *Renew. Energy*2009, *34*, 755–761.
11. Muneer Asif, M.; Kelley, R. Life cycle assessment: A case study of a dwelling home in Scotland.*Build. Environ.* 2007, *42*, 1391–1394.
12. Huberman, N.; Pearlmutter, D. A life-cycle energy analysis of building materials in the Negev desert. *Energy Build.* 2008, *40*, 837–848.
13. Thormark, C. A low energy building in a life cycle e its embodied energy, energy need for operation and recycling potential. *Build. Environ.* 2002, *37*, 429–435.
14. Blengini, G.A. Life cycle of buildings, demolition and recycling potential: A case study in Turin, Italy. *Build. Environ.* 2009, *44*, 319–330.
15. Chen, T.Y.; Burnett, J.; Chau, C.K. Analysis of embodied energy use in the residential building of Hong Kong. *Energy* 2001, *26*, 323–340.
16. Yildiz, A.; Güngör, A. Energy and exergy analyses of space heating in buildings. *Appl. Energy*2009, *86*, 1939–1948.
17. Arena, A.P.; Rosa, C. Life cycle assessment of energy and environmental implications of the implementation of conservation technologies in school buildings in Mendoza–Argentina. *Build. Environ.* 2003, *38*, 359–368.
18. De Miranda, F. Inventario de ciclo de Vida da Geracao Hidrelectrica no Brasil—Usina de Itaipu: Primeira Aproximacao 2003. Universidade de Sao Paulo. Available online: http://www.maternatura.org.br/hidreletricas/biblioteca_docs/ICV%20Itaipu.pdf (accessed on 7 August 2012).
19. Sartori, I.H. Energy use in the life cycle of conventional and low-energy buildings: A review article. *Energy Build.* 2007, *39*, 249–257.
20. Ortiz, O.; Castells, F.; Sonnermann, G. Sustainability based on LCM of residential dwellings: A case study in Catalonia, Spain. *Build. Environ.* 2009, *44*, 584–594.
21. Ortiz, O.; Castells, F.; Sonnermann, G. Sustainability Operational energy in the life cycle of residential dwellings: The experience of Spain and Colombia. *Appl. Energy* 2010, *87*, 673–680.
22. Chung, W.S.; Tohno, S.; Shim, S.Y. An estimation of energy and GHG emission intensity caused by energy consumption in Korea: An energy IO approach. *Appl. Energy* 2009, *86*, 1902–1914.
23. Thormark, C. The effect of material choice on the total energy need and recycling potential of a building. *Build. Environ.* 2006, *41*, 1019–1026.
24. Stephan, A.; Stephan, L. Life cycle energy and cost analysis of embodied, operational and user-transport energy reduction measures for residential buildings. *Appl. Energy* 2016, *161*, 445–464.
25. Koesling, M.; Ruge, G.; Fystro, G.; Torp, T.; Hansen, S. Embodied and operational energy in buildings on 20 Norwegian dairyfarms—Introducing the building construction approach to agriculture. *Energy Build.* 2015, *108*, 330–345.
26. Guan, L.; Walmsely, M.; Chen, G. Life Cycle Energy Analysis of Eight Residential Houses in Brisbane, Australia. *Procedia Eng.* 2015, *121*, 653–661.
27. Brown, N.; Olsson, S.; Malmqvist, T. Embodied greenhouse gas emissions from refurbishment of

residential building stock to achieve a 50% operational energy reduction. *Build. Environ.* 2014,*79*, 46–56.

28. Zhu, J.; Chew, D.; Lv, S.; Wu, W. Optimization method for building envelope design to minimize carbon emissions of building operational energy consumption using orthogonal experimental design (OED). *Habitat Int.* 2013, *37*, 148–154.
29. Pinky Devi, L.; Sivakumar, P. A case study on life cycle energy use of residential buildingin Southern India. *Energy Build.* 2014, *80*, 247–259.
30. Praseeda, K.I.; Reddy, B.V.V.; Mani, M. Embodied and operational energy of urban residential buildings in India. *Energy Build.* 2016, *110*, 211–219.
31. Stephan, A.; Crawford, R.H.; Myttenaere, K. Towards a comprehensive life cycle energy analysis framework for residential buildings. *Energy Build.* 2012, *55*, 592–600.
32. Jollands, M.; Setunge, S.; Iftekhar, A.; Haque, N.; Islama, H. Life cycle assessment and life cycle cost implications of wall assemblages designs. *Energy Build.* 2014, *84*, 33–45.
33. Iddon, C.R.; Firth, S.K. Embodied and operational energy for new-build housing: A case study of construction methods in the UK. *Energy Build.* 2013, *67*, 479–488.
34. Ibn-Mohammed, T.; Greenough, R.; Taylor, S.; Ozawa-Meida, L.; Acquaye, A. Operational *vs.* embodied emissions in buildings—A review of current trends. *Energy Build.* 2013, *66*, 232–245.
35. ISO 14040. *Environmental Management—Life Cycle Assessment—Principles and Framework, CEN*; ISO: Brussels, Belgium, 2006.
36. ISO 14044. *Environmental Management—Life Cycle Assessment—Requirements and Guidelines, CEN*; ISO: Brussels, Belgium, 2006.
37. CIPECAD. *CYPE Ingenieros S.A.*; CIPECAD: Madrid, Spain, 2015.
38. CTE_DB-HE. Código Técnico de la Edificación. Documento Básico de Ahorro de Energía. Available online: http://161.111.13.202/apache2-default/cte/CTE_DB-HE.pdf (accessed on 19 April 2016).
39. Galán-Marín, C.; Rivera-Gómez, C.; Petric-Gray, J. Effect of animal fibres reinforcement on stabilized earth mechanical properties. *J. Biobased Mater. Bioenergy* 2010, *4*, 121–128.
40. Barber, A.; Pellow, G. *Life Cycle Assessment: New Zealand Merino Industry, Merino Wool Total Energy Use and Carbon Dioxide Emissions*; The Agribusiness Group: Canterbury, New Zealand, 2006.
41. Resurreccion, E.P.; Colosi, L.M.; White, M.A.; Clarens, A.F. Comparison of algae cultivation methods for bioenergy production using a combined life cycle assessment and life cycle costing approach. *Bioresource Technol.* 2012, *126*, 298–306. [PubMed]
42. *ECOINVENT Database*; The Ecoinvent Center, A Competence Centre of ETH; PSI; Empa & ART: Switzerland; Available online: http://www.ecoinvent.org (accessed on 19 April 2016).
43. Garcia-Martinez, A. Life Cycle Assessment (LCA) of Buildings. Methodological Proposal for the Development of Environmental Declarations of Dwellings in Andalusia. Doctoral Dissertation, Universidad de Sevilla, Sevilla, Spain, 2010. Available online: http://fondosdigitales.us.es/media/thesis/1546/Q_Tesis_AGM.pdf (accessed on 19 April 2016).
44. Kellenberger, D.; Althaus, H.-J.; Jungbluth, N.; Künniger, T.; Lehmann, M.; Thalmann, P. *Life Cycle Inventories of Building Products. Data v1.1. Ecoinvent Report No 7*; Swiss Centre for Life Cycle Inventories: Dübendorf, Switzerland, 2004.
45. *SPAHOUSEC PROJECT Análisis del Consumo Energético del Sector Residencial en España INFORME FINAL*; IDAE Secretaría General Departamento de Planificación y Estudios: Madrid, Spain, 2011; pp. 1–76. Available online: http://www.idae.es/uploads/documentos/documentos_informe_spahousec_acc_f68291a3.pdf (accessed on 19 April 2016).

CHAPTER 17

STRENGTHENING OF MASONRY COLUMNS WITH BFRCM OR WITH STEEL WIRES: AN EXPERIMENTAL STUDY

Marinella Fossetti[1] and Giovanni Minafò[2]

[1]Facoltà di Ingegneria e Architettura, University of Enna "Kore", 94100 Enna, Italy

[2]Dipartimento di Ingegneria Civile, Ambientale, Aerospaziale, dei Materiali (DICAM), University of Palermo, 90128 Palermo, Italy

ABSTRACT

Nowadays, innovative materials are more frequently adopted for strengthening historical constructions and masonry structures. The target of these techniques is to improve the structural efficiency with retrofitting methods while having a reduced aesthetical impact. In particular, the use of basalt fiber together with a cementitious matrix emerges as a new technique. This kind of fiber is obtained by basalt rock without other components, and consequently it could be considered a natural material, compatible with masonry. Another innovative technique for strengthening masonry columns consists of applying steel wires in the correspondence of mortar joints. Both techniques have been recently proposed and some aspects of their structural performances are still open. This paper presents the results of an experimental study on the compressive behavior of clay brick masonry columns reinforced either with Basalt Fiber–Reinforced Cementitious Matrix (BFRCM) or with steel wire collaring. Uniaxial compressive tests were performed on eight retrofitted columns and four control specimens until failure. Two masonry grades were considered by varying the mix used for the mortar. Results are presented and discussed in terms of axial stress-strain curves, failure modes and crack patterns of tested specimens. Comparisons with unreinforced columns show the capability of these techniques in increasing ductility with limited strength enhancements.

INTRODUCTION

The use of Fiber-Reinforced Polymer (FRP) composites as wraps for upgrading existing masonry columns has become increasingly popular in recent years. These wrapping techniques allow obtaining large increases of the strength and deformation capacity, and are effective for columns with poor structural features. FRP wrapping allows an outstanding combination of properties, such as ease of handling, speed of installation and the use of a material with a high strength-to-weight ratio. Despite these advantages, some critical issues are involved in this method, including ensuring a durable bond between FRP and masonry and the stress concentration near the column's edges.

For these reasons, several research works have been carried out in the recent past with the aim of investigating the compressive behavior of FRP-confined masonry columns. One of the first studies on the effect of external FRP wrapping on masonry columns was carried out by Corradi *et al.* [1]. Researchers performed compressive tests on FRP-confined clay brick columns and proposed an adaptation of the model of Campione and Miraglia [2] for masonry members. Similarly, Aiello *et al.* [3] performed compressive tests on masonry columns built with clay or calcareous blocks, and reinforced

Citation: Marinella Fossetti and Giovanni Minafò, Strengthening of Masonry Columns with BFRCM or with Steel Wires: An Experimental Study, doi:10.3390/fib4020015.

with external FRP wrapping or internal FRP bars. Based on obtained results, they proposed a calibration of some design expressions for obtaining the strength of confined masonry columns. Di Ludovico *et al.* [4] presented the results of an experimental investigation on the compressive behavior of tuff or clay brick masonry columns confined with carbon or glass FRP and proposed new expressions for evaluating the strength enhancement in these members. Faella *et al.* [5] tested 54 natural stone and artificial brick masonry columns under uniaxial compression and investigated the effects of three different kinds of composite systems applied for confinement. Researchers observed several failure modes in the various tests and values of the confined-to-unconfined strength ratio ranging between 1.22 and 3.94.

As general findings, most of the experimental studies pointed out that the strength increase is generally higher in the case of weak masonry columns and for stiffer wrapping layers, confirming the influence of the basic parameters on the behavior of confined members; the stress concentration at the corners induces brittle failure, and it does not allow the column to reach the maximum confining stress; and the bond between masonry and FRP depends on the kind of resin and tissue adopted.

From a theoretical point of view, recent studies investigated the strength increase due to FRP confinement of masonry columns. Koksal *et al.* [6] developed a Drucker-Prager–type constitutive model for the nonlinear finite element analysis of FRP-confined masonry columns, while Lignola *et al.*[7] summarized previous literature studies and proposed a theoretical approach based on the Mohr-Coulomb strength criterion for determining the strength increase due to FRP confinement. Despite the advantages of this method highlighted by this large amount of research works, some drawbacks were spotted when organic resins were used to bind the fibers, such as poor behavior at high temperatures, inapplicability on wet surfaces, high costs and problems of compatibility between materials for long-term effects.

To solve these problems, different techniques were developed to be used as alternative methods for enhancing the structural performances of poor masonry columns. Among these, the use of wraps with Basalt Fiber–Reinforced Cementitious Matrix (BFRCM) allows obtaining an external surface that is easy to finish, and with a material compatible with that of the column. In fact, basalt fibers are obtained from basalt rocks through a melting process and contain no other chemical components in the manufacturing process, and thus are low in cost. Basalt fibers show mechanical features comparable to that of glass fibers and exhibit good resistance to chemical and high temperature exposure [8]. The ultimate strain of basalt is higher compared to other common FRP materials and thus it is interesting to use this advantage in column strengthening to enhance the seismic performance. There is little research concerning the application of basalt fibers in civil engineering and their strengthening efficiency on masonry elements. Yilmaz *et al.* [9] investigated the efficacy of external confinement with open-grid basalt-reinforced mortar on the axial behavior of masonry columns by testing five specimens. Their results showed that the open-grid BFRCM provided a fairly small compressive strength gain but, on the other hand, basalt jacket–reinforced mortar improves energy dissipation of the masonry columns notably. Similarly, Carloni *et al.* [10] performed an experimental investigation on the compressive behavior of polyparaphenylene benzobisoxazole (PBO) FRCM–reinforced masonry columns, finding small increments of strength but a noticeable increase of ultimate strain.

Another alternative method is the application of steel wires in the correspondence of mortar joints, to provide overlapping hoops at every course over the entire height of the column. The main advantages of this technique are related to the low costs and the reduced aesthetical impact, which makes this method suitable for buildings of historical and/or architectural interest. The idea was initially proposed by Jurina [11,12], who performed an experimental investigation, showing the capabilities of the technique in enhancing the axial capacity of the column.

Borri *et al.* [13,14] developed the technique proposed in [11] by using prestressed high-strength steel cords, applied with mortar or with resin to masonry columns with various cross-section shapes. Strength gains of up to three times were recorded, with higher values for octagonal sections, while square and rectangular specimens proved to be more sensitive to stress concentrations at the corners.

This paper presents the results of an experimental comparative analysis on the compressive behavior of confined masonry columns reinforced with different techniques. In particular, four clay brick masonry columns wrapped with BFRCM (series BF) and four specimens strengthened by high-strength steel wires placed in the joints (series SW) were tested until failure.

All the columns were manufactured with binding mortars typical of existing masonry buildings, which require structural retrofitting. In particular, two masonry strengths were targeted by varying the mortar grade and composition: series M1 for low-strength mortar, and series M3 for medium-strength mortar. Adopted nomenclature follows the definition of mortar classes of Eurocode 6 [15]. In addition to the reinforced specimens, two unconfined clay brick columns were prepared as control specimens (denoted as U) for each series, M1 and M3, of specimens.

In this paper, after describing the geometry and construction of test specimens, the characterization of constituent materials is presented by discussing the results of bending tests on mortar samples and compressive tests on clay brick and mortar cubes. The preparation of test specimens is also described in detail, by examining the construction stages of the specimens. Results of the compressive tests are finally presented and discussed by analyzing the axial stress-strain curves, strength and ultimate strain gains and failure modes. Results of performed tests allow the examination of the efficiency of the investigated reinforcing techniques, BFRCM wrapping and steel wire collaring, in a comparative manner. Results obtained could be useful to clarify the structural performances of these strengthening techniques, which could be efficient and economical alternatives to traditional FRP wrapping.

MATERIALS AND METHODS

Specimens

All specimens were about 960 mm in height and with square cross-section shape of 230 mm × 230 mm (Figure 1). The identification of unreinforced and reinforced specimens is reported in the first column of Table 1 and Table 2, respectively. In particular, the first two letters identify the mortar grade, while the third and the four letters identify the strengthening technique (U for unreinforced specimens, BF for basalt fiber–reinforced specimens, SW for columns strengthened with steel wires). For example, the specimen M1_BF is a column made with binding mortar M1 and wrapped by BFRCM. A final code number (1 or 2) has been used for identical specimens to distinguish the columns.

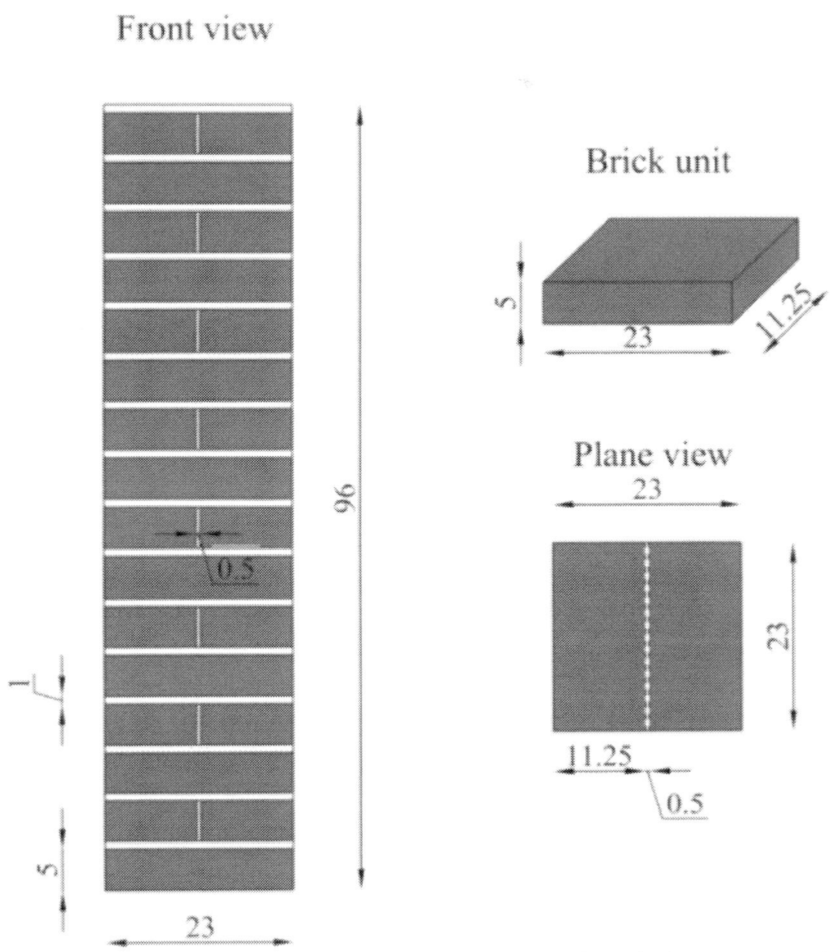

Figure 1: Geometry of test specimens (units in cm).

Table 1: Details of unconfined brick masonry columns (control specimens) and key results

Specimen	Mortar	Reinforcement	Maximum Stress f_{co} (N/mm^2)	Ultimate Stress f_{uo} (N/mm^2)	Axial Strain at Maximum Stress ε_{co}	Ultimate Axial Strain ε_{uo}	Absorbed Energy E_o (MJ/m^3)
M1_U_1	M1	NONE	4.31	3.74	0.0145	0.0169	0.0415
M1_U_2	M1	NONE	5.05	4.20	0.0139	0.0167	0.0491
	Average value		**4.68**	**3.97**	**0.0142**	**0.0168**	**0.0453**
M3_U_1	M3	NONE	9.14	7.53	0.0120	0.0132	0.0685
M3_U_2	M3	NONE	9.70	8.23	0.0128	0.0147	0.0739
	Average value		**9.42**	**7.88**	**0.0124**	**0.0139**	**0.0712**

Table 2: Details of confined brick masonry columns and key results

Specimen	Mortar	Reinforcement	Normalized Maximum Stress f_{cc}/f_{co}	Normalized Ultimate Stress f_{cu}/f_{uo}	Normalized Axial Strain at Maximum Stress $\varepsilon_{cc}/\varepsilon_{co}$	Normalized Ultimate axial Strain $\varepsilon_{cu}/\varepsilon_{uo}$	Normalized Absorbed Energy E/E_o
M1_BF_1	M1	basalt fiber mesh	1.81	1.72	1.5842	1.6053	2.8046
M1_BF_2	M1	basalt fiber mesh	1.67	1.66	1.4093	1.4709	2.5259
M1_SW_1	M1	steel wires	1.30	1.30	2.4869	2.3753	3.9057
M1_SW_2	M1	steel wires	1.35	1.32	2.4176	2.7062	4.3391
M3_BF_1	M3	basalt fiber mesh	1.19	1.19	1.0575	1.0373	1.3959
M3_BF_2	M3	basalt fiber mesh	1.09	1.09	1.2579	1.2669	1.5027
M3_SW_1	M3	steel wires	1.02	1.03	1.3004	1.3217	1.5743
M3_SW_2	M3	steel wires	1.33	1.32	1.6178	1.6373	2.3014

Materials

In order to characterize the mechanical behavior of the clay brick units, 10 cubes with size equal to 50 mm were cut from the bricks and tested under uniaxial compression (Figure 2a). An average value of the cube compressive strength of about 42.3 MPa was recorded.

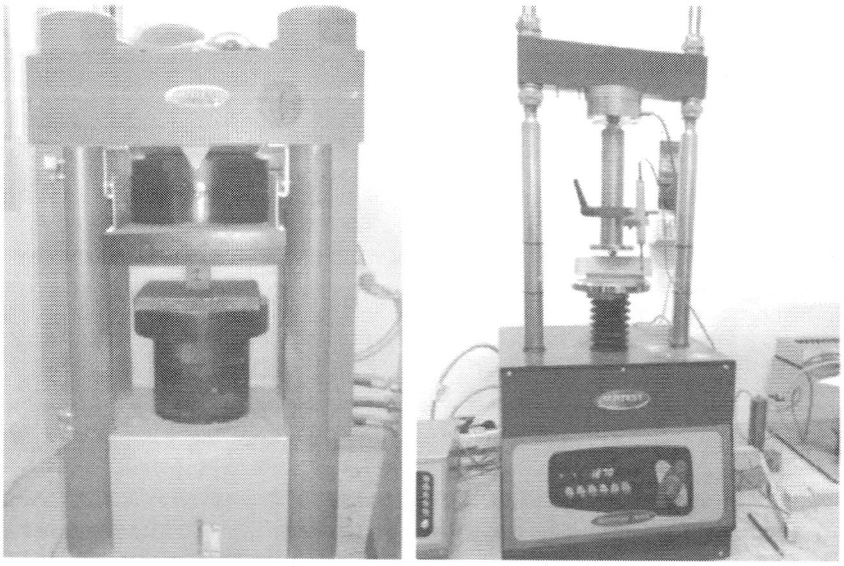

(a) (b)

Figure 2: Mechanical characterization of constituent materials: (**a**) compressive test on a clay brick cube; (**b**) three-point bending test on a mortar prism.

Mortar of M1 series was manufactured by assembling hydraulic lime and sand with a volume ratio of 1:3. The mortar used for the M3 mortar was composed of Portland Cement (14% in volume), hydraulic lime (14% in volume) and sand (72% in volume). Water was added until workability was achieved. The mechanical properties of the two mortar mixes were obtained by means of three-point bending tests on six standard 40 mm × 40 mm × 160 mm prisms (Figure 2b) and uniaxial compressive tests on six standard 40 mm × 40 mm × 40 mm cubes for each mortar grade series. The average 28-day tensile-flexural strength was 0.44 MPa for the M1 series and 3.13 MPa for the M3 series. The average 28-day compressive strength was equal to 0.55 MPa for M1 series and 4.54 MPa for M3 series.

The tissue adopted for reinforcing the specimens consists of an open mesh basalt fabric, with mesh openings having sizes equal to 6 mm × 6 mm. Its mechanical properties were deduced from the technical data provided by the producer, due to the difficulties in arranging a test set-up with the available machine. The tensile strength of the basalt fiber net was equal to 60 kN/m, ultimate strain was 1.8% and elastic modulus equal to 89 GPa.

Steel wires had a diameter equal to 1.8 mm. Also in this case, mechanical properties were provided by the producer. In particular, tensile strength was equal to 2845 MPa, Young modulus equal to 190 GPa and ultimate strain equal to 2.6%.

Preparation of Test Specimens

Columns were made up by assembling two brick units for the course, arranged orthogonally to the units in the upper and lower courses, and connected with mortar joints with a thickness of 10 mm. Specimens were reinforced after manufacturing and after a curing time of about two weeks.

The vertical edges of BF specimens were preliminary rounded with a corner radius of 20 mm for limiting stress concentration at the corners of the external wrap during the test. The basalt fabric was applied after initially treating the external surface with a high-strength cement-based mortar, reinforced with glass fibers (Figure 3a). A single layer of fabric was installed, with an overlapping length equal to the half perimeter of the section. This overlapping zone between the beginning and the end of the tissue plays a fundamental role on the efficacy of the method, since it allows the transfer of circumferential stresses along the external jacket. Finally, an additional cover of mortar was laid on the external surface after specimens were wrapped with the basalt fiber mesh, ensuring complete filling of voids and adhesion of the fabric. The overall thickness of the external reinforcing layer was equal to about 10 mm.

(a) (b)

Figure 3. Preparation of specimens: (**a**) basalt-reinforced specimen; (**b**) specimen reinforced with steel wires.

It should be noted that in this case continuous wrapping was allowed along the member's height due to the small scale of the specimens. When full-scale columns are reinforced in realistic conditions, a certain angle of the fibers with the horizontal axis (more or less like a helix) is necessary. It is clear that this angle influences the maximum available confinement pressure and the adoption of an in-elevation efficiency coefficient of confinement should be considered in calculations. This fact highlights the need for future studies on full-scale columns.

An easy procedure was followed to reinforce specimens with steel wires. Horizontal mortar joints of specimens were flattened mechanically and four high-strength steel wires with a diameter equal to 1.8 mm were introduced by collaring the joint (Figure 3b). Tightening of the wires was made by hand-pinch closing. Afterwards, mortar joints were filled with the above-mentioned fiber-reinforced mortar. Finally, every specimen was capped with a self-leveling mortar on the top surface, in order to warrant its plane regularity in contact to the plate of the testing machine.

Test Set-Up and Instrumentation

Compression tests on all columns were carried out with a hydraulic actuator with maximum load capacity of 2000 kN, which enables tests in displacement-controlled mode. An electronic control unit, together with a user interface via personal computer, regulated the test type and the equipment.

To measure the applied displacements of the samples, a linear voltage displacement transducer (LVDT) was placed between the press plates. Moreover, two transducers were installed in the axial and orthogonal direction at the middle height of each specimen. Figure 4 shows a typical test set-up. All specimens were loaded up to failure with small increments of displacement (0.5 mm/s).

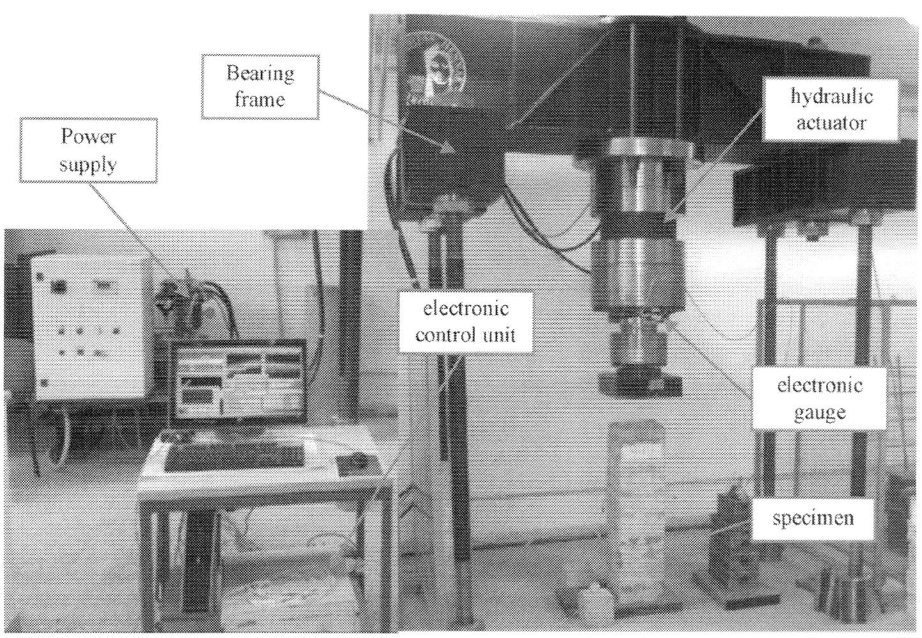

Figure 4: Test equipment and set-up.

RESULTS AND DISCUSSION

Unreinforced Columns

The monotonic stress-strain curves of the control specimens are shown in Figure 5. Axial stress has been calculated by dividing the axial load by the cross-section area. The axial strain is obtained from the displacement transducer placed between the press plates.

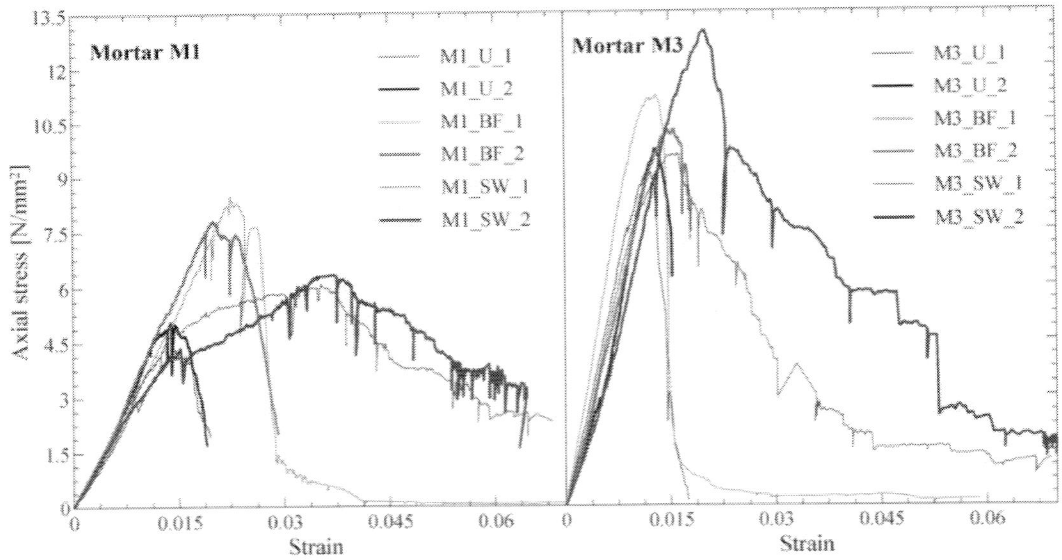

Figure 5: Axial stress-strain curves of unconfined and confined clay brick masonry columns.

Figure 5 plots the stress-strain curves for unreinforced clay brick masonry columns under axial compression characterized by an ascending branch up to the peak and by a rapidly descending post-peak branch until collapse (brittle failure). The peak stress f_{co} and the corresponding strain value ε_{co} are reported in Table 1 for each control specimen. Similarly, the ultimate or failure values are defined as (f_{uo}, ε_{uo}). The conventional ultimate strains were fixed as the strain values at 85% of the maximum stress. Also, the energy absorption capacity (E_o) measured for all specimens as $\Sigma(d\sigma \cdot d\varepsilon)$ is reported inTable 1.

It is worth noting that the use of high-grade binding mortars significantly influences the mean value of the maximum stress, ultimate stress and absorbed energy. For these parameters, increases of 101%, 98% and 57% were recorded by adopting the mortar grade M3 instead of M1.

Reinforced Columns

Table 2 shows the peak and ultimate stress, peak and ultimate strain, and absorbed energy values normalized to those relative to unreinforced specimens. They prove that strengthening by BFRCM wraps or by steel wires can enhance the performance of brick masonry columns with a substantial gain in strength, ductility and energy absorption capacity.

In particular, comparing the key results between reinforced and unreinforced columns (control specimens), the mean strength, ultimate strain and absorbed energy for series M1 increased respectively by 53%, 104% and 239%. In contrast, for series M3, the mean strength, ultimate strain and absorbed energy increased respectively by 16%, 32% and 69%.

In more detail, regarding series M1, the mean values of maximum stress, ultimate strain and absorbed energy increased respectively by 74%, 54% and 167% for BF specimens and by 33%, 154% and 312% for SW specimens. Similarly, for series M3 columns, they increased respectively by 14%, 15% and 45% for BF specimens and by 18%, 48% and 94% for SW specimens.

Concerning the overall effect of the two investigated techniques, an average increase of maximum stress of 44% was achieved for BF specimens, while it was equal to 25% for SW specimens. With reference to the ultimate strain and absorbed energy, observed increases were equal to 35% and 106% for BF specimens and 101% and 203% for SW specimens.

Failure Modes and Crack Patterns at Failure

In order to describe and compare the main damage patterns and the failure mechanisms for unconfined and confined columns, Figure 6 shows the failure mode observed for some specimens.

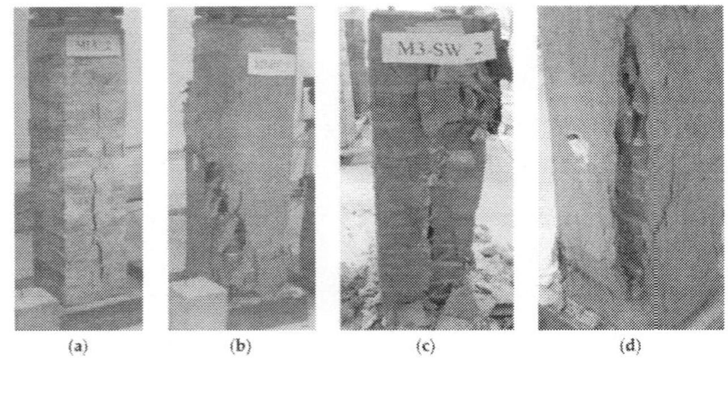

Figure 6: Failure mode for clay brick masonry columns: **(a)** unconfined; **(b)** confined with BFRCM wraps; **(c)** reinforced by steel wire collaring. **(d)** Tensile fracture of the basalt fiber mesh.

Unconfined specimens were characterized by brittle failure. In fact, the stress-strain curves of Figure 5 were characterized by an almost linear ascending branch up to the peak, followed by a rapidly descending post-peak branch until collapse.

Generally, BF specimens behaved monolithically in the first stage, up to a strain value equal to about one-third of the peak strain. During this phase, no crack was observed in the external mortar layer and this load stage corresponded to an initial linear ascending branch of the stress-strain curves of Figure 5. Afterwards, first cracking of the external layer occurred at the corners and the trend of the stress-strain curves assumed a non-linear shape up to the peak load. Failure of BFRCM-confined specimens occurred with the propagation of a unique fracture plane inclined at about 45 degrees across the mortar joints. Moreover, failure mechanisms occurred suddenly, when the jacket could not restrain the unstable expansion of the clay brick masonry core. In all cases the failure mechanisms were due to the jacket collapse at the corner.

Alternately, the behavior of the SW specimens was characterized by a slow damage process. First cracking occurred for a strain value similar to that recorded for unreinforced specimens. In particular, multiple sub-vertical cracks were observed in the brick units, while mortar joints were restrained by the wires. Progressive rupture of steel wires was also observed in the correspondence of knots and the failure of specimens occurred when all wires failed in one or two mortar joints.

CONCLUSIONS

This paper presented the results of an experimental investigation on the compressive behavior of clay brick masonry columns reinforced with BFRCM wraps or with steel collars in the mortar joints, varying the mortar grade. From obtained results and for the limits of investigated variables, the following main conclusions can be drawn:

- -BFRCM jacketing proved to be more effective in terms of strength enhancement, while significant increases of the ultimate strain and absorbed energy were observed mainly in specimens reinforced with steel wires;
- -the efficiency of both strengthening techniques depended on the mortar grade of the masonry, both being more effective for low-grade mortar masonry;
- -failure of the BFRCM layer was observed at the corners, highlighting the importance of ensuring an adequate corner radius to avoid stress concentration;
- -failure of the specimens reinforced with steel wires occurred due to the opening of the wires' knots, stressing the need for future studies on possible new joints to close the steel wires.

Further investigations should be addressed in the future in order to highlight the role of scale effect on full-size members and to draw more general conclusions.

ACKNOWLEDGMENTS

This work has been developed within the activities of the educational Project K.I.S.E.D.A.—*Kore International School of Earthquake engineering and Dynamic Analysis* financed by the Italian National Operative Programme—PON "R&C" 2007–2013 (Code: PONa3_00374/F1) to the University of Enna "Kore". Materials used for strengthening the specimens discussed in this paper have been funded by Mapei, Spa.

REFERENCES

1. Corradi, M.; Grazini, A.; Borri, A. Confinement of brick masonry columns with FRP materials.*Compos. Sci. Technol.* 2007, *67*, 1772–1783.
2. Campione, G.; Miraglia, N. Strength and strain capacities of concrete compression members reinforced with FRP. *Cem. Concr. Compos.* 2003, *23*, 31–41.
3. Aiello, M.A.; Micelli, F.; Valente, L. FRP confinement of square masonry columns. *J. Comp. Constr.* 2009, *14*, 583–596.
4. Di Ludovico, M.; D'Ambra, C.; Prota, A.; Manfredi, G. FRP Confinement of tuff and clay brick columns: Experimental study and assessment of analytical models. *J. Comp. Constr.* 2010, *14*, 583–596.
5. Faella, E.; Martinelli, E.; Paciello, S.; Camorani, G.; Aiello, M.A.; Micelli, F.; Nigro, E. Masonry columns confined by composite materials: Experimental investigation. *Comp. Part B Eng.* 2011,*42*, 692–704.
6. Köksal, H.O.; Aktan, S.; Kuruşçu, A.O. Elastoplastic Finite-Element Analysis of FRP-Confined Masonry Columns. *J. Comp. Constr.* 2012, *16*, 407–417.
7. Lignola, G.P.; Angiuli, R.; Prota, A.; Aiello, M.A. FRP confinement of masonry: Analytical modeling. *Mater. Struct.* 2014, *47*, 2101–2115.
8. Matthys, S.; Toutanji, H.; Audenaert, K.; Taerwe, L. Axial load behavior of large-scale columns confined with fiber-reinforced polymer composites. *ACI Struct. J.* 2005, *102*, 258–267.
9. Yilmaz, I.; Mezrea, P.; Ispir, M.; Binbir, E.; Bal, I.; Ilki, A. External confinement of Brick Mansory Columns with Open—Grid Basalt Reinforced Mortar. In Proceedings of the Fourth Asia-Pacific Conference on FRP in Structures (APFIS 2013), Melbourne, Australia, 11–13 December 2013.
10. Carloni, C.; Mazzotti, C.; Savoia, M.; Subramaniam, K.V. Confinement of Masonry Columns with PBO FRCM Composites. *Key Eng. Mater.* 2015, *624*, 644–651.
11. Jurina, L. Prove a collasso su colonne cerchiate in muratura. In Proceedings of the IF CRASC'09 Conference, Naples, Italy, 2–4 December 2009.
12. Jurina, L. Tecniche di cerchiatura di colonne in *muratura. L'Edilizia-Structural* 2010, *164*, 38–49. (In Italian)
13. Borri, A.; Castori, G.; Corradi, M. Masonry confined by steel fiber-composite wraps. *Materials*2011, *4*, 311–326.
14. Borri, A.; Castori, G.; Corradi, M. Strengthening of fair face masonry columns with steel hooping. *Mater. Struct.* 2014, *47*, 2117–2130.
15. European Committee for Standardisation (CEN). EN 1996-1-1: Eurocode 6: Design of Masonry Structures—Part 1-1: General Rules for Reinforced and Unreinforced Masonry Structures. Brussels, 2005. Available online: http://eurocodes.jrc.ec.europa.eu/showpage.php?id=136(accessed on 9 May 2016).

INDEX